International Conference on Big Data Analytics, Data Mining and Computational Intelligence (BigDaCI2025), Connected Smart Cities (CSC2025), and e-Health (EH2025)

Held at the Multi-Conference on Computer Science and Information Systems (MCCSIS 2025)

Lisbon, Portugal
23-25 July 2025

ISBN: 979-8-3313-2768-2

Printed from e-media with permission by:

Curran Associates, Inc.
57 Morehouse Lane
Red Hook, NY 12571

Printed with permission by Curran Associates, Inc. (2026)

For permission requests, please contact International Association for Development of the Information Society (IADIS) at the address below.

International Association for Development of the Information Society (IADIS)
Rua Sao Sebastiao Da Pedreira
No. 100, 30, 1050-209
Lisbon, Portugal

Phone: 351 21 3151373
Fax: 351 21 3151244

secretariat@iadis.org

Additional copies of this publication are available from:

Curran Associates, Inc.
57 Morehouse Lane
Red Hook, NY 12571 USA
Phone: 845-758-0400
Fax: 845-758-2633
Email: curran@proceedings.com
Web: www.proceedings.com

INTERNATIONAL CONFERENCES ON

BIG DATA ANALYTICS, DATA MINING AND COMPUTATIONAL INTELLIGENCE 2025

CONNECTED SMART CITIES 2025

and

E-HEALTH 2025

part of the

MULTI CONFERENCE ON COMPUTER SCIENCE AND INFORMATION SYSTEMS 2025

PROCEEDINGS OF THE INTERNATIONAL CONFERENCES ON

BIG DATA ANALYTICS, DATA MINING AND COMPUTATIONAL INTELLIGENCE 2025

CONNECTED SMART CITIES 2025

and

E-HEALTH 2025

Lisbon, Portugal

JULY 23 - 25, 2025

Organised by

iadis

international association for development of the information society

As a member of Crossref (a non-profit membership organization for scholarly publishing working with the purpose to make content easy to find, link, cite and assess) each published paper in this book of proceedings will be allocated a DOI (Digital Object Identifier) number for its fast and easy citation and indexation.

Volume Editors:
Ajith Abraham, Guo Chao Peng and Pedro Isaías

Computer Science and Information Systems Series Editors:
Piet Kommers and Pedro Isaías

Associate Editor: Luís Rodrigues

TABLE OF CONTENTS

CONNECTED SMART CITIES

E-HEALTH

SHORT PAPERS

BIG DATA ANALYTICS, DATA MINING AND COMPUTATIONAL INTELLIGENCE

CONNECTED SMART CITIES

E-HEALTH

REFLECTION PAPERS

E-HEALTH

POSTERS

BIG DATA ANALYTICS, DATA MINING AND COMPUTATIONAL INTELLIGENCE

E-HEALTH

FOREWORD

These proceedings contain the papers and posters of the 10th International Conference on Big Data Analytics, Data Mining and Computational Intelligence (BigDaCI 2025), the 11th International Conference on Connected Smart Cities (CSC 2025) and of the 17th International Conference on e-Health (EH 2025), which were organised by the International Association for Development of the Information Society, from 23-25 July, 2025. These conferences are part of the 19th Multi Conference on Computer Science and Information Systems 2025, 23-25 July, which had a total of 608 submissions.

The aim of the Big Data Analytics, Data Mining and Computational Intelligence (BigDaCI'25) Conference is to serve as a forum to present current and future work as well as to exchange research ideas in this field. BigDaCI'25 invites authors to submit their original and unpublished work that demonstrate current research using big data analytics, computational intelligence and other intelligent computing techniques and their applications in science, technology, business and commerce. Submissions were accepted under the following areas and topics:
- Big Data Algorithms and Architectures
- Computational Intelligent Frameworks for Big Data Processing
- Data Mining Topics and Applications
- Big Data Applications
- Multi-Agent Systems: Models, Architectures and Applications

The Connected Smart Cities (CSC 2025) conference aims to provide a premier and exciting forum for the exchange of scientific ideas, research findings, and industrial practices among academics and practitioners, who are from different disciplines (e.g. information systems, computer science, telecommunication engineering, urban studies, and business and economics) but have common interests in smart cities. It welcomes all forms of relevant research, including conceptual, theoretical, quantitative, qualitative, and mixed-method studies. Papers with an interdisciplinary nature are particularly encouraging. In order to offer wide-ranging discussion opportunities, the conference seeks the most recent and relevant research on the following main topics:
- Wireless Connectivity in Smart Cities;
- New Business Concerns in Smart Cities;
- Data and Software Applications in Smart Cities;
- User-Centric Smart Cities;
- Security, Privacy and Identity Issues in Smart Cities;
- Greenness and Sustainability of Smart Cities;
- Smart City as an Integrated Ecosystem.

The use of ICTs (Information and Communication Technologies) in Healthcare Services is the main mechanism to improve efficiency and effectiveness. Nowadays ICTs are being developed to achieve the following objectives:
– To integrate heterogeneous systems;
– To develop frameworks to make all data meaningful, accessible and available everywhere and permanently;
– To develop AIDC (Automatic Identification and Data Collectors) systems;
– To develop intelligent systems to support clinical and management decisions;

The use of these technologies also improves the quality of patient care and reduces clinical risk. At the same time, the patient will be part of the healthcare process, having more information about diseases and access to his/her electronic health record.

The e-Health (EH) 2025 conference aims to draw together information systems, practitioners and management experts from all quadrants involved in developing computer technology to improve healthcare quality.

Submissions were accepted under the following 3 main areas in the field of e-Health within specific topics:
-Research Issues
-Management Issues
-Applications

These conferences received 168 submissions from more than 23 countries. Each submission has been anonymously reviewed by an average of four independent reviewers, to ensure that accepted submissions were of a high standard. Consequently, only 25 full papers were approved which means an acceptance rate of 15%. A few more papers were accepted as short papers and reflection paper, posters and doctoral paper. An extended version of the best papers may be published in the IADIS International Journal on Computer Science and Information Systems (ISSN: 1646-3692), IADIS International Journal on WWW/Internet (ISSN: 1645-7641), and also in other selected journals.

This volume has taken shape as a result of the contributions from a number of individuals. We are grateful to all authors who have submitted their papers to enrich the conference proceedings. We wish to thank all members of the organizing committee, delegates, invitees and guests whose contribution and involvement are crucial for the success of the conference. We are especially grateful to Professor Dr. Piet Kommers, UNESCO Professor of Educational Technology (The Netherlands), for serving as the keynote speaker at MCCSIS 2025.

Last but not least, we hope that everybody enjoyed the presentations, and we invite all participants for next year's edition of these conferences.

Ajith Abraham, Bennett University, India
BigDaCI 2025 Program Chair

Guo Chao (Alex) Peng, Sun Yat-Sen University, China
CSC 2025 Program Chair

Pedro Isaias, Universidade Aberta (Portuguese Open University), Portugal
EH 2025 Program Chair

Piet Kommers, UNESCO Professor of Educational Technology, The Netherlands
Pedro Isaias, Universidade Aberta (Portuguese Open University), Portugal
MCCSIS 2025 General Conference Co-Chairs

July 2025

PROGRAM COMMITTEE

BIG DATA ANALYTICS, DATA MINING AND COMPUTATIONAL INTELLIGENCE 2025 PROGRAM CHAIR

Ajith Abraham, Bennett University, India

CONNECTED SMART CITIES 2025 PROGRAM CHAIR

Guo Chao (Alex) Peng, Sun Yat-Sen University, Guangzhou, China

E-HEALTH 2025 PROGRAM CHAIR

Pedro Isaias, Universidade Aberta (Portuguese Open University), Portugal

MCCSIS 2025 GENERAL CONFERENCE CO-CHAIRS

Piet Kommers, UNESCO Professor of Educational Technology, The Netherlands
Pedro Isaias, Universidade Aberta (Portuguese Open University), Portugal

BIG DATA ANALYTICS, DATA MINING AND COMPUTATIONAL INTELLIGENCE 2025

COMMITTEE MEMBERS

Abhishek Shukla, R.D.Engineering College Technical Campus, India
Adel M. Alimi, University of Sfax, Tunisia
Adina Florea, University Politehnica of Bucharest, Romania
Aleksandar Lazarevic, AIDA Insights, USA
Alessandro Giuliani, University of Cagliari, Italy
Alfredo Cuzzocrea, Icar-cnr & University of Calabria, Italy
Álvaro García-Barragán, Universidad Politécnica de Madrid, Spain
Anand Nayyar, Duy Tan University, Vietnam
Andrei Olaru, University Politehnica of Bucharest, Romania
Andres Castillo, Pontificial University of Salamanca, Spain
Angel Moreno, University of Alcala De Henares, Spain
Archil Maysuradze, Lomonosov Moscow State University, Russia
Arnulfo Alanis, Tecnologico Nacional de México, campus Tijuana, Mexico
Audrone Lupeikiene, VU Institute of Data Science and Digital Technology, Lithuania
Carson K. Leung, University of Manitoba, Canada
Christos Kalloniatis, University of the Aegean, Greece
Christos Makris, University of Patras, Greece
Daniel Lemire, LICEF Research Center, Canada

Dariusz Krol, Wrocław University of Science and Technology, Poland
Domenico Talia, Università della Calabria, Italy
Eva Sciacca, Istituto Nazionale di Astrofisica, Italy
Fani Liapi, University of Bedfordshire, United Kingdom
Filippo Sciarrone, Universitas Mercatorum, Italy
Gerard Murray, 1729 Distillery, Australia
Gianni Costa, Icar-cnr, Italy
Horia-Nicolai Teodorescu, Technical University of Iasi, Romania
Igor Bernik, University of Maribor, Slovenia
I-Hsien Ting, Kaohsiung University, Taiwan
Javier Carbo Rubiera, Univ. Carlos III de Madrid, Spain
Jayesh Soni, Florida International University, USA
Jorge Adolfo Ramirez Uresti, Tecnologico de Monterrey, Mexico
Jose Molina Lopez, Universidad Carlos III de Madrid, Spain
Jose Alfredo Ferreira Costa, Federal University, Brasil
Kauser Ahmed, Vit University, India
Ki Chan, Hong Kong University of Science and Technology, Hong Kong
Laurentiu Vasiliu, Peracton ltd., Ireland
Marco Villani, University of Modena and Reggio Emilia, Italy
Maria de los Angeles Constantino-Gonzalez, Tecnologico de Monterrey, Mexico
Maricela Quintana, Tecnológico de Monterrey, Mexico
Marius Silaghi, Florida Institute of Technology, USA
Michael Vassilakopoulos, University of Thessaly, Greece
Mohamed Ben Halima, Higher Institute of Computer Sciences and Multimedia, Tunisia
Mutlu Mete, University of North Texas, USA
Panayiotis Bozanis, University of Thessaly, Greece
Paula Miranda, School of Technology, Setubal Polytechnic University, Portugal
Pavel Osipov, Riga Technical University, Latvia
Prabhat Mahanti, University of New Brunswick, Canada
Rafael Leon, Universidad Politécnica de Madrid, Spain
Riccardo Ortale, ICAR-CNR, Italy
Silvia Parusheva, University of Economics - Varna, Bulgaria
Sviatoslav Braynov, University of Illinois, USA
Tibebe Beshah, Addis Ababa University, Ethiopia
Tomonobu Ozaki, Nihon University, Japan
Victor Lopez, Universidad Tecnologica de Panama, Panama
Victor Manuel Landassuri-Moreno, Universidad Autónoma del Estado de México, Mexico
Vincent Hilaire, Université de Technologie de Belfort-Montbeliard, France
Viorel Negru, West University of Timisoara, Romania
Wiem Chebil, University of Monastir and University of Sousse, Tunisia
Yolanda Escudero Martin, Universidad de Alcala, Spain
Zahid Akhtar, State University of New York Polytechnic Institute, USA

CONNECTED SMART CITIES 2024

COMMITTEE MEMBERS

Alessandro Bocci, University of Pisa, Italy
Andreas Menychtas, University of Piraeus, Greece
Carlos Cruz, Universidad de Alcalá, Spain
Chafik Mahaya, University of Biskra, Algeria
Davide Careglio, Universitat Politècnica de Catalunya, Spain

Felipe Araya, Universidad Tecnica Federico Santa Maria, Chile
Giovanni Merlino, University of Messina, Italy
Hanna Kavalionak, ISTI-CNR, Italy
Irfan Al-Anbagi, University of Regina, Canada
Jacopo Massa, University of Pisa, Italy
Javier Faulin, Public University of Navarra, Spain
Kamran Soomro, University of West England, United Kingdom
Leonardo Sierra, Universidad de la Frontera, Chile
Luca Ferrucci, University of Pisa, Italy
Mailyn Moreno, Facultad de Informática, Universidad Complutense d, Spain
Mani Krishna, University of Massachusetts Amherst, USA
Marcos Tostado-Véliz, Universidad de Jaén, España
Nabil Badr, Higher Institute of Public Health, St Joseph Unive, USA
Nishant Saurabh, University of Utrecht, Netherlands
Noureddine Zemmouri, University of Biskra, Algeria
Patrizo Dazzi, University of Pisa, Italy
Peter Mechant, Ghent University, Belgium
Pierfrancesco Bellini, University of Florence, Italy
Rabinarayan Mahapatra, Texas A&M University, USA
Ramin Yahyapour, GWDG/University Goettingen, Germany
Raquel Martinez Espana, Universidad de Murcia, Spain
Riccardo Bettati, Texas A&M University, USA
Rosaldo Rossetti, University of Porto, Portugal
Sandro Leuchter, Mannheim University of Applied Sciences, Germany
Valerio Besozzi, University of Pisa, Italy
Xiao Huang, Emory University, United States
Zahra Najafabadi, University of Innsbruck, Austria
Zhu Ye, Cleveland State University, USA

E-HEALTH 2025

COMMITTEE MEMBERS

Adnan Jouned, Medical University of Vienna, Austria
Ahmed Aljabbary, Dallah Hospital - Alnakheel, Saudi Arabia
Ehsan Bitaraf, Iran University of Medical Sciences, Iran
Elizabeth Krupinski, Emory University, United States
Enzo Pasquale Scilingo, University of Pisa, Italy
Eric Campo, LAAS-CNRS, France
Fatemeh Sarani Rad, North Dakota State University, USA
Frederic Bousefsaf, Université de Lorraine, France
Hamit Soyel, Queen Mary, United Kingdom
Jesuk Ko, Universidad Mayor de San Andres (UMSA), Bolivia
José Luis Bayo Monton, Universitat Politècnica de València, Spain
Karla Felix Navarro, University of Montreal, Canada
Larbi Boubchir, Liasd - University of Paris 8, France
Laurent Billonnet, Ensil - Université de Limoges, France
Lenka Lhotska, Czech Technical University in Prague, Czech Republic
Malina Jordanova, Bulgarian Academy of Sciences, Bulgaria
Mariusz Duplaga, Jagiellonian University Medical College, Poland
Martin Beer, Sheffield Hallam University, United Kingdom
Maryam Jafarpour, Medical University of Vienna, Austria
Maurice Mars, University of KwaZulu-Natal, South Africa

KEYNOTE LECTURE

A.I. FOR STUDENTS' METACOGNITIVE SKILLS AND CREATIVE THINKING

Prof. Dr. Piet Kommers
UNESCO Professor of Educational Technology,
The Netherlands

ABSTRACT

Similar to the advancements made in ICT three decades ago, the emergence of AI today signals to educators that the human learning process may be at risk. This keynote lecture will provide an additional perspective on this issue. It is crucial to recognize that the WWW supplies all the information that AI currently presents. The main difference lies in its rhetoric concerning establishing a specific viewpoint. In other words, AI now acts as a powerful catalyst for refining criteria for student assessments, concentrating on discovering unique creative solutions rather than merely elaborating on existing ones. The foundation of creative thinking is metacognition: the awareness of what one already knows, one's preferred thinking style, and an evolving intuition about how learning builds on prior knowledge. This lecture will outline a roadmap for innovative didactic, instructional, and curricular approaches in the future. Previous research into conceptual representations, ontologies, and metaphoric analogies has supported education in expanding the fields of knowledge elicitation and conceptual thinking. Once AI is integrated, we will witness a growing need among young people to navigate complexities, such as sources of information that do not rely on empirical and validated scientific evidence. This keynote will highlight new aspirations that have emerged and how to seamlessly incorporate them into 'standing education' at this very moment. Topics such as teacher education, student selection, and assessment will be discussed.

Full Papers

OBJECT SEGMENTATION BASED DEEP LEARNING MODELS: A REVIEW AND A BENCHMARKING STUDY

Rajaa Miftah, Mostafa Hanoune and Mohssine Bentaib
LTIM laboratory. Faculty of Science Ben M'sik
Hassan II University of Casablanca
Casablanca, Morocco

ABSTRACT

The Object segmentation, a fundamental task in computer vision, involves delineating pixel-level boundaries of objects within images, enabling precise understanding and analysis. Over the past decade, deep learning models have significantly advanced object segmentation, with various architectures designed to address the complexity of this task. In this paper, we present a comprehensive review and benchmarking analysis of deep learning models for object segmentation. Leveraging the COCO dataset, we evaluate models across multiple performance metrics including mask Average Precision (mask AP) and AP at IoU thresholds. Key models such as EVA and Mask Frozen-DETR, utilizing different backbone architectures, are examined for their efficacy in segmentation tasks. The discussion emphasizes the importance of backbone architecture in shaping model performance, highlighting the nuanced interplay between architecture design and segmentation accuracy. This study serves as a valuable resource for researchers and practitioners seeking to understand the landscape of object segmentation models and make informed decisions in model selection and deployment for diverse applications.

KEYWORDS

Deep Learning, Object Segmentation, Object Detection, EVA, Mask Frozen-DETR, Computer Vision

1. INTRODUCTION

In the dynamic field of computer vision, the tasks of identifying, classifying, and segmenting objects within digital images are foundational yet pose considerable challenges. These processes are crucial for a variety of applications that range from autonomous driving and medical image analysis to video surveillance and augmented reality. Among these, object segmentation through deep learning models has seen notable advancements, significantly enhancing the performance and efficiency of visual recognition systems.

Object segmentation, which involves determining the exact pixel-level boundaries of objects in an image, represents a more complex challenge than object detection, which merely identifies and locates objects within bounding boxes, or classification, which labels objects without indicating their position or size. Furthermore, object localization, which specifies the position of an object, typically offers less detailed information than segmentation. Given its complexity and broad applicability, object segmentation has been the focus of extensive research efforts, leading to the development of a wide array of models aimed at addressing increasingly complex tasks.

This paper seeks to provide a critical review and benchmarking of the most effective deep learning models for object segmentation, offering an exhaustive overview of current technologies and methodologies. The subsequent sections include a detailed exploration of the evolution of object segmentation models, highlighting significant developments and the technologies that have influenced the current landscape. We examine architectures that have demonstrated efficacy in segmentation, such as FCN, U-Net, Mask R-CNN, and newer innovations like Deep Lab and YOLACT.

Further, we present a comparative analysis of these models based on their reported performance using the COCO dataset. This analysis is structured to offer a clear overview of how each model performs under various conditions, assisting in understanding the trade-offs involved in model selection depending on specific application needs. The discussion extends to the implications of our findings, considering practical aspects of deploying these models in real-world scenarios. Challenges such as computational demands, real- time processing requirements, and the balance between accuracy and operational speed are addressed.

The paper concludes with a summary of the key insights, potential areas for future research, and the broader impact of object segmentation technologies on advancing the field of computer vision. Through this comprehensive review and benchmark study, we aim to serve as a definitive resource for researchers and practitioners looking to select and utilize the most appropriate models for their specific object segmentation needs.

2. BACKGROUND

The landscape of deep learning models for object segmentation is rich and varied, encompassing a broad range of approaches that reflect the ongoing innovation within the field. This section provides an overview of significant models that have shaped the current state of the art in object segmentation. Each model is discussed in the context of its architecture, performance characteristics, and the specific innovations it brings to handling the complex task of segmenting objects at the pixel level.

EVA, introduced by Fang, Yuxin et al. (Fang et al., 2023), explores the limits of masked visual representation learning at scale. This innovative framework delves into the complexities of masked representation learning, offering valuable insights into the capabilities and limitations of current methodologies. By pushing these boundaries, EVA contributes to the advancement of computer vision research, providing a deeper understanding of visual representation learning.

Mask R-CNN, developed by He et al. (2020), is an extension of the Faster R-CNN, adding a branch for predicting segmentation masks on each Region of Interest (RoI), in parallel with the existing branch for classification and bounding box regression. This model marked a significant milestone in object segmentation, offering robust detection and precise segmentation, even under challenging conditions (see Figure 1).

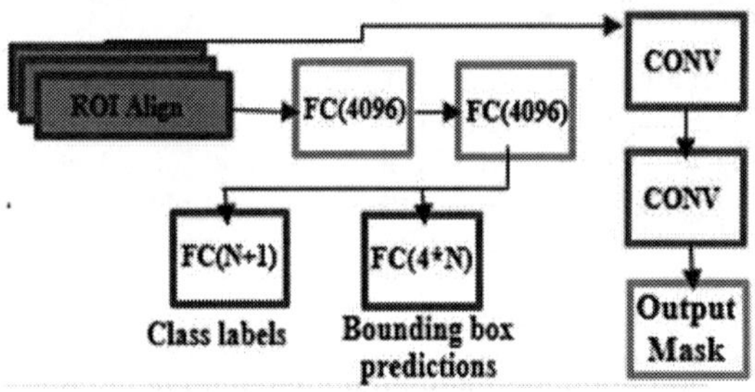

Figure 1. Mask R-CNN Architecture

BlendMask, introduced by Chen et al. (2020), builds upon the foundations of Mask R-CNN and advances the concept using a blending mechanism to generate masks. This approach provides an efficient and flexible method for instance segmentation.

CenterMask leverages the VoVNetV2 backbone to enhance the efficiency of instance segmentation by integrating a spatial attention-guided mask (SAG-Mask) with an anchor-free detection method. This combination, developed by Lee and Park (2020), delivers a balance between speed and accuracy.

SOLO and its successor SOLOv2, introduced by Wang et al. (2020), provide a direct instance segmentation method that predicts categories and instance masks without bounding boxes. SOLOv2 improved upon this with a dynamic segmentation head and faster speeds, representing a significant shift towards simplifying the instance segmentation pipeline.

DetectoRS, proposed by Qiao et al. (2021), integrates Recursive Feature Pyramid and Switchable Atrous Convolution into detector architectures, notably enhancing the representational capabilities and accuracy of the model on complex scenes.

QueryInst, by Fang et al. (2021), utilizes a set-based global reasoning strategy which formulates instance segmentation as a set prediction problem, optimizing the segmentation outcomes with fewer parameters and increasing inference speeds.

Mask2Former, introduced by Cheng et al. (2022), is designed for semantic and panoptic segmentation that excels in predicting masks from a set of learnable mask queries, showcasing superior performance across multiple segmentation tasks (see Figure 2).

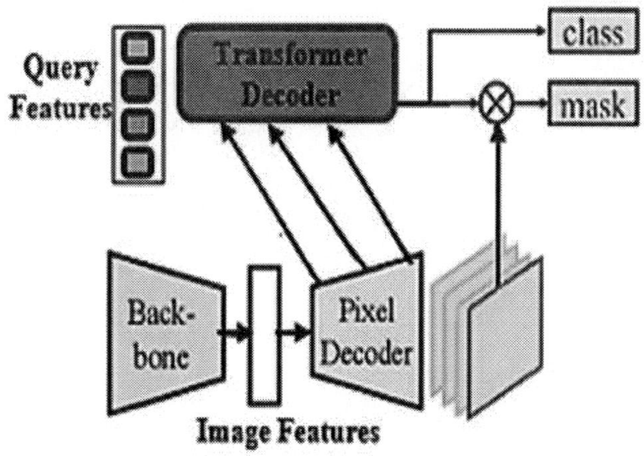

Figure 2. Original Mask2Former Architecture

3. COMPARISON

In this section, we present a comparative analysis of the various deep learning models discussed in the Background section for object segmentation, based on existing works and tests conducted on the COCO dataset.

The comparison presented in this table offers valuable insights into the relative strengths and weaknesses of each model with respect to object segmentation performance. The comparison is structured around performance metrics including mask Average Precision (mask AP), AP at IoU threshold of 0.5 (AP50), AP at IoU threshold of 0.75 (AP75), AP for small objects (APS), AP for medium-sized objects (APM), and AP for large objects (APL). The table below (Table 1) summarizes the performance of each model various deep learning models discussed in the Background.

Table 1. Detailed comparisons on multiple popular segmentation models on the coco dataset

Model Ref	Journal	Model	Backbone	mask AP	AP50	AP75	APS	APM	APL
(Fang et al., 2023)	CVPR 2023	EVA	ViT	55.5	80.0	-	36.3	58.0	72.4
(Liang & Yuan, 2023)		Mask Frozen-DETR	FocalNet-L	55.3	79.3	61.4	37.8	58.4	70.4
(Yan et al., 2023)	CVPR 2023	UNINEXT-H	-	51.8	76.2	56.7	33.3	55.9	67.5
(Cheng et al., 2022)	CVPR 2022	Mask2Former	Swin-L	50.5	74.9	54.9	29.1	53.8	71.2
(Fang et al., 2021)	ICCV 2021	QueryInst	ResNet-101-FPN	49.1	74.2	53.8	31.5	51.8	63.2
(Qiao et al., 2021)	CVPR 2021	DetectoRS	Cascade-ResNeXt	48.5	72.0	53.3	31.6	50.9	61.5
(Qiao et al., 2021)	CVPR 2021	DetectoRS	Cascade-ResNeXt	47.1	71.1	51.6	30.3	49.5	59.6
(X. Wang, Zhang, et al., 2020)	NeurIPS 2020	SOLOv2	ResNet-DCN-FPN	41.7	63.2	45.1	18.0	45.0	61.6
(Chen et al., 2020)	CVPR 2020	BlendMask	DCN-ResNet	41.3	63.1	44.6	22.7	44.1	54.5
(Lee & Park, 2020)	arXiv 2019	CenterMask + VoVNetV2-99	-	40.6	62.3	44.1	20.1	42.8	57.0
(X. Wang, Kong, et al., 2020)	ECCV 2020	SOLO	DCN-FPN	40.4	62.7	43.3	17.6	43.3	58.9
(Cao, Cholakkal, et al., 2020)	CVPR 2020	D2Det	ResNet	40.2	61.5	43.7	21.7	43.0	54.0
(Ke et al., 2021)	CVPR 2021	BCNet	FPN-ResNet	39.6	61.2	42.7	22.3	42.3	51.0
(Cai & Vasconcelos, 2018)	ICCV 2019	Cascade R-CNN	Cascade-FPN-ResNet	39.5	61.4	42.9	21.2	42.5	52.1
(Fan et al., 2020)	ECCV 2020	CPMask	-	39.2	60.8	42.2	22.2	41.8	50.1
(Xie et al., 2022)		PolarMask++	DCN-ResNeXt	38.7	64.1	40.0	22.2	40.2	52.0
(Cao, Anwer, et al., 2020)	ECCV 2020	SipMask	ResNet	38.1	60.2	40.8	17.8	40.8	54.3
(Ying et al., 2019)		EmbedMask	FPN-ResNet	37.7	59.1	40.3	17.9	40.4	53
(He et al., 2020)	ICCV 2017	Mask R-CNN	FPN-ResNeXt	37.1	60.0	39.4	16.9	39.9	53.5
(S. Wang, Gong, et al., 2020)		RDSNet	-	36.4	57.9	39.0	16.4	39.5	51.6
(Xie et al., 2020)	CVPR 2020	PolarMask	FPN-ResNeXt	32.9	55.4	33.8	15.5	35.1	46.3
(Xie et al., 2020)	CVPR 2020	PolarMask	FPN-ResNet	30.4	51.9	31	13.4	32.4	42.8

Average Precision (AP):
AP % AP at IoU=.50:.05:.95 APIoU=.50 % AP at IoU=.50 APIoU=.75 % AP at IoU=.75

AP Across Scales:
APsmall % AP for small objects: area < 322 APmedium AP for medium objects: 322 < area < 962 APlarge AP for large objects: area >962

4. RESULTS

This section evaluates the performance of deep learning models for object segmentation using key metrics from Table I, which reflect segmentation accuracy, detection performance, and localization precision. Before presenting individual bar charts, we briefly outline the importance of each metric:

- Mask Average Precision (mask AP): This metric evaluates segmentation accuracy, reflecting the precision of object boundaries and pixel-level segmentation.
- AP at IoU Threshold of 0.5 (AP50) and 0.75 (AP75): These metrics assess a model's ability to detect and localize objects accurately at moderate and strict IoU thresholds, respectively.
- AP Across Scales (APsmall, APmedium, APlarge): These metrics provide insights into a model's performance across objects of different sizes, highlighting its effectiveness in segmenting small, medium, and large objects.

Now, let's proceed to the individual bar charts for each metric to visualize the performance of the models in more detail.

1) *Mask Average Precision (mask AP) chart*: This chart illustrates the mask Average Precision scores for each model. Among the models evaluated, EVA demonstrates the highest mask AP of 55.5%, indicating its superior segmentation accuracy compared to other models. Mask Frozen-DETR follows closely behind with a mask AP of 55.3%, showcasing its competitive performance in accurately delineating object boundaries (see Figure 3).

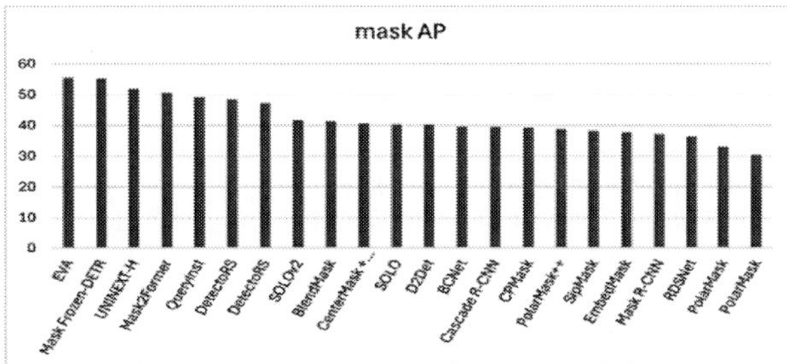

Figure 3. Models Comparison based Mask AP

2) *AP at IoU Threshold of 0.5 (AP50) Chart*: The AP50 scores chart reveals that EVA achieves the highest performance with an AP50 score of 80.0%, demonstrating its exceptional ability to detect objects with moderate overlap. Mask Frozen-DETR also performs impressively with an AP50 score of 79.3%, positioning it as a strong contender in object detection tasks (see Figure 4).

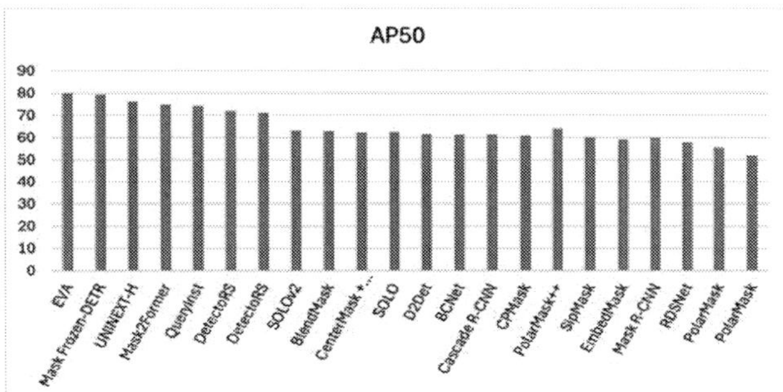

Figure 4. Models Comparison based Mask AP

3) *AP at IoU Threshold of 0.75 (AP75) Chart*: In terms of AP75 scores, Mask Frozen-DETR emerges as the top performer with a score of 61.4%. This indicates its proficiency in accurately localizing objects with high overlap thresholds. With a notable difference, UNINEXT-H also demonstrates strong performance in precise object localization under stricter IoU criteria., achieving a score of 56.7% (see Figure 5).

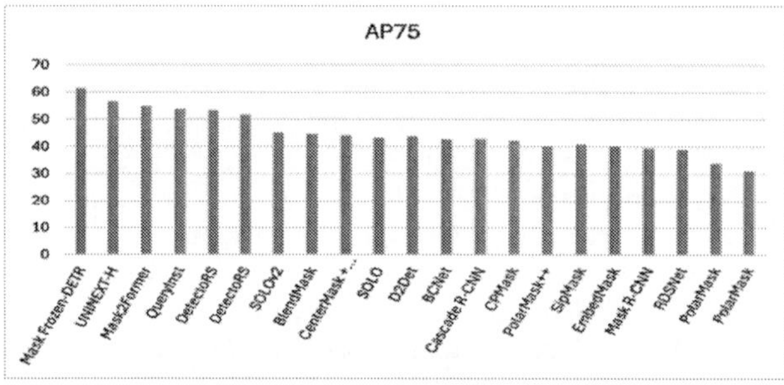

Figure 5. Models Comparison based AP75

4) *AP for Small Objects (APS) Chart*: For the APS metric, Mask Frozen-DETR stands out as the best performer with a score of 37.8%, showcasing its superior ability to segment small objects accurately. Closely following is EVA, with a competitive score of 36.3%. These results illustrate the effectiveness of these models in handling challenges associated with smaller object segmentation (see Figure 6).

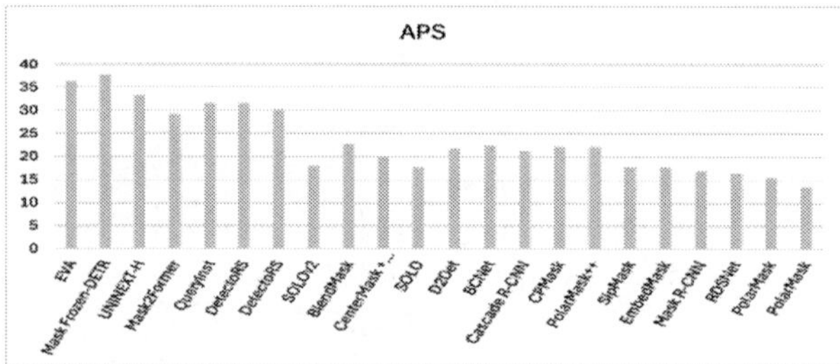

Figure 6. Models Comparison based AP small

5) *AP for Medium-Sized Objects (APM) Chart*: In the APM metric, Mask Frozen-DETR leads narrowly with a score of 58.4%, demonstrating its effectiveness in segmenting medium-sized objects. EVA follows closely with an almost matching score of 58.0%, underscoring its comparable proficiency in this category. This tight competition highlights the nuanced differences in model performance within medium object segmentation (see Figure 7).

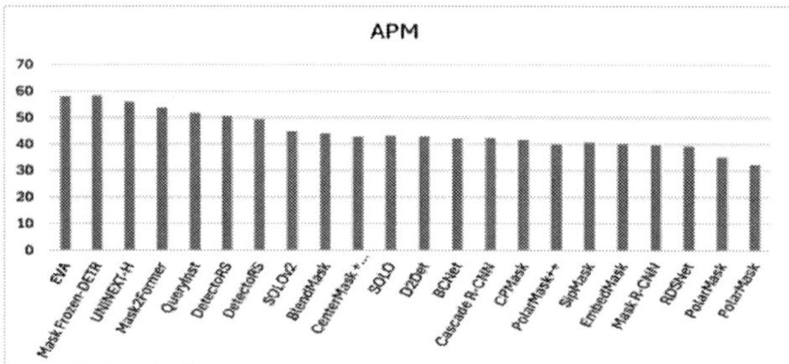

Figure 7. Models Comparison based AP medium.

6) *AP for Large Objects (APL) Chart:* EVA leads in segmenting large objects with a remarkable score of 72.4%, which highlights its superior capability in handling larger object boundaries effectively. Mask2Former follows with a score of 71.2%, showcasing its strong performance in this category. Mask Frozen-DETR also demonstrates notable capability with a score of 70.4%, making it another strong contender in the segmentation of large objects. This grouping illustrates the competitive nature of these models when dealing with larger-scale segmentation tasks (see Figure 8).

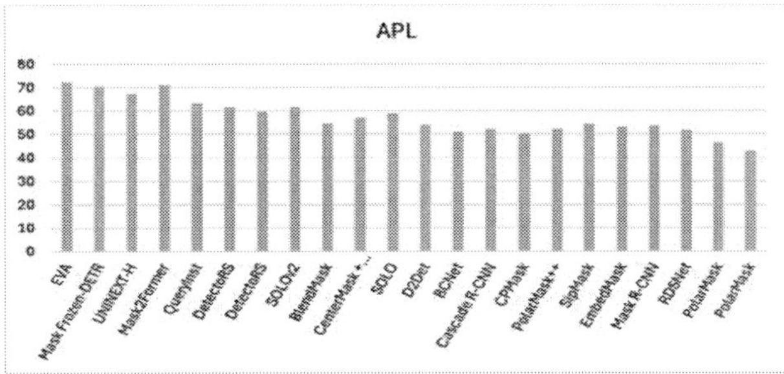

Figure 8. Models Comparison based AP large

5. DISCUSSION

In this paper, an analysis of the performance of segmentation models on the COCO dataset clearly demonstrates that the choice of backbone architecture plays a pivotal role in determining the effectiveness of a model in object segmentation tasks. Notably, models such as EVA, which employs the Vision Transformer (ViT) architecture as its backbone, and Mask Frozen-DETR, which utilizes a FocalNet-L backbone, are particularly noteworthy.

Both EVA and Mask Frozen-DETR perform strongly, each leveraging distinct backbone architectures. EVA, built on the ViT backbone, uses self-attention to capture global context, representing images as token sequences. This enables effective extraction of high-level semantic features, leading to accurate object segmentation despite moving away from traditional convolutional designs.

In contrast, Mask Frozen-DETR uses the FocalNet-L backbone from the FocalNet series, which incorporates focal attention to highlight important regions and suppress irrelevant features. Its depth and capacity allow for capturing fine details and context, contributing to strong performance in AP50 and AP75 scores.

The comparison between EVA and Mask Frozen-DETR highlights the importance of choosing the right backbone for segmentation tasks. EVA leverages the ViT backbone for strong global context and semantic understanding, while Mask Frozen-DETR uses FocalNet-L to enhance detail and discriminative features. This underscores how backbone choice significantly impacts performance and should be carefully matched to task requirements.

6. CONCLUSION

In conclusion, this comprehensive study offers valuable insights into the intricate landscape of deep learning models for object segmentation, highlighting significant advancements and key performance metrics. Through a detailed comparative analysis, models such as EVA and Mask Frozen-DETR emerge as frontrunners, showcasing exemplary performance under various evaluation criteria.

The importance of backbone architecture in influencing model effectiveness is underscored, emphasizing the need for careful consideration in model design. Moving forward, continued research and innovation in object segmentation hold promise for addressing evolving challenges and advancing the capabilities of computer vision systems.

This study contributes to fostering a deeper understanding of object segmentation methodologies and their practical implications, paving the way for future advancements in the field.

REFERENCES

Cao, J., Cholakkal, H., Anwer, R. M., Khan, F. S., Pang, Y. and Shao, L. (2020). , "D2DET: Towards high quality object detection and instance segmentation," in Proceedings of the IEEE Computer Society Conference on Computer Vision and Pattern Recognition, pp. 11482–11491, doi: 10.1109/CVPR42600.2020.01150.

Cao, J., Anwer, R. M .,Cholakkal, H., Khan, F. S., Pang, Y. and Shao, L. (2020). "SipMask: Spatial Information Preservation for Fast Image and Video Instance Segmentation," in Lecture Notes in Computer Science (including subseries Lecture Notes in Artificial Intelligence and Lecture Notes in Bioinformatics), vol. 12359 LNCS, pp. 1–18, doi: 10.1007/978-3-030-58568- 6_1.

Chen, H., Sun, K., Tian, Z., Shen,, C., Huang, Y. and Yan, Y. (2020). "Blendmask: Top-down meets bottom-up for instance segmentation," in Proceedings of the IEEE Computer Society Conference on Computer Vision and Pattern Recognition, pp. 8570–8578, doi: 10.1109/CVPR42600.2020.00860.

Cheng, B., Schwing, A. G. and Kirillov, A. (2021). "Per-Pixel Classification is Not All You Need for Semantic Segmentation," in Advances in Neural Information Processing Systems, vol. 22, no. NeurIPS, pp. 17864–17875.

Cheng, B., Misra, I., Schwing, A. G., Kirillov, A. and Girdhar, R. (2022). "Masked-attention Mask Transformer for Universal Image Segmentation," in Proceedings of the IEEE Computer Society Conference on Computer Vision and Pattern Recognition, vol. 2022-June, pp. 1280–1289, doi: 10.1109/CVPR52688.2022.00135.

Fan, Q., Ke, L., Pei, W., Tang, C. K . and Tai, Y. W. (2020). "Commonality-Parsing Network Across Shape and Appearance for Partially Supervised Instance Segmentation," in Lecture Notes in Computer Science (including subseries Lecture Notes in Artificial Intelligence and Lecture Notes in Bioinformatics), 2020, vol. 12353 LNCS, pp. 379–396, doi: 10.1007/978-3-030- 58598-3_23.

Fang, Y. et al. (2023). "EVA: Exploring the Limits of Masked Visual Representation Learning at Scale," in Proceedings of the IEEE Computer Society Conference on Computer Vision and Pattern Recognition, 2023, vol. 2023-June, pp. 19358–19369, doi: 10.1109/CVPR52729.2023.01855.

Fang, Y. et al. (2021). "QueryInst: Instances as Queries," Iccv, pp. 6910–6919, 2021, [Online]. Available: http://arxiv.org/abs/2105.01928.

Fang, H. S., Sun, J., Wang, R., Gou, M., Li, Y. L. and Lu, C. (2019). "InstaBoost: Boosting instance segmentation via probability map guided copy-pasting," in Proceedings of the IEEE International Conference on Computer Vision, 2019, vol. 2019-Octob, pp. 682–691, doi: 10.1109/ICCV.2019.00077.

He, K., Gkioxari, G., Dollár, P. and Girshick, R. (2020). "Mask R-CNN," IEEE Trans. Pattern Anal. Mach. Intell., vol. 42, no. 2, pp. 386–397, 2020, doi: 10.1109/TPAMI.2018.2844175.

Ke, L., Tai, Y. W., and Tang, C. K. (2021). "Deep Occlusion-Aware Instance Segmentation with Overlapping BiLayers," in Proceedings of the IEEE Computer Society Conference on Computer Vision and Pattern Recognition, 2021, pp. 4018–4027, doi: 10.1109/CVPR46437.2021.00401.

Lee, Y. and Park, J. (2020). "CenterMask: Real-time anchor-free instance segmentation," in Proceedings of the IEEE Computer Society Conference on Computer Vision and Pattern Recognition, 2020, pp. 13903–13912, doi: 10.1109/CVPR42600.2020.01392.

Liang, Z. and Yuan, Y. (2023). "Mask Frozen-DETR: High Quality Instance Segmentation with One GPU," 2023.

Qiao, S., Chen, L. C. and Yuille, A. (2021). "DetectoRS: Detecting Objects with Recursive Feature Pyramid and Switchable Atrous Convolution," in Proceedings of the IEEE Computer Society Conference on Computer Vision and Pattern Recognition, 2021, pp. 10208–10219, doi: 10.1109/CVPR46437.2021.01008.

Wang, X., Zhang, R., Kong, T., Li, L. and Shen, C. (2020). "SOLOv2: Dynamic and fast instance segmentation," in Advances in Neural Information Processing Systems, 2020, vol. 2020- Decem, no. NeurIPS, pp. 1–17.

Wang, X., Kong, T., Shen, C., Jiang, Y., and Li, L. (2020). "SOLO: Segmenting Objects by Locations," in Lecture Notes in Computer Science (including subseries Lecture Notes in Artificial Intelligence and Lecture Notes in Bioinformatics), 2020, vol. 12363 LNCS, pp. 649–665, doi: 10.1007/978-3-030-58523-5_38.

Wang, S., Gong, Y., Xing, J., Huang, L., Huang, C. and Hu, W. (2015). "RDSNet: A new deep architecture for reciprocal object detection and instance segmentation," in AAAI 2020 - 34th AAAI Conference on Artificial Intelligence, 2020, no. Girshick 2015, pp. 12208–12215, doi: 10.1609/aaai.v34i07.6902.

Xie, E., Wang, W., Ding, M., Zhang, R. and Luo, P. (2022). "PolarMask++: Enhanced Polar Representation for Single-Shot Instance Segmentation and beyond," IEEE Trans. Pattern Anal. Mach. Intell., vol. 44, no. 9, pp. 5385–5400, 2022, doi: 10.1109/TPAMI.2021.3080324.

Xie, E. et al. (2020). "PolarMask: Single shot instance segmentation with polar representation," in Proceedings of the IEEE Computer Society Conference on Computer Vision and Pattern Recognition, 2020, vol. 1, pp. 12190–12199, doi: 10.1109/CVPR42600.2020.01221.

Yan, B. et al., (2023). "Universal Instance Perception as Object Discovery and Retrieval," 2023, pp. 15325–15336, doi: 10.1109/cvpr52729.2023.01471.

Ying, H., Huang, Z., Liu, S., Shao, T. and Zhou, K. (2019). "EmbedMask: Embedding Coupling for One-stage Instance Segmentation," 2019, [Online]. Available: http://arxiv.org/abs/1912.01954.

PREDICT AND ESTIMATE THE CURRENT STOCK PRICES WITH POPULAR MACHINE LEARNING ALGORITHMS

Ying Bai[1] and Dali Wang[2]
[1]Johnson C. Smith University
100 Beatties Ford Rd., Charlotte NC 28216 USA
[2]Christopher Newport University
One Avenue of the Arts, Newport News VA 23606 USA

ABSTRACT

It is a challenging task to correctly and accurately predict and estimate the stock prices to get the maximum profit in the current market, and it is also critically important for all financial institutions to obtain the optimal stock prices under the current fluctuation situation. In this study, we use different machine learning (ML) algorithms, such as random forest (RF), support vector machine (SVM), decision regression trees (DRT), and nonlinear regression (NR), to easily and correctly predict and estimate the current and future possible stock prices based on the historical stock data. By using some appropriate pre-data-processing techniques, the current stock prices could be accurately and quickly estimated via those models. In this research, different ML algorithms are designed and built to help decision makers working in the financial institutions to easily and conveniently predict the current stock prices. A comparison among different ML algorithms is performed to get the optimal one for the stock prices predictions. The minimum training and checking RMSE values for RF and DRT models can be accurate to 0.010 and 0.012.

KEYWORDS

Machine Learning, Deep Learning, Random Forest, Support Vector Machine, Regression Trees, Predict Current Stock Prices, AI Applications in Financial Implementations

1. INTRODUCTION

As the fast development of AI technologies, such as fuzzy inference systems, machine learning and deep learning, today various AI related algorithms have been widely implemented in financial fields to estimate and predict the stock values, currency exchanging rates, bonus analyses and all other related applications (Chen & Chiang et al, 2019).

Most of researches is concentrated on stock predictions or estimations based on neural networks, machine learning, and deep learning studies. Different and various machine learning algorithms accompanied with some sophisticated additions are applied on stock analyses and predications to improve the accuracy of prediction on stock markets. L. S. Chong et al. reported to use Ensemble of Deep Neural Networks to performance prediction for stock markets (Chong & Lim et al, 2020). L. Yu developed an algorithm based on deep learning and neural networks to improve the analyses for economic and financial data (Yu, 2022). V. Polepally et al. and Y. K. Pardeshi and P. P. Kale reported to use machine learning and deep learning algorithms to improve the prediction accuracy for current stock prices (Polepally et al, 2021, Pardeshi and Kale, 2021). H. J. Singh, et al. and Y. Lin et al. developed a novel multivariate recurrent neural network and a new convolutional neural network with long short-term memory combined model to estimate the current stock prices and their tendency (Singh & Ghosh, 2022, Lin, 2021). K. Singh et al. performed a comparative studies and analysis for different stock price prediction techniques developed in recent years (Singh & Thapliyal, 2022). S. Roy and S. Tanveer developed an algorithm to forecast stock price by using DeepNet method (Roy & Tanveer, 2023). Instead of using any traditional machine learning model, M. Tarsi et al. utilized a Long Short-Term Memory (LSTM), which is a variation of machine learning model, to predict the stock price (Tarsi & Douzi et al, 2023). Mandee, A et al. utilized an explainable artificial intelligence XAI to predict stock market trends (Mandee et al, 2022).

K. Chinprasatsak et al. reported to use neural network for forecasting high and low price on foreign exchange market (Chinprasatsak, et al, 2020). A. Alamsyah and W. H. Aprillia and P. Aggarwal and A. K. Sahani performed some studies in comparisons to the foreign currency prediction performance with neural network algorithms (Alamsyah & Aprillia, 2022, Aggarwal & Sahani, 2020). T. A. Bui et al. and G. Singh et al. reported to use neural networks and CNN-RNN based hybrid machine learning model to predict the currency exchange rate (Bui et al, 2020, Singh et al, 2022). E. Sarmas et al. performed a comparison study among different machine learning classification methods used for currency exchange rate trends (Sarmas et al, 2022). A. L. C. Tak and R. Logeswaran also developed a foreign currency prediction method based on machine learning technique (Tak & Logeswaran, 2022).

To correctly and accurately predict and estimate the current stock prices to get the maximum profit via different AI methods, some correct AI models are necessary with popular algorithms, such as RF, SVM, DRT and NR. Combining with some appropriate pre-data-processing techniques, the current stock prices could be accurately and quickly estimated via those models. Unlike other similar reported studies, in which only a single ML model is adopted, in this research, multiple different popular ML models are adopted and built to help decision makers working in the financial institutions to easily and conveniently predict the current stock prices.

Stock prices are changed at any moment, and they may be varied significantly day by day, month by month and year by year. Due to the heavy complicity and unforeseen variations in the current market, to correctly and accurately predict the stock prices needs the following factors and operational steps to be taken:

1) The changing or variation of the stock prices can be considered as a periodic function, and this period could be 3 months, 6 month or longer, which depends on the target period on each research. In our case, we used 3 months as a period.
2) Based on assumption above, we utilized the Google Stock dataset to train and check our different target ML models.

This study is divided into 5 sections; after this Introduction, an introduction to two Google Stock datasets used as the training and checking for AI models is provided in section 2. Various ML algorithms are discussed in section 3. The experiment training studies and predicting results are given in section 4. The conclusion and future works are provided in section 5.

2. THE GOOGLE STOCK DATASET

Two Google Stock datasets (Kaggle, 2012), one contained 5-year stock transaction records from Jan. 3, 2012, to Dec. 30, 2016, and the other included 1-month stock transaction records from Jan. 3, 2017, to Jan. 31, 2017, are utilized in this study. The first one is used as the training and checking data for ML models, and the second works as the testing and validation purpose for those models.

Each dataset contained six columns, Date, Open, High, Low, Volume and Close, with both 5-year and 1-month stock price records. Each related column can be mapped to the Opening price, Highest price, Lowest price, transaction Volume and Closing price. For our study, we only need four of them; Open, High, Low and Close. In fact, we use the first three columns, Open, High, Low as inputs and the Close column as the output.

A critical key issue in using those data to train, check and test our ML models is the data preprocessing. As everybody knows, the stock prices are changed or varied at every moment, and the changed amounts are significantly different with a relatively wider range, or even dynamically, for a period of time. This provided a challenging issue when using ML models to estimate the output or the closing price due to the significant variations in the price values. In the worst case, the ML models could not perform their fitting functions properly due to significant variations in the input values with big different price values for different time period.

To effectively solve this key issue, we need to preprocess those data, exactly to perform a normalization job for those data to enable them to be used in our model training and checking. In summary, we only take care of those relatively changing values on the prices, but not for the absolute changing values, which is good enough for us since we only pay our attention to the changing values in trends or tendency.

3. MACHINE LEARNING ALGORITHM

As mentioned in the abstract section, different popular ML algorithms are adopted and studied in this study to analyze and compare the detection functions among different ML algorithms. Now let's have a quick discussion about those algorithms.

3.1 Random Forest Algorithm

The random forest algorithm is similar to ensemble learning method that is composed of a set of decision trees. An example algorithm of ensemble method is called bagging. In fact, random forest is an extension of the bagging method since it utilizes both bagging and feature randomness to create an uncorrelated forest of decision trees. The so-called feature randomness (feature bagging) is to generate a random subset of features from the training dataset, which ensures low correlation among decision trees. For a regression task (our case), all decision trees will be averaged to yield the predicted values.

The bagging algorithm is used to train a random forest model. Given a training set X = x1, ..., xn with responses Y = y1, ..., yn, bagging repeatedly (B times) selects a random sample with replacement of the training set and fits trees to these samples (Wikipedia, 2025):

For b = 1, ..., B:
  Sample, with replacement, n training examples from X, Y; call these Xb, Yb.
  Train a classification or regression tree fb on Xb, Yb.

After training, predictions for unseen samples x' can be made by averaging the predictions from all the individual regression trees on x':

$$\hat{f} = \frac{1}{B}\sum_{b=1}^{B} f_b(x') \quad (3.1)$$

Equation (3.1) is used to get an average of all decision bagging results for regression. Figure 1 shows an illustration for RF structure.

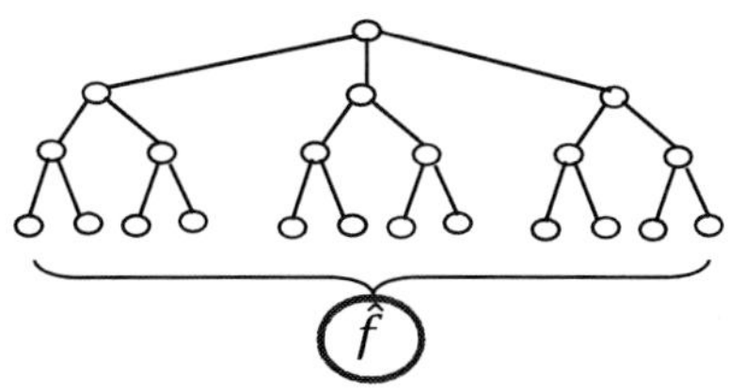

Figure 1. A random Forest structure

3.2 Support Vector Machine Algorithm

The support vector machine (SVM) is a kind of supervised learning algorithm. The objective of the SVM algorithm is to find a hyperplane that, to the best degree possible, separates data points of one class from those of another class. Best is defined as the hyperplane with the largest margin between the two classes, represented by plus versus minus in the Fig. 2. Margin means the maximal width of the slab parallel to the hyperplane that has no interior data points. Only for linearly separable problems can the algorithm find such a hyperplane, for most practical problems the algorithm maximizes the soft margin allowing a small number of misclassifications.

In fact, a so-called hyperplane is to use a single line or an intersection line to present a single plan or one dimension in a multiple dimensional space or Euler space in mathematics. Regularly the SVM is used for classification problems, but it can also be used in regression issues. The SVM is mainly used in unsupervised learning, such as clustering process. Let's use a simple example to illustrate how to use SVM to perform clustering process.

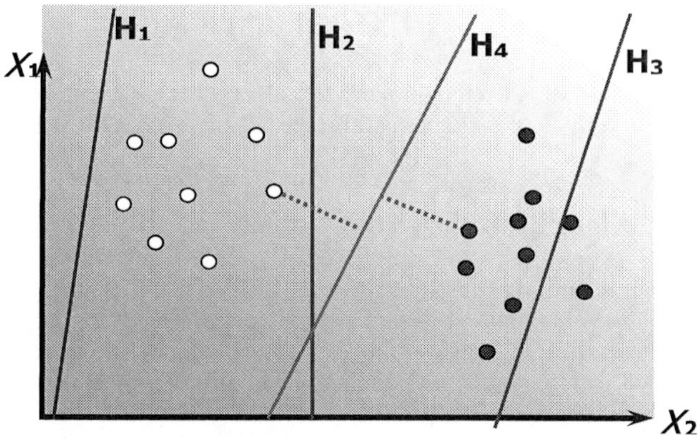

Figure 2. An illustration of SVM

In Figure 2, two classes are represented by two groups of white and black data points, each of them can be considered as a p dimensional vector (a list of p-numbers), and we want to know whether we can separate such points with a $(p-1)$-dimensional hyperplane. This is called a linear classifier. Each hyperplane can be mapped to a single line, H_i, $i = 1 \sim 4$. The purpose of SVM is to find or build a hyperplane in between two classes of data sets to indicate which class it belongs to. The challenge is to train the machine to understand structure from data and mapping with the right class label, for the best result, the hyperplane has the largest distance or largest margin to the nearest training data points of any class.

Suppose we have a training dataset of **n** points, (x_1, y_1), … (x_n, y_n), and these data points belong to two classes. The condition is, $y_i = 1$ if related x_i belong to one class, otherwise $y_i = -1$. We want to find the **maximum-margin hyperplane** that divides the group of points x_i for which $y_i = 1$ from the group of points for which $y_i = -1$, which is defined so that the distance between the hyperplane and the nearest point x_i from either group is maximized.

A hyperplane can be written as the set of points x_i satisfying

$$w^T x_i - b = 0 \tag{3.2}$$

where $\boldsymbol{w}$ is the normal and normalized vector to the hyperplane.

As shown in Fig. 3, two parallel hyperplanes (H2 and H3) that separate the two classes of data are built, so that the distance between them is as large as possible. The region bounded by these two hyperplanes are called the margin, and the maximum-margin hyperplane (H1) is the hyperplane that lies halfway between them. With a normalized or standardized dataset, these hyperplanes can be described by the equations.

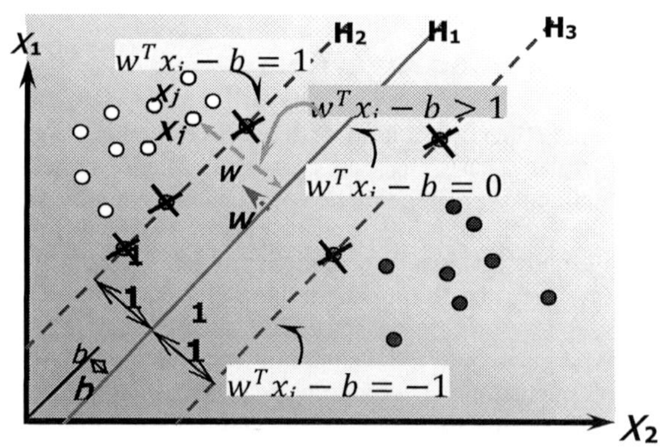

Figure 3. An illustration of hyperplanes with SVM

$$w^T x_i - b = 1 \tag{3.3}$$

(H_2 - points on or above this boundary is of one class, with label 1) and

$$w^T x_i - b = -1 \tag{3.4}$$

(H_3 - points on or below this boundary is of the other class, with label −1)

The points on two boundary hyperplanes are called support vectors, which are indicated by **x** marks. An example point x_j is shown in Figure 3. The vertical distance between the point x_j and the hyperplane $w^T x_i - b = 0$ is $w^T x_j - b$, which is greater than 1. Thus, that point belongs to the top class with $y_i = 1$. The $\boldsymbol{w}$ is a normalized vector with magnitude of 1 and it only provides a direction that is vertical to the hyperplane $w^T x_i - b = 0$.

3.3 Nonlinear Regression Algorithm

Nonlinear regression is a statistical technique that helps describe nonlinear relationships in experimental data. Nonlinear regression models are generally assumed to be parametric, where the model is described as a nonlinear equation. Typically, machine learning methods are used for non-parametric nonlinear regression.

Parametric nonlinear regression models the dependent variable (called the response) as a function of a combination of nonlinear parameters and one or more independent variables (called predictors). The model can be single response variable or multiple response variables.

The parameters can take the form of an exponential, Gaussian, trigonometric, power, or any other nonlinear function. To determine the nonlinear parameter estimates, an iterative algorithm is typically used.

$$y = f(x, \beta) + \varepsilon \tag{3.5}$$

relates a vector of independent variables, x, and its associated observed dependent variables, y, with some error ε. The function f is nonlinear in the components of the vector of parameters β, but otherwise arbitrary.

Popular algorithms for fitting a nonlinear regression include:

Gauss-Newton algorithm
Gradient descent algorithm
Levenberg-Marquardt algorithm

Different ML algorithms discussed above are involved in some MATLAB Toolboxes, such as Statistics and Machine Learning Toolbox and Deep Learning Toolbox. Each algorithm can be triggered by a related function. Table 1 shows all functions related to each algorithm used in MATLAB.

Table 1. Map between algorithms & functions

Algorithm	Function
Random Forest	fitrensemble() Treebagger()
Decision Regression Tree	fitrtree()
SVM	fitrsvm()
Nonlinear Regression	fitnlm()

Table 2. Errors among algorithms

Algorithm	Max	Min	RMSE
Random Forest	0.0425	0.0003	0.0202
	0.0231	0.0002	0.0101
Decision Regression Tree	0.0269	0.0005	0.0121
SVM	0.2613	0.0044	0.1298
Nonlinear Regression	0.0864	0.0015	0.0335

4. EXPERIMENTAL RESULTS

By using related functions in MATLAB Toolboxes, we performed training, checking and validation functions to different ML models with Google Stock dataset to predict the current stock prices.

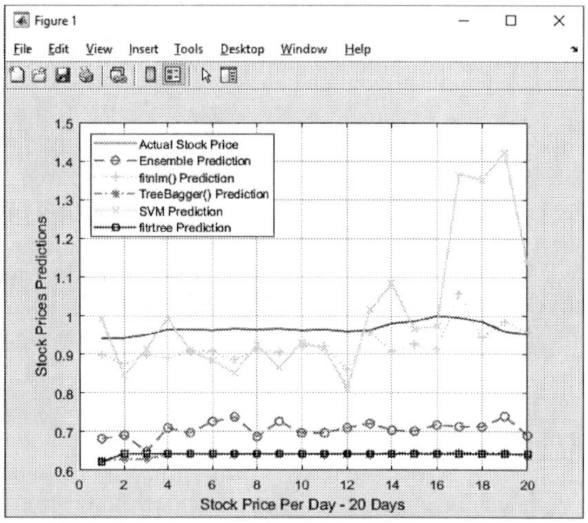

Figure 4. The stock prediction results

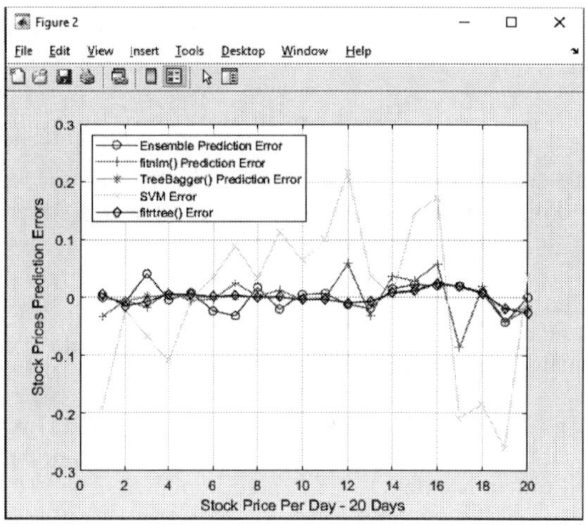

Figure 5. The comparison of prediction errors

We used 1,258 stock records in one Google Stock dataset that contained 5-year stock transaction records from Jan. 3, 2012, to Dec. 30, 2016, as the training data, and 20 stock records in another Google Stock dataset that included 1-month stock transaction records from Jan. 3, 2017, to Jan. 31, 2017, as the testing data to above ML models. The training and testing results are shown in Figure 4.

A comparison of prediction results among those ML algorithms is also performed and the comparison result is shown in Figure 5. The errors between the predicted and actual stock prices are shown in Table 2, where the maximum errors, the minimum errors and RMSE are displayed based each algorithm.

It can be found from Figures 4 and 5 as well as Table 2, the random forest and decision regression tree algorithms provided better prediction results with smaller RMSE values. The nonlinear regression algorithm provided a satisfactory prediction result. But the SVM algorithm gives a worse prediction result.

These comparison results based these ML algorithms are only limited to this case study and may not be true when these algorithms are used for other applications.

5. CONCLUSION

Five popular ML algorithms are analyzed and studied for our stock prediction applications. Among those algorithms, the random forest and decision regression tree provided better prediction results with smaller RMSE values compared with all other methods. The nonlinear regression algorithm gives a satisfactory prediction result. The SVM provided a worse prediction result.

REFERENCES

Aggarwal, P. and Sahani, A. K. (2020), "Comparison of Neural Networks for Foreign Exchange Rate Prediction," IEEE 15th International Conference on Industrial and Information Systems (ICIIS), RUPNAGAR, India, pp. 415-419, doi: 10.1109/ICIIS51140.2020.9342733.

Alamsyah and Aprillia, W. H. (2022), "Comparison of Predictions Foreign Currency Exchange Rates (USD, JPY, EUR, GBP, CHF, CAD) to Rupiah (IDR) using Artificial Neural Network and Long Short-Term Memory," 6th International Conference on Information Technology, Information Systems and Electrical Engineering (ICITISEE), Yogyakarta, Indonesia, pp. 585-590, doi: 10.1109/ICITISEE57756.2022.10057683.

Biswal, A. (2023), Stock Price Prediction Using Machine Learning: An Easy Guide, lesson 21 of 34, Simplilearn.

Bui, T. A., Pashchenko, F. F., Pashchenko, A. F. and Kudinov, Y. I. (2020), "Using Neural Networks to Predict Exchange Rates," 13th International Conference "Management of large-scale system development" (MLSD), Moscow, Russia, pp. 1-5, doi: 10.1109/MLSD49919.2020.9247748.

Chen, C. T., Chiang, L. K., Huang, Y. C. and Huang, S. H. (2019), "Forecasting Interaction of Exchange Rates Between Fiat Currencies and Cryptocurrencies Based on Deep Relation Networks," IEEE International Conference on Agents (ICA), Jinan, China, 2019, pp. 69-72, doi: 10.1109/AGENTS.2019.8929155.

Chinprasatsak, K., Niparnan, N. and Sudsang,A. (2020), "Neural Network for Forecasting High Price and Low Price on Foreign Exchange Market," 17th International Conference on Electrical Engineering/Electronics, Computer, Telecommunications and Information Technology (ECTI-CON), Phuket, Thailand, pp. 461-465, doi: 10.1109/ECTI-CON49241.2020.9158133.

Chong, L. S., Lim, K. M. and Lee, C. P. (2020), "Stock Market Prediction using Ensemble of Deep Neural Networks," 2020 IEEE 2nd International Conference on Artificial Intelligence in Engineering and Technology (IICAIET), Kota Kinabalu, Malaysia, pp. 1-5, doi: 10.1109/IICAIET49801.2020.9257864.

Guo, Y. (2020), "Stock Price Prediction Based on LSTM Neural Network: The Effectiveness of News Sentiment Analysis," 2nd International Conference on Economic Management and Model Engineering (ICEMME), Chongqing, China, pp. 1018-1024, doi: 10.1109/ICEMME51517.2020.00206.

Lamba, V., Hooda, S., Solanki, V., Ahuja, R., Ahuja, S. and Kaur, A. (2021), "Comparative Analysis of Artificial Neural Networks for Predicting Nifty 50 value in The Indian Stock Market," 5th International Conference on Information Systems and Computer Networks (ISCON), Mathura, India, pp. 1-5, doi: 10.1109/ISCON52037.2021.9702400.

Lin, Y., Zhong, Q., Huang,Q., Li, M. and Ma, F. (2021), "A new convolutional neural network and long short-term memory combined model for stock index prediction," 14th International Congress on Image and Signal Processing, BioMedical Engineering and Informatics (CISP-BMEI), Shanghai, China, pp. 1-6, doi: 10.1109/CISP-BMEI53629.2021.9624337.

Mandee, A., Agarwal, A., Bhatia, A., Malhi, P., Kaler and Pannu, H. S. (2022), "Machine Learning Based Explainable Financial Forecasting," 4th International Conference on Computer Communication and the Internet (ICCCI), Chiba, Japan, pp. 34-38, doi: 10.1109/ICCCI55554.2022.9850272.

Moghaddam, H., Moghaddam, M. H. and Esfandyari, M. (2016), "Stock market index prediction using artificial neural network", Journal of Economics, Finance and Administrative Science, pp.1-12.

Pardeshi, Y. K. and Kale, P. P. (2021), "Technical Analysis Indicators in Stock Market Using Machine Learning: A Comparative Analysis," 12th International Conference on Computing Communication and Networking Technologies (ICCCNT), Kharagpur, India, pp. 1-6, doi: 10.1109/ICCCNT51525.2021.9580172.

Polepally, V., Nandini Reddy, N. S., Sindhuja, M., Anjali, N. and Reddy, K. J. (2021), "A Deep Learning Approach for Prediction of Stock Price Based on Neural Network Models: LSTM and GRU," 12th International Conference on Computing Communication and Networking Technologies (ICCCNT), Kharagpur, India, pp. 1-4, doi: 10.1109/ICCCNT51525.2021.9579782.

Roy, S. and Tanveer,S. (2023), "Stock Price Forcasting using DeepNet," International Conference on Computational Intelligence, Communication Technology and Networking (CICTN), Ghaziabad, India, pp. 418-420, doi: 10.1109/CICTN57981.2023.10141100.

Sarmas, E. et al., (2022), "Comparison of Machine Learning Classifiers for Exchange Rate Trend Forecasting," 13th International Conference on Information, Intelligence, Systems & Applications (IISA), Corfu, Greece, pp. 1-7, doi: 10.1109/IISA56318.2022.9904380.

Singh, A., Bhardwaj, G., Srivastava, A. P., Bindra, A., Chaudhary, P. and Ritika, (2022), "Application of Neural Network to Technical Analysis of Stock Market Prediction," 3rd International Conference on Intelligent Engineering and Management (ICIEM), London, United Kingdom, pp. 302-306, doi: 10.1109/ICIEM54221.2022.9853162.

Singh, H. J., Ghosh, K. and Sharma, S. (2022), "A Novel Multivariate Recurrent Neural Network based Analysis Model for Predicting Stock Prices," 2nd International Conference on Innovative Sustainable Computational Technologies (CISCT), Dehradun, India, pp. 1-6, doi: 10.1109/CISCT55310.2022.10046540.

Singh, K., Thapliyal, M. P. and Barthwal, V. (2022), "A comparative analysis of stock price prediction techniques," 2022 7th International Conference on Computing, Communication and Security (ICCCS), Seoul, Korea, Republic of, pp.1-4, doi: 10.1109/ICCCS55188.2022.10079696.

Singh, G. et al., (2022), "CNN-RNN based Hybrid Machine Learning Model to Predict the Currency Exchange Rate: USD to INR," 2nd International Conference on Advance Computing and Innovative Technologies in Engineering (ICACITE), Greater Noida, India, pp. 1668-1672, doi: 10.1109/ICACITE53722.2022.9823844.

Tak, A, L. C. and Logeswaran,R. (2022), "Foreign Currency Exchange Market Prediction using Machine Learning Techniques," IEEE International Conference on Distributed Computing and Electrical Circuits and Electronics (ICDCECE), Ballari, India, pp. 1-5, doi: 10.1109/ICDCECE53908.2022.9792832.

Tarsi, M., Douzi, S. and Marzak, A. (2023), "Predicting stock price using LSTM and Social Media dataset," 3rd International Conference on Innovative Research in Applied Science, Engineering and Technology (IRASET), Mohammedia, Morocco, pp. 1-4, doi: 10.1109/IRASET57153.2023.10152930.

Thu, H. G. T., Thanh, T. N. and Quy, T. L. (2021), "A Neighborhood Deep Neural Network Model using Sliding Window for Stock Price Prediction," IEEE International Conference on Big Data and Smart Computing (BigComp), Jeju Island, Korea (South), pp. 69-74, doi: 10.1109/BigComp51126.2021.00022.

Yu, L. (2022), "Economic and Financial Data Analysis System Based on Deep Learning and Neural Network Algorithm," International Conference on Knowledge Engineering and Communication Systems (ICKES), Chickballapur, India, pp. 1-5, doi: 10.1109/ICKECS56523.2022.10060240.

https://www.kaggle.com/datasets/vaibhavsxn/google-stock-prices-training-and-test-data, 2012.

https://en.wikipedia.org/wiki/Random_forest, 2025.

ANALYSIS OF THE CORRELATION BETWEEN MORTALITY RATES AND THE HEALTHCARE WORKFORCE IN SOUTH KOREA

Chanyoo Kim and Emanuel Santos
Concordia International School Hanoi
Van Tri Golf Club, Tho Da, Kim No, Dong Anh, Hanoi, Vietnam

ABSTRACT

A country's mortality rate depends on numerous factors that influence or mitigate death rates. While death is inevitable, various national achievements–ranging from advancements in education and economic stability to improved sanitation and nutrition–can prolong life expectancy and lower mortality rates. Among these, a robust healthcare workforce, characterized by optimal doctor density, plays a vital role by ensuring accessible medical services and addressing conditions that may lead to preventable death. For instance, a country with a well-established doctor-to-patient ratio would be better equipped to provide timely treatment and offer preventive care, ensuring that healthcare access is evenly distributed across the population and ultimately reducing the risk of preventable deaths.

This paper explores whether there is a correlation between doctor density and mortality rates in South Korea, specifically focusing on four administrative divisions: Seoul, Gwangju, Gangwon-do, and Jeollanam-do.

Data were collected from KOSIS on population, deaths by cause, and doctors by specialties. The data from 2010 to 2020 were cleaned and analyzed, particularly focusing on circulatory system-related deaths and external factors of mortality. Visualizations, observations, and analyses derived from the data revealed a mix of insightful correlations between the doctor densities of the selected administrative divisions in South Korea and their mortality rates.

KEYWORDS

Mortality Rates, Doctor Density, South Korea, Data Analysis

1. INTRODUCTION

With growing demands from aging populations, rising chronic disease rates, and advances in medical technology that have the potential to streamline healthcare delivery, many countries are increasingly investing in the healthcare and medical sector (World Health Organization, 2023). South Korea is among the leaders in this trend, spending nearly 10% of its GDP on healthcare (*South Korea Health Spending as Percent of GDP - Data, Chart*, n.d.).

While increasing the number of healthcare professionals is often regarded as a key strategy for improving healthcare access and quality (Research, n.d.), it is becoming increasingly clear that this alone does not fully address the complexities of modern healthcare challenges. This highlights the need to examine to what extent increasing the number of doctors actually impacts mortality rates–a key indicator of healthcare quality–and to identify other potentially more influential factors in improving healthcare outcomes.

This study aims to explore the correlation between doctor density and mortality rates, with a focus on four administrative divisions in South Korea: Seoul, Gwangju, Gangwon-do, and Jeollanam-do. Specifically, the analysis centers on South Korea's two leading causes of deaths and examines the relationship between the corresponding mortality rates and the doctor-to-patient ratios of relevant types of doctors. In particular, the study explores the correlation between circulatory system-related deaths and thoracic surgeons, and between deaths caused by external factors–such as accidents, injuries, violence, and infections–and emergency medicine doctors.

The paper is structured as follows: Section 2 will discuss data collection; Section 3 will cover data cleaning and transformation; Section 4 will present data analysis, including key visualizations; and Section 5 will address the findings and limitations of the study, as well as provide a conclusion, summarizing the study and suggesting further research opportunities and implications.

2. DATA COLLECTION

All data used in this investigation were collected from the Korean Statistical Information Service (KOSIS) (*KOSIS KOrean Statistical Information Service*, n.d.), a statistical database portal managed by Statistics Korea (KOSTAT)–the official government agency in South Korea responsible for producing and managing national statistics across fields such as the economy, environment, and society. Three datasets were obtained: Population (*KOSIS*, 2023a), Number of Deaths by Causes (*KOSIS*, 2023b), and Number of Doctors by Specialty (*KOSIS*, 2023c).

The population dataset (2000–2022) was downloaded as an excel file on August 21, 2024, containing 20 rows and 34 columns. Each row represents an administrative division of South Korea (cities and provinces), while each column represents a year, divided into sub-columns by gender: male, female, and total (sum of males and females).

The deaths dataset (2010–2020) was downloaded on September 4, 2024, also as an excel file. It initially had 1,081 rows and 14 columns, detailing the number of deaths by gender, administrative division, cause, and year. The first column listed 19 causes of death, followed by columns for gender and administrative division, with subsequent columns indicating yearly data.

The doctors by specialty dataset (2010–2020) was also downloaded on September 4, 2024. It originally consisted of 56 rows and 295 columns, detailing the number of doctors by gender, administrative division, and specialty in a given year. The first two columns specified region and gender, while the remaining columns captured yearly counts of doctors, covering a total of 28 areas of expertise.

3. DATA CLEANING/TRANSFORMATION

All three datasets were cleaned and transformed to facilitate analysis. Some common formatting issues included misaligned column headings, blank cells meant to contain essential information, such as administrative division names, and inconsistencies in row headings across datasets.

For the population dataset, preprocessing involved creating accurate column headings and removing unnecessary rows and columns by using a customized Unpivot and Split Function. Additionally, text inconsistencies were standardized. For instance, "By Administrative District" was replaced with "Region".

The deaths by cause dataset was similarly cleaned by removing unnecessary rows/columns and populating blank cells intended for text entries, such as "total" or "male," with the appropriate values. Terminology was also standardized to align with the other datasets.

For the doctors by specialties dataset, text was translated from Korean to English to ensure consistency with other datasets. Proper column headings were created using a Concatenate Function and Split Function, and an Unpivot Function was also applied to reshape the original dataset into a format suitable for analysis.

Once cleaning and transformation were completed, the datasets were joined by matching the variables *administrative division* and *year* across the three datasets: population, deaths by cause, and doctors by specialties.

4. DATA ANALYSIS

The Data Visualizations and Analysis were done with Tableau (Tableau Software, 2024), a commonly used business intelligence and data visualization tool. All Tableau visualizations are interactive and available upon request.

The population data, deaths by cause data, and doctors by specialties data were used to generate scatter plots and map visualizations. Scatter plots were chosen to highlight clear trends between key variables in an easily-observable manner, as well as the strength of those trends. Maps provide a general overview of the country, allowing for comparisons between different administrative divisions and their neighboring ones.

Through scatter plots, other visualizations, and identification of trends across and between different datasets, unexpected results were discovered between different variables.

4.1 Doctor Density for Each Administrative Division

This section examines the doctor-to-population ratio for each administrative division in South Korea, comparing them to identify which areas have a high supply of doctors relative to their population. A map of the country was created to provide an overview, allowing for a clear comparison of each administrative division and identification of neighboring or nearby areas. The map was generated using the dataset of doctors by specialty, covering the years 2010 to 2020, though specific doctor specialties are not addressed yet.

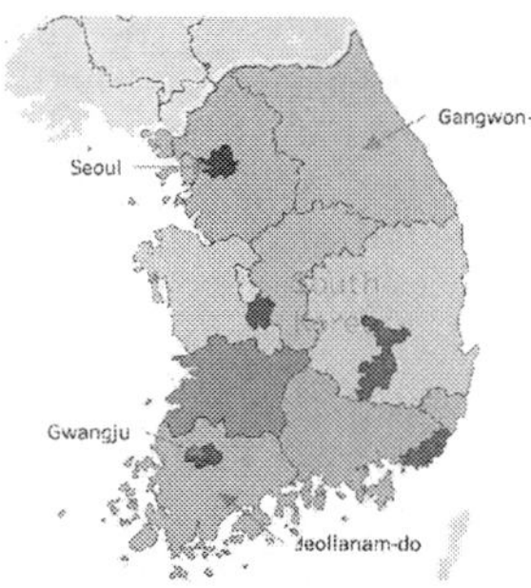

Figure 1. Mean doctor density from 2010 to 2020 of each administrative division

Figure 1 presents a map of the country, with each administrative division shaded in a gradient of blue–the darker the color, the higher the doctor-to-population ratio. As observable in Figure 2, Seoul has the highest doctor density, followed by Gwangju. Seoul has a doctor-to-population ratio of approximately 1.9 doctors per 1,000 people, while Gwangju has 1.6 doctors per 1,000 people. Two of the areas with the lowest doctor densities are the provinces Jeollanam-do and Gangwon-do, with ratios of 1.09 and 1.07 doctors per 1,000 people, respectively. It is also worth noting that Seoul and Gangwon-do are relatively close to each other, in terms of location, and that Gwangju is a city within the province Jeollanam-do. This allows differences in mortality rates, particularly between neighbouring regions, to be attributed to these discrepancies in doctor densities. Specifically, Seoul and Gwangju have well-developed healthcare systems with a high doctor-to-patient ratio, as shown in Figure 1. While Gwangju is not as urbanized as Seoul is, it still maintains a balance between urban areas and rural regions. In contrast, Gangwon-do and Jeollanam-do are predominantly rural, characterized by limited healthcare facilities and lower doctor densities (Hoon, n.d.). These characteristics can help illustrate the relationship between the doctor densities and specific causes of deaths, as well as any underlying factors that might influence their correlation. As a result, the administrative divisions selected to be the main focus of analysis were Seoul, Gwangju, Gangwon-do, and Jeollanam-do.

4.2 Burden of Causes of Death in Total Mortality

This section explores the burden of each cause of death in the total mortality data to find out which cause contributes most to the overall trendline between death rates and number of doctors, which will be discussed later in depth. The horizontal and vertical axis of the graph are time and number of deaths, respectively. The graph was created using the dataset for deaths by cause, which ranges from 2010 to 2020.

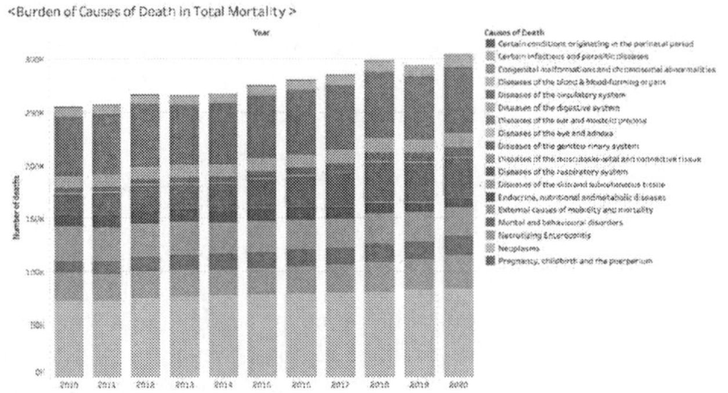

Figure 2. Burden of causes of death in total mortality of all administrative divisions

As shown in Figure 2, the number of deaths per year tends to increase over time from 2010 to 2020; there were approximately 255,000 deaths in 2010 and 305,000 deaths in 2020. It is also evident that the three leading causes of deaths are neoplasms, the light pink portion; circulatory system diseases, the dark green portion; and external causes of morbidity and mortality, the gray portion. While neoplasms were the leading cause of death, they were excluded from analysis in this article because their trendlines showed high p-values across all four administrative divisions when compared to the number of radiation oncologists. Therefore, circulatory system diseases and external causes of morbidity and mortality were chosen as the main focus of analysis to explore potential correlations between mortality rates and the density of doctors.

4.3 Trends Between Deaths Data and Doctors Data

Before investigating specific correlations between two leading causes of death–diseases of the circulatory system, and external causes–and the densities of specific types of doctors that address these causes in the chosen four administrative divisions, a general relationship between the sum of all deaths and the sum of all doctors in all 16 administrative divisions was looked into first.

4.3.1 Total Mortality Rates and Healthcare Workforce Distribution Across All Administrative Divisions

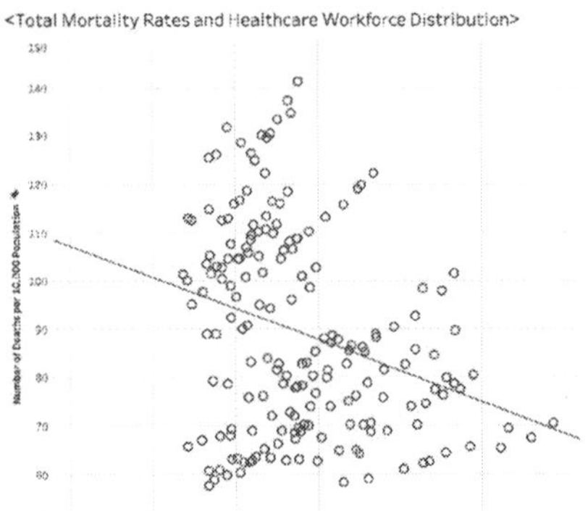

Figure 3. Total mortality rates and doctor density (2010–2020)

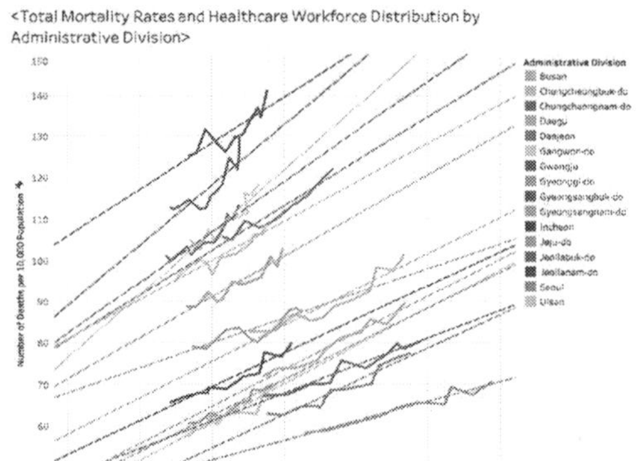

Figure 4. Total mortality rates and doctor density (2010–2020) by administrative division

Figure 3 illustrates the relationship between the doctor density (on the x-axis) and the death rate (on the y-axis). Each dot on the scatter plot represents the number of deaths per 10,000 people and number of doctors per 10,000 people in a specific administrative division for a given year between 2010 and 2020. Specific administrative divisions, death causes, doctor types, and time (year) were not identified because the sole purpose of Figure 3 was to find a general relationship between mortality rates and doctor density. Observations from Figure 3 show that there is barely any correlation between the doctor densities and death rates. The p-value of the linear trendline is less than 0.0001, indicating a statistically significant relationship between the two variables. However, the low r-squared value (0.085) suggests that the number of doctors merely accounts for the variation in the number of deaths. This weak explanatory power, along with the large spread of data points shown in the graph, indicates that there is no correlation between the two variables. However, it does appear that as the density of doctors increases, the range of the death ratio decreases. The administrative divisions were then identified by allocating each one a color, and by doing so, a novel relationship was found.

As illustrated in Figure 4, when each administrative division was identified and analyzed separately, the correlation between the number of deaths and number of doctors in each division was positive, with a significant r-squared value for each. In particular, the lowest r-squared value was 0.69 in Gyeongsangbuk-do, and the highest was 0.95 in Daegu. Unlike the trendline with a weak r-squared value shown in Figure 3, when administrative divisions were not identified, the r-squared values were consistently high when data values were compared within the same divisions. This illustrates that as of now, it is better to assume that the number of deaths and number of doctors for a given number of people in a population has a positive correlation. Paths were created connecting data points within each administrative division in chronological order with respect to time. The paths reveal that, generally, as time progresses, both mortality rates and doctor density increase, which also strengthens the trend displayed in figure 2 that the number of deaths tends to rise over time. This highlights that, despite rising investments in the healthcare workforce with more doctors across all administrative divisions, death rates continue to climb—likely due to South Korea's aging population, as it has the second lowest birth rates in the world, just 0.01 more births per woman than that of Taiwan (CIA, 2023).

4.3.2 Deaths by Circulatory System Diseases and Doctors Specializing in Thoracic Surgery

This section investigates the correlation between the number of deaths caused by circulatory system diseases and the number of doctors specializing in thoracic surgery, operations on organs in the chest. While cardiology may relate more to this particular cause of death, the dataset did not include information on the number of cardiologists; hence, the best alternative was chosen instead, which were thoracic surgeons (eSkyCityDEVRG, 2023). Figure 5 displays the relationship between the two variables for each year from 2010 to 2020 in four administrative divisions: Seoul, Gwangju, Gangwon-do, and Jeollanam-do.

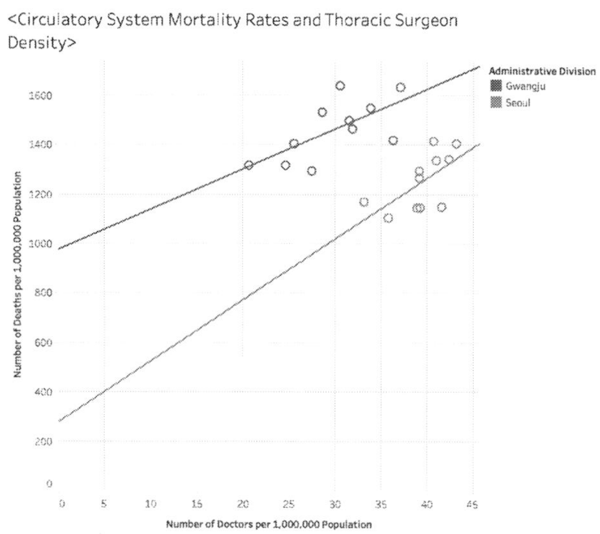

Figure 5. Circulatory system-related mortality rates and thoracic surgeon density for each year between 2010 and 2020 in Seoul and Gwangju

Figure 5 above shows the relationship between the number of thoracic surgeons per million people (x-axis) and circulatory system-related deaths per million people (y-axis) in Seoul and Gwangju. These two administrative divisions were analyzed using linear trendlines due to their low p-values–0.037 for Seoul and 0.025 for Gwangju–indicating statistical significance. Both trendlines show a positive correlation, with r-squared of approximately 0.4 for Seoul and 0.45 for Gwangju. suggesting thoracic surgeon numbers account for around 40% and 45% of the variation in circulatory system-related deaths.

The trend may stem from South Korea's aging population, where 19% were aged 65 and over in 2020 and the median age rose from 36.6 in 2010 to 42.8 in 2020 (*South Korea Demographics 2024 (Population, Age, Sex, Trends) - Worldometer*, 2024). As the population becomes more vulnerable to circulatory diseases, mortality rates may remain high despite an increase in the number of thoracic surgeons. Alternatively, this also suggests that increasing the number of doctors alone may not significantly reduce mortality rates, as around 60% of the variation is attributable to other factors.

Contrarily, the p-values for all trendlines of Gangwon-do and Jeollanam-do exceeded 0,05, suggesting any observed trends may have resulted from chance. Hence, in-depth analysis was not conducted for these two regions. This constraint will be further discussed in the limitations section.

4.3.3 Deaths by External Causes of Morbidity and Mortality and Doctors Specializing in Emergency Medicine (EM)

This section examines the relationship between the number of deaths caused by external factors of morbidity and mortality (such as accidents, injuries, violence, infections, and other external causes) and the number of EM doctors. EM doctors are trained to gain the skills required to care for people with life-threatening or urgent health care needs, making it a good variable for analysis (Swiner, 2023). Figure 6 presents the relationship between the two variables for each year from 2010 to 2020 in four administrative divisions: Seoul, Gwangju, Gangwon-do, and Jeollanam-do.

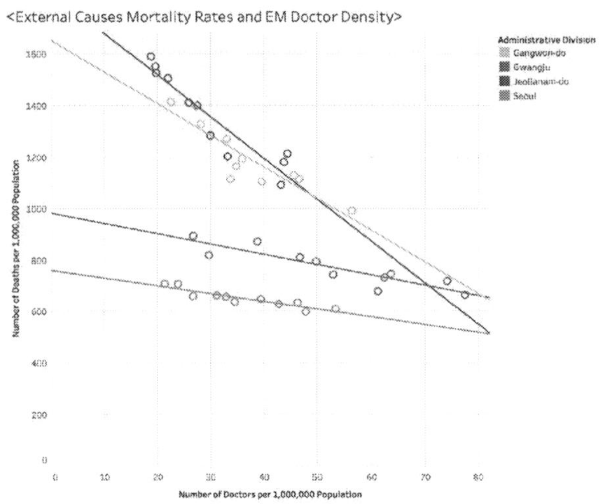

Figure 6. External causes mortality rates and emergency medicine (EM) doctor density for each year between 2010 and 2020 in Seoul, Gwangju, Gangwon-do, and Jeollanam-do

Figure 6 illustrates the relationship between the number of EM doctors per million people (x-axis) and deaths caused by external factors of morbidity and mortality per million people (y-axis). Unlike previous graphs, there is a negative correlation between the two variables. All trend lines have a p-value below 0.05 and a r-squared value above 0.8, making the data statistically significant.

The negative relationship may reflect the role of EM doctors, who provide immediate care for urgent cases–so higher supply of doctors enables more timely treatment, potentially reducing deaths. The difference in slopes of the regression lines among different regions may be a result of the disparities in overall healthcare qualities. For instance, Seoul, a highly urbanized area with advanced medical infrastructure, would experience a less dramatic effect when the number of EM doctors increases, as other healthcare variables are already optimized. This is visualized in the relatively flat line, represented in orange. In contrast, Jeollanam-do, an aging and relatively less developed province, would experience a more significant impact from increased EM doctor density, as it has fewer supporting healthcare resources or factors that already contribute to reducing mortality rates. This is represented by a dynamic trendline, represented in blue.

5. CONCLUSION

5.1 Discussion & Limitations

5.1.1 Discussion

Overall, no clear relationship between doctor density and mortality rates in South Korea was found, as some correlations were positive while others were negative. When comparing the total number of doctors and total number of deaths, there was no correlation between the two variables. When each administrative division was identified and graphed accordingly, it was possible to assume that there was, in fact, a positive correlation between the two variables within each division. In addition, when the leading causes of death were observed, along with the most related types of doctors, there was a combination of a seemingly positive linear correlation and a negative linear correlation between the specific type of death cause and doctor specialty. Such a mix of patterns suggests that there is no obvious or consistent correlation between doctor density and mortality rates in South Korea, at least when considering the scale of this study and the limitations that exist. These findings suggest that reducing total deaths requires considering and analyzing multiple factors beyond just the number of doctors. In other words, simply increasing the number of doctors does not necessarily lower mortality rates. These additional factors could include the demographics of the population; availability of hospitals and medical tools, which relates to socioeconomic factors; environmental conditions like air quality; lifestyle and behavior such as obesity and cigarette consumption; and many more.

5.1.2 Limitations

While the project has yielded some insightful conclusions, there are clearly several limitations that could have hindered reaching more extensive and accurate findings. There are two major limitations of this study that could be addressed in future research to reach more decisive conclusions. The first limitation is that the project only accounts for data from 2010 to 2020. Thus, when one cause of death was analyzed against one type of doctor, there were only 11 data points in total for each administrative division, resulting in weak correlations. The lack of data may have also contributed to the high p-values of the trendlines for the relationship between density of thoracic surgeons and the rate of circulatory system-related deaths in Gangwon-do and Jeollanam-do, limiting the opportunity to draw meaningful conclusions. The range of data could also possibly indicate that the findings of this study are not up-to-date, especially if other factors had also been considered to reach more insightful conclusions. The second limitation is that only the doctor-to-population ratio has been investigated as a possible cause of mortality rates. As discussed throughout the article, there are possibly many other underlying causes that would have contributed to the difference in death rates; therefore, also taking these factors into consideration would have yielded more accurate results.

Overall, these difficulties serve as valuable lessons for future studies, and addressing these issues will lead to more productive and in-depth analyses, ultimately allowing for more accurate conclusions.

5.2 Conclusion & Future Work

The primary objective of this study is to see whether mortality rates correlate with doctor density in South Korea, specifically focusing on the correlation between circulatory system-related deaths and thoracic surgeons, as well as external causes of mortality and emergency medicine doctors. Data from four administrative divisions with distinct characteristics–Seoul, Gwangju, Gangwon-do, and Jeollanam-do–were analyzed in particular.

As recognized previously in the Limitations section, there remains significant room for further research and improvement upon the findings of this paper. Future studies could explore additional underlying variables that may explain trends in mortality rates and assess the long-term significance of the healthcare sector.

Furthermore, examining the implications of the recent 2024 South Korean medical crisis–where thousands of residents and interns resigned in response to new government policies aimed at increasing medical student admission quotas–could provide valuable insights into how systemic disruptions may affect healthcare outcomes (*Doctors on Strike: Why Are South Korea's Government and Medical Community at Odds?*, 2024). This addition could broaden the scope of the study, lead to more conclusive and well-explained findings through deeper analysis, and highlight the interplay of various factors beyond merely the number of doctors.

REFERENCES

CIA. (2023). Total Fertility Rate - The World Factbook. Www.cia.gov. https://www.cia.gov/the-world-factbook/field/total-fertility-rate/country-comparison/

Doctors on Strike: Why are South Korea's Government and Medical Community at Odds? (2024). Wilson Center. https://www.wilsoncenter.org/blog-post/doctors-strike-why-are-south-koreas-government-and-medical-community-odds

eSkyCityDEVRG. (2023, March 27). Cardiovascular and Thoracic Surgery - Valley Health Physician Alliance. Valley Health Physician Alliance. https://valleyhealthphysicians.com/services/cardiovascular-and-thoracic-surgery/

Hoon, M. K. (n.d.). Rural South Korea faces stark health care inequality: report. https://asianews.network/rural-south-korea-faces-stark-health-care-inequality-report/

KOSIS. (2023). Kosis.kr. https://kosis.kr/statHtml/statHtml.do?orgId=101&tblId=DT_1B040A3&language=en&conn_path=I3

KOSIS. (2023b). Kosis.kr. https://kosis.kr/statHtml/statHtml.do?orgId=101&tblId=DT_1B34E09&language=en&conn_path=I3

KOSIS. (2023c). Kosis.kr. https://kosis.kr/statHtml/statHtml.do?orgId=117&tblId=DT_117110_A009&conn_path=I3

KOSIS KOrean Statistical Information Service. (n.d.). Kosis.kr.
https://kosis.kr/eng/statisticsList/statisticsListIndex.do?menuId=M_01_01&vwcd=MT_ETITLE&parmTabId=M_01_01

Research, B. D.-B. B. D.-B. is the communications manager for the C. for H. P. for P. C. and O. (n.d.). More primary care physicians leads to longer life spans. News Center. https://med.stanford.edu/news/all-news/2019/02/more-primary-care-physicians-lead-to-longer-life-spans.html

South Korea Demographics 2024 (Population, Age, Sex, Trends) - Worldometer. (2024). Worldometers.info. https://www.worldometers.info/demographics/south-korea-demographics

South Korea Health spending as percent of GDP - data, chart. (n.d.). TheGlobalEconomy.com.https://www.theglobaleconomy.com/South-Korea/Health_spending_as_percent_of_GDP/

Swiner, N. (2023, July 7). What Is an Emergency Medicine Doctor? WebMD. https://www.webmd.com/a-to-z-guides/what-is-emergency-medicine-doctor

Tableau Software. (2024). Tableau Desktop (Version 2024.2.2) [Computer software]. https://www.tableau.com/products/desktop

World Health Organization. (2023). Global Health Expenditure Database. Apps.who.int. https://apps.who.int/nha/database

DEEP NEURAL NETWORKS FOR IMAGE CLASSIFICATION: A COMPARATIVE STUDY AND BENCHMARKING

Rajaa Miftah, Mostafa Hanoune and Mohssine Bentaib
LTIM laboratory. Faculty of Science Ben M'sik
Hassan II University of Casablanca
Casablanca, Morocco

ABSTRACT

The landscape of deep neural networks for image classification has undergone multiple significant changes due to advancements in model architectures and learning strategies. In this paper, we thoroughly review, compare, and benchmark deep neural network models designed explicitly for the ImageNet dataset to elaborate on the three aspects of their performance: compositional ability, computational efficiency, and energy consumption. Our comparative analysis includes classic convolutional neural networks and transformer-based architectures, namely the Vision Transformers. The objective of the paper is to present a nuanced view of the state-of-the-art in the field of image classification, allowing the reader to understand the compromise between model complexity and performance. This paper integrates recent researches and supplies comparative benchmarking. Thus, in essence, this paper will be a significant resource for establishing and implementing the most effective and efficient image classification solutions for academics and industry. It is essential and timely to respond to the scientific relevance of the topic and pave the way for upcoming improvements in the field of computer vision with such a fast pace of development as now.

KEYWORDS

Image Classification, Deep Neural Networks, Deep Learning, Computer Vision, Visual Recognition, Spatial-Temporal Features

1. INTRODUCTION

Over the past years, the field of computer vision has made impressive advancements thanks to the development of deep learning, with deep neural networks making significant contributions. One of the key concerns in computer vision is image classification, one with broad applications in several industries such as health, autonomous driving, and organization of consumer images. Due to the capacity of deep neural networks to learn diverse feature hierarchies directly from image pixels, image classification has become more efficient and relatively accurate than before.

The ImageNet Large scale visual recognition challenge (ILSVRC) was first introduced in 2010 and has since provided a standard benchmark to assess image classification results. Researchers have developed and fine-tuned various deep neural networks based on (ILSVRC) constraints, aiming to achieve higher accuracy while managing computational complexity and model size. From traditional convolutional neural networks (CNNs) to more recent transformer-based models such as Vision Transformers (ViTs), the landscape of DNNs for image classification continues to evolve rapidly.

In this paper, we embark on a comprehensive exploration of the landscape of deep neural networks (DNNs) for image classification, focusing particularly on the performance evaluation and comparative analysis of state-of-the-art models on the ImageNet dataset. Our study encompasses a diverse array of architectures, each offering unique strengths and trade-offs in terms of accuracy, efficiency, and scalability. By conducting a rigorous benchmarking analysis, we seek to elucidate the relative merits of different DNN models, providing researchers and practitioners with valuable insights for informed decision-making in model selection and deployment.

Our detailed review covers a broad spectrum of model types, from traditional convolutional neural networks to cutting-edge transformer-based designs. This diversity allows us to identify key trends and performance patterns that can inform future innovations. Moreover, we delve into the implications of model complexity, examining how the trade-offs between computational demand and performance can affect practical deployment in real-world applications.

Through this endeavor, we aim to contribute to the ongoing discourse surrounding the development and optimization of DNNs for image classification. By dissecting the performance nuances of these models and providing a clear benchmarking framework, we aspire to advance the state-of-the-art in computer vision research and applications, enabling the community to build more effective and efficient systems.

2. BACKGROUND

The field of image classification changed dramatically with the advent and evolution of deep neural networks (DNNs). The milestones in this journey highlight a shift from traditional machine learning methods to more sophisticated and robust neural architectures, influencing a wide range of applications in computer vision.

Initial approaches in image classification relied heavily on hand-engineered features and linear classifiers. Seminal works like Lowe's Scale-Invariant Feature Transform (SIFT) (Lowe, 2004) and Dalal and Triggs' Histograms of Oriented Gradients (HOG) (Dalal & Triggs, 2005) set the stage for the first wave of advancements in this area.

The landscape has witnessed transformative changes with the introduction of CNNs, particularly after Krizhevsky et al. presented AlexNet (B. A. Krizhevsky et al., 2012), which significantly outperformed traditional methods in the ImageNet challenge. This success spurred further developments, leading to more sophisticated models such as ZFNet (Zeiler & Fergus, 2014), VGG (Simonyan & Zisserman, 2015), and ResNet, each introducing deeper and more complex architectures.

The introduction of attention mechanisms and transformers marked another paradigm shift. The transformer, originally proposed by Vaswani et al. (Vaswani et al., 2017) for natural language processing, was adapted for image classification with models like Vision Transformer (ViT) (Dosovitskiy et al., 2021). These models have shown remarkable success, challenging the supremacy of CNNs in certain scenarios.

More recently, the integration of CNNs and transformers has led to hybrid architectures, which combine the locality and inductive biases of CNNs with the global reception fields of transformers. Notable examples include CoAtNet (Z. Dai et al., 2021) which merges these paradigms to achieve enhanced performance across various benchmarks.

The ImageNet dataset (Deng et al., 2009) remains a central benchmark in this field, facilitating the comparison and evaluation of different models. Other datasets like CIFAR-10/100 (A. Krizhevsky, 2009) and Google's JFT-300M (Sun et al., 2017) also play critical roles in training and testing the limits of these advanced models.

Despite the advancements, challenges such as model interpretability, computational efficiency, and data bias persist. Addressing these issues is crucial for the next wave of innovations in image classification.

2.1 Notable Architectures in Image Classification

EVA: Enhanced Visual Attention (EVA) model, introduced by Li et al., explores the limits of masked visual representation learning at scale (A. Krizhevsky et al., 2017). EVA employs innovative techniques for capturing intricate visual patterns and semantics, contributing to improved performance in various computer vision tasks.

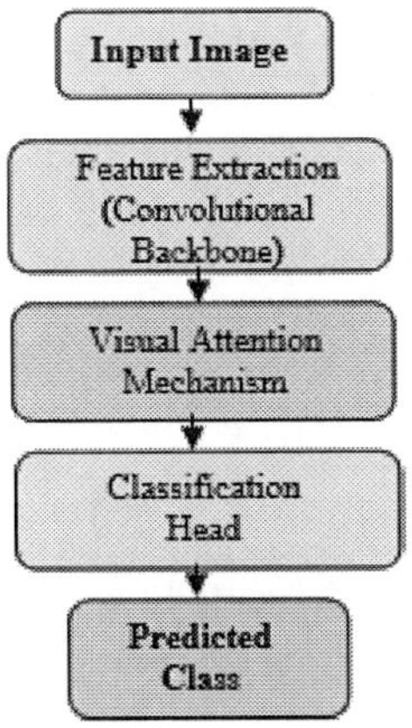

Figure 1. EVA Architecture for Image Classification

Group DETR v2: A variant of the Detection Transformer (DETR) model, Group DETR v2, extends the capabilities of object detection through group-based attention mechanisms (He et al., 2016). This approach enhances the model's ability to capture contextual information and relationships between objects in complex scenes.

DETA: Contrastive Captioners are Image-Text Foundation Models (DETA), proposed by Yu et al. (2022), leverage contrastive learning for image classification tasks. DETA exploits the complementary information between images and captions, leading to enhanced feature representations for classification.

GLIP: Global-Local Interaction Protocol (GLIP) model, introduced by Li et al. (2022), enhances feature representation through global and local interaction mechanisms. GLIP effectively integrates global context with local details, facilitating a more comprehensive understanding of visual content.

DyHead: Dynamic Head model (DyHead), presented by X. Dai & Liu (2021), dynamically adjusts the number of detection heads for improved object detection. DyHead optimizes computational resources by adapting the model's capacity based on the complexity of the input data.

CenterNet++: CenterNet++, developed by Duan et al., introduces an efficient and accurate approach for object detection through keypoint estimation (Duan et al., 2021). CenterNet++ achieves robust object localization by directly predicting keypoint coordinates, eliminating the need for complex post-processing steps.

YOLOv4-P6 CSP-P6: YOLOv4-P6 CSP-P6, an enhanced version of YOLOv4, incorporates cross-stage partial connections and a panoptic segmentation module for object detection. YOLOv4-P6 CSP-P6 achieves real-time performance with improved accuracy (Alexey Bochkovskiy, Chien-Yao Wang, 2020), making it suitable for various deployment scenarios.

2.2 Transfer Learning and Pre-Trained Models

Transfer learning has emerged as a powerful technique for leveraging pre-trained models on large-scale image datasets for video analysis tasks. By fine-tuning pre-trained CNNs on video-specific datasets, researchers can harness the knowledge encoded in these models to accelerate training and improve performance on object detection and related tasks.

Pre-trained CNN models, such as those trained on the ImageNet dataset, have been widely adopted as feature extractors for video analysis tasks (Zeiler & Fergus, 2014). Fine-tuning these models on video-specific datasets has been shown to yield state-of-the-art results in object detection, action recognition, and video captioning.

In summary, deep learning foundations, including CNNs, RNNs, and their hybrid architectures, form the bedrock of video analysis methodologies. By understanding the capabilities and limitations of these models, researchers can design more effective algorithms for object detection and other video understanding tasks.

3. TEMPORAL CONTEXT MODELING IN VIDEO UNDERSTANDING

Understanding the temporal dynamics inherent in video sequences is essential for accurate and robust video analysis tasks, including object detection. Temporal context modeling aims to capture long-range dependencies and temporal coherence within video data, enabling more effective interpretation and prediction of visual content.

3.1 Temporal Convolutional Networks

Temporal Convolutional Networks (TCNs) have gained prominence for their ability to capture temporal dependencies in sequential data efficiently. TCNs extend the concept of convolutional layers to the temporal domain, allowing for the extraction of temporal features across frames in a video sequence. Unlike traditional recurrent architectures, TCNs do not suffer from the vanishing gradient problem, making them well-suited for capturing long-range dependencies in video data.

The work of Bai et al (2018). introduced the Temporal Convolutional Network architecture, demonstrating its effectiveness in capturing long-range dependencies for sequence modeling tasks. TCNs have since been applied to various video understanding tasks, including action recognition and video captioning, showcasing their versatility and efficacy in capturing temporal context.

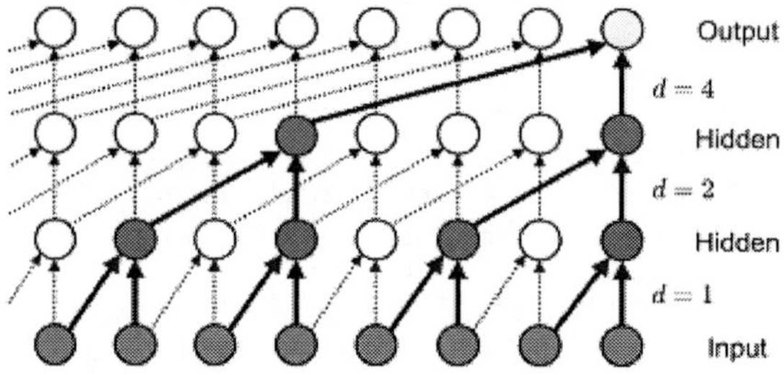

Figure 2. TCNs architecture

3.2 Recurrent Neural Networks with Attention Mechanisms

Recurrent Neural Networks (RNNs) with attention mechanisms have emerged as powerful tools for capturing temporal context in video sequences. By dynamically attending to relevant temporal frames, attention mechanisms enable RNNs to focus on salient information while filtering out irrelevant noise (Concha et al., 2018). This selective attention mechanism allows RNNs to effectively model long-range dependencies and capture complex temporal patterns within video data.

3.3 Spatiotemporal Feature Aggregation Networks

Spatiotemporal Feature Aggregation Networks aim to integrate spatial and temporal information explicitly for enhanced video understanding (Girdhar et al., 2017). These networks leverage both spatial and temporal convolutions to capture spatial-temporal features simultaneously, facilitating more robust representation learning for dynamic visual content.

The work of Feichtenhofer et al. (2019) introduced the SlowFast network architecture, which incorporates two pathways for processing video data: a slow pathway for capturing spatial information and a fast pathway for modeling temporal dynamics. By aggregating features from both pathways, the SlowFast network achieves state-of-the-art performance in action recognition tasks, demonstrating the efficacy of spatiotemporal feature aggregation for video understanding.

Temporal context modeling is essential for video object detection. Approaches such as Temporal CNNs, attention-enhanced RNNs, and Spatiotemporal feature fusion capture dynamic dependencies, boosting detection accuracy and robustness.

4. FUSION OF MULTI-MODAL DATA MODALITIES

To enhance object detection in video, multi-modal fusion—combining visual, textual, and motion data—offers complementary cues that boost performance and adaptability across diverse conditions.

4.1 Visual and Motion Information Fusion

Combining visual and motion cues, such as optical flow, enhances object detection by adding temporal context. Fusing these features helps distinguish static from moving objects, improving accuracy in dynamic scenes.

Recent advancements in this area include the work of Carreira & Zisserman (2017), which introduced the I3D (Inflated 3D ConvNet) architecture for action recognition. I3D leverages both RGB frames and optical flow sequences to capture spatial and temporal dynamics simultaneously, achieving state-of-the-art performance in action recognition tasks.

4.2 Textual Context Integration

incorporating textual context like scene descriptions or object labels enriches video understanding and boosts object detection accuracy by complementing visual features with semantic cues.

One notable approach in this domain is the work of Anne Hendricks et al. (2017), which introduced dense captioning networks for generating dense captions describing regions of interest in images and videos. By jointly modeling visual and textual information, dense captioning networks enable fine-grained localization and detailed description of objects, enhancing object detection performance.

4.3 Sensor Fusion for Environmental Awareness

Beyond visual and textual modalities, sensor data from various sources, such as LiDAR (Reutebuch et al., 2005) or radar, can be fused with visual information to enhance environmental awareness and object detection capabilities. Sensor fusion enables object detection systems to adapt to diverse environmental conditions, such as low-light scenarios or adverse weather conditions, by leveraging complementary sensory inputs.

The work of Chen et al. (2017) exemplifies this approach by proposing a fusion framework that integrates LiDAR data with visual information for 3D object detection in autonomous driving scenarios. By combining depth information from LiDAR with visual features extracted from camera images, the proposed framework achieves robust object detection performance in challenging real-world environment.

5. ADDRESSING CHALLENGES IN TRAINING DEEP MODELS FOR VIDEO ANALYSIS

Training deep models for video analysis poses several challenges that need to be addressed to ensure optimal performance and generalization. This section explores key challenges in training deep models for video analysis and discusses recent advancements and strategies to mitigate these challenges.

5.1 Data Augmentation and Domain Adaptation

A key challenge in video model training is limited labeled data. Data augmentation (e.g., temporal jittering, frame subsampling) and domain adaptation (e.g., adversarial training, fine-tuning) help improve generalization and robustness in real-world scenarios.

Recent work by Wang et al. (2016) has demonstrated the effectiveness of data augmentation and domain adaptation techniques in improving object detection and action recognition performance in video analysis tasks.

5.2 Model Compression and Efficient Inference

Deep video models are often computationally intensive. Compression techniques like pruning, quantization, and distillation reduce model size and cost with minimal performance loss. Efficient inference and hardware acceleration further enable real-time deployment on resource-limited devices.

The work of Han et al. (2016) and Bolukbasi et al. (2017) has investigated various model compression and efficient inference techniques for video analysis, demonstrating significant reductions in model size and computational complexity while maintaining competitive performance.

5.3 Robustness to Environmental Variability

Deep models for video analysis should exhibit robustness to environmental variability, including changes in lighting conditions, occlusions, and camera viewpoints. Adversarial training, robust optimization techniques, and ensemble learning approaches can enhance model robustness and resilience to environmental variations by exposing models to diverse training scenarios and perturbations.

Recent research by Liang et al. (2018) has addressed the challenge of robustness to environmental variability in video analysis, proposing novel training strategies and regularization techniques to improve model generalization and performance in challenging real-world conditions.

In summary, tackling challenges in video model training requires data augmentation, domain adaptation, model compression, efficient inference, and robustness to variability. Advancements in these areas enable more robust, efficient, and generalizable models for video analysis.

6. FUTURE RESEARCH DIRECTIONS

As the field of object detection in video sequences using deep learning continues to evolve, several promising avenues for future research emerge. This section outlines key areas for exploration and provides references to recent works that contribute to these directions:

6.1 Interpretable Models

Developing interpretable models is key to building trust in object detection systems. Future research should explore attention mechanisms, saliency maps, and feature visualization to clarify model decisions and performance factors.

Recent works contributing to interpretable deep learning models include the study by Alvarez-Melis and Jaakkola (2018), who proposed self-explaining neural networks for robust interpretability.

6.2 Continual Learning and Adaptation

Object detection models should possess the ability to adapt and learn continuously from streaming video data to accommodate evolving environments and emerging object classes. Future research should investigate techniques for continual learning, online adaptation, and domain generalization to ensure model robustness and adaptability over time.

6.3 Efficient Deployment for Real-time Inference

Efficient deployment of deep learning models for real-time object detection in video sequences is essential for practical applications. Future research should focus on developing lightweight model architectures, efficient inference algorithms, and hardware-accelerated implementations to enable real-time performance on resource-constrained devices.

Recent works contributing to efficient deployment include the study by Tan and Le (2019), who proposed EfficientNet for rethinking model scaling for convolutional neural networks to achieve efficient inference.

6.4 Semantic Understanding and Contextual Reasoning

Improving semantic understanding and context reasoning can enhance object detection in complex scenes. Future work should integrate context, segmentation, and scene understanding into detection models.

Recent works contributing to semantic understanding and contextual reasoning include the work of Zhu et al. (2019), who introduced Deformable ConvNets v2 for improved semantic understanding and contextual reasoning in object detection.

Integrating recent findings into future directions can advance video object detection and support the development of more robust, interpretable models for dynamic visual environments.

7. CONCLUSION

In conclusion, this paper has provided a comprehensive exploration of deep learning methodologies for enhancing object detection accuracy and efficiency in dynamic video environments by leveraging temporal context. By reviewing foundational principles and investigating novel approaches for capturing temporal dependencies, we have highlighted the importance of leveraging temporal context for object detection in video sequences.

Through synthesizing existing literature and proposing innovative insights, this paper offers a roadmap for researchers and practitioners to navigate the complex landscape of object detection in video sequences using deep learning techniques. We underscore the significance of interpreting model decisions, adapting to evolving environments, efficiently deploying models for real-time inference, and enhancing semantic understanding and contextual reasoning.

Looking ahead, future research directions include developing interpretable models, enabling continual learning and adaptation, efficient deployment for real-time inference, and enhancing semantic understanding and contextual reasoning capabilities. By addressing these future research directions, researchers can advance the state-of-the-art in video understanding and facilitate the development of robust and efficient deep learning systems for dynamic visual environments.

REFERENCES

Alvarez-Melis, D., & Jaakkola, T. S. (2018). Towards robust interpretability with self-explaining neural networks. *Advances in Neural Information Processing Systems*, *2018-Decem*(NeurIPS), 7775–7784.

Bai, S., Kolter, J. Z., & Koltun, V. (2018). *An Empirical Evaluation of Generic Convolutional and Recurrent Networks for Sequence Modeling*.

Bochkovskiy, A., Wang, C.-Y., & Liao, H.-Y. M. (2020). *YOLOv4: Optimal Speed and Accuracy of Object Detection*.

Bolukbasi, T., Wang, J., Dekel, O., & Saligrama, V. (2017). Adaptive neural networks for efficient inference. *34th International Conference on Machine Learning, ICML 2017*, *2*, 812–821.

Carreira, J., & Zisserman, A. (2017). Quo Vadis, action recognition? A new model and the kinetics dataset. *Proceedings - 30th IEEE Conference on Computer Vision and Pattern Recognition, CVPR 2017*, *2017-Janua*, 4724–4733. https://doi.org/10.1109/CVPR.2017.502

Chen, X., Ma, H., Wan, J., Li, B., & Xia, T. (2017). Multi-View 3D Object Detection Network for Autonomous Driving. *In Proceedings of the IEEE Conference on Computer Vision and Pattern Recognition*, 1907–1915.

Concha, D. T., Maia, H. D. A., Pedrini, H., Tacon, H., Brito, A. D. S., Chaves, H. D. L., & Vieira, M. B. (2018). Multi-stream Convolutional Neural Networks for Action Recognition in Video Sequences Based on Adaptive Visual Rhythms. *Proceedings - 17th IEEE International Conference on Machine Learning and Applications, ICMLA 2018*, *May 2019*, 473–480. https://doi.org/10.1109/ICMLA.2018.00077

Dai, X., & Liu, M. (2021). *Dynamic Head : Unifying Object Detection Heads with Attentions*.

Dai, Z., Liu, H., Le, Q. V., & Tan, M. (2021). CoAtNet: Marrying Convolution and Attention for All Data Sizes. *Advances in Neural Information Processing Systems*, *5*, 3965–3977.

Dalal, N., & Triggs, B. (2005). Histograms of Oriented Gradients for Human Detection. *IEEE Computer Society Conference on Computer Vision and Pattern Recognition (CVPR'05)*, v–vi. https://doi.org/10.1007/978-3-319-24947-6

Deng, J., Dong, W., Socher, R., Li, L.-J., Li, K., & Fei-Fei, L. (2009). ImageNet: A Large-Scale Hierarchical Image Database. *IEEE Computer Society Conference on Computer Vision and Pattern Recognition.*

Dosovitskiy, A., Beyer, L., Kolesnikov, A., Weissenborn, D., Zhai, X., Unterthiner, T., Dehghani, M., Minderer, M., Heigold, G., Gelly, S., Uszkoreit, J., & Houlsby, N. (2021). an Image Is Worth 16X16 Words: Transformers for Image Recognition At Scale. *ICLR 2021 - 9th International Conference on Learning Representations.*

Duan, K., Bai, S., Xie, L., Qi, H., Huang, Q., & Tian, Q. (2021). *CenterNet ++ for Object Detection. 1*, 1–13.

Feichtenhofer, C., Fan, H., Malik, J., & He, K. (2019). Slowfast networks for video recognition. *Proceedings of the IEEE International Conference on Computer Vision, 2019-Octob*, 6201–6210. https://doi.org/10.1109/ICCV.2019.00630

Girdhar, R., Ramanan, D., & Gupta, A. (2017). ActionVLAD : Learning spatio-temporal aggregation for action classification. *Proceedings of the IEEE Conference on Computer Vision and Pattern Recognition (CVPR), typically 25*, 971–980.

Han, S., Liu, X., Mao, H., Pu, J., Pedram, A., Horowitz, M. A., & Dally, W. J. (2016). EIE: Efficient Inference Engine on Compressed Deep Neural Network. *Proceedings - 2016 43rd International Symposium on Computer Architecture, ISCA 2016, 16*, 243–254. https://doi.org/10.1109/ISCA.2016.30

He, K., Zhang, X., Ren, S., & Sun, J. (2016). Deep Residual Learning for Image Recognition. *IEEE Conference on Computer Vision and Pattern Recognition (.*

Hendricks, L. A., Wang, O., Shechtman, E., Sivic, J., Darrell, T., & Russell, B. (2017). Localizing Moments in Video with Natural Language. *In Proceedings of the IEEE International Conference on Computer Vision*, 5803–5812.

Krizhevsky, A. (2009). *Learning Multiple Layers of Features from Tiny Images.*

Krizhevsky, A., Sutskever, I., & Hinton, G. E. (2017). ImageNet classification with deep convolutional neural networks. *Communications of the ACM, 60*(6), 84–90. https://doi.org/10.1145/3065386

Krizhevsky, B. A., Sutskever, I., & Hinton, G. E. (2012). *ImageNet Classification with Deep Convolutional Neural Networks.* https://doi.org/10.1145/3065386

Li, L. H., Zhang, P., Zhang, H., Yang, J., Li, C., Zhong, Y., Wang, L., Yuan, L., Zhang, L., Hwang, J. N., Chang, K. W., & Gao, J. (2022). Grounded Language-Image Pre-training. *Proceedings of the IEEE Computer Society Conference on Computer Vision and Pattern Recognition, 2022-June*, 10955–10965. https://doi.org/10.1109/CVPR52688.2022.01069

Liang, S., Li, Y., & Srikant, R. (2018). Enhancing the reliability of out-of-distribution image detection in neural networks. *6th International Conference on Learning Representations, ICLR 2018 - Conference Track Proceedings*, 1–15.

Lowe, D. G. (2004). Distinctive Image Features from Scale-Invariant Keypoints. *International Journal of Computer Vision*, 27–38.

Reutebuch, S. E., Andersen, H. E., & McGaughey, R. J. (2005). Light detection and ranging (LIDAR): An emerging tool for multiple resource inventory. In *Journal of Forestry* (Vol. 103, Issue 6, pp. 286–292).

Simonyan, K., & Zisserman, A. (2015). Very deep convolutional networks for large-scale image recognition. *3rd International Conference on Learning Representations, ICLR 2015 - Conference Track Proceedings*, 1–14.

Sun, C., Shrivastava, A., Singh, S., & Gupta, A. (2017). Revisiting Unreasonable Effectiveness of Data in Deep Learning Era. *Proceedings of the IEEE International Conference on Computer Vision, 2017-Octob*, 843–852. https://doi.org/10.1109/ICCV.2017.97

Tan, M., & Le, Q. V. (2019). EfficientNet: Rethinking model scaling for convolutional neural networks. *36th International Conference on Machine Learning, ICML 2019, 2019-June*, 10691–10700.

Vaswani, A., Shazeer, N., Parmar, N., Uszkoreit, J., Jones, L., Gomez, A. N., Kaiser, Ł., & Polosukhin, I. (2017). Attention is all you need. *Advances in Neural Information Processing Systems, 2017-Decem*(Nips), 5999–6009.

Wang, L., Xiong, Y., Wang, Z., Qiao, Y., Lin, D., Tang, X., & van Gool, L. (2016). Temporal segment networks: Towards good practices for deep action recognition. *Lecture Notes in Computer Science (Including Subseries Lecture Notes in Artificial Intelligence and Lecture Notes in Bioinformatics), 9912 LNCS*, 20–36. https://doi.org/10.1007/978-3-319-46484-8_2

Yu, J., Wang, Z., Vasudevan, V., Yeung, L., Seyedhosseini, M., & Wu, Y. (2022). *CoCa: Contrastive Captioners are Image-Text Foundation Models. 1*, 1–19. http://arxiv.org/abs/2205.01917

Zeiler, M. D., & Fergus, R. (2014). Visualizing and Understanding Convolutional Networks. *In European Conference on Computer Vision*, 818–833.

Zhu, X., Hu, H., Lin, S., & Dai, J. (2019). Deformable convnets V2: More deformable, better results. *Proceedings of the IEEE Computer Society Conference on Computer Vision and Pattern Recognition, 2019-June*, 9300–9308. https://doi.org/10.1109/CVPR.2019.00953

A FOCUSED APPROACH FOR SPEECH EMOTION RECOGNITION IN REAL-WORLD ENVIRONMENTS

Giuseppe Lentini, Paolo Ranzi, Isabella Della Torre, Ismaela Avellino, Andrea Buccoliero and Antonio Colangelo
GPI S.p.A, Trento, Italy

ABSTRACT

Speech Emotion Recognition (SER) systems often suffer performance drops when deployed in real-world conditions due to noise, speaker variability, and uncontrolled environments. In this work, we propose a novel and focused approach for SER that explicitly addresses these challenges. Our method combines targeted data selection, robust preprocessing techniques, and a lightweight transformer architecture optimised for generalisation across diverse datasets. The techniques used in this study led to a 13% increase in test set accuracy compared to prior studies. Additionally, following the EmoBOX benchmark, we evaluated our model on ten publicly available SER datasets and demonstrated that it achieves state-of-the-art results in cross-corpus settings. In particular, our system demonstrates robust performance even in noisy, spontaneous, and multilingual scenarios, highlighting its potential for real-world deployment and achieving results comparable to human-level performance. The proposed framework represents a step forward in building practical and scalable emotion-aware technologies. Such improvements in SER models pave the way for increasingly reliable applications in telemedicine, mental health monitoring, and other emotion-aware technologies, bringing us closer to human-level interaction and understanding.

KEYWORDS

Deep Learning, Transformers, wav2vec 2.0, Speech Emotion Recognition, Real-World Environment, Telemedicine

1. INTRODUCTION

Human speech is one of the most powerful and sophisticated communication devices. By varying tone, pitch, rhythm, and volume, the voice can express our state of mind more accurately than posture or facial expression. Paul Ekman (1992), an early researcher of emotions, listed six universal emotions that are shared: joy, sadness, fear, anger, surprise, and disgust. They are universally recognised across cultures and can be accessed through voice modulations, which makes the voice a uniquely powerful medium for emotional recognition. In the past several years, the increasing availability of vocal data and advancements in artificial intelligence (AI) has provided an opportunity for speech emotion recognition (SER) systems. Such systems aim to automate the recognition of emotions in human language using advanced signal processing, machine learning, and deep neural networks. SER systems analyse vocal traits such as pitch, intensity, and spectral characteristics that allows determining a speaker's emotional states with remarkable accuracy. Recent research on SER has demonstrated promising results within controlled environments. We made an initial data release using a SER model to detect postpartum depression by voice examination (Fanos et al., 2023). The experiment not only stressed the value of existing SER procedures, but also found important drawbacks that led to further evolution. Our study aims to improve the robustness of the model along with its applicability and generalisability in real-world settings. Specifically, this study addresses the serious issue of performance degradation observed in cross-corpus validation scenarios, where models with good performance in a specific dataset tend to perform badly in others. Our study is introducing a new methodology aimed specifically to address this generalisation gap in an explicit manner so as to provide more uniform performance across a variety of datasets and closer proximity to real-world settings. Ultimately, our goal is to develop SER models that can be reliably used in sensitive applications such as telemedicine, remote mental health monitoring, and emotion-aware healthcare tools, providing clinicians and patients with accessible, non-invasive, and real-time insights into emotional states, and helping improve patient outcomes and quality of care.

2. RELATED STUDIES

Early SER systems were based on classical machine learning techniques like SVMs and Random Forests. Although these techniques were effective in controlled conditions, they would tend to fail with the speech data variability resulting from the speaker variability and background noise. Deep learning has significantly improved SER with the help of neural networks, especially CNNs and LSTMs. CNN-based frameworks have been efficient in feature extraction and classification, showing prowess in emotion recognition (Dal Rì et al., 2023). For improving performance further, hybrid models such as the Attention-Guided 3D-CNN-LSTM have been investigated, extracting spatiotemporal relations in speech signals (Atila & Sengur, 2021). Among the most widely utilised features, MFCCs have been most utilised owing to their capability of representing pertinent spectral characteristics of speech signals (Bakhshi, 2020). Prior studies by Costantini et al. (2022) have investigated subsets of acoustic features for machine learning-based SER, where the effectiveness of MFCCs fused with prosodic and spectral features has been examined. Recent research has also explored new feature extraction techniques to enhance classification performance. For example, the research of Chattopadhyay et al. (2020) proposed a hybrid feature fusion technique to enhance emotion recognition by combining spectral and prosodic features, with better performance than conventional MFCC-based systems. More recently, Transformer-based architectures have emerged as state-of-the-art solutions in speech processing. Models like Wav2Vec 2.0 (Baevski et al., 2020) leverage self-supervised learning and attention mechanisms to achieve superior results in SER, demonstrating remarkable improvements in feature representation and generalisation (Wang et al., 2021).

3. METHODOLOGY

3.1 Data Selection

For this study, we selected the Emozionalmente dataset as the primary set since it provides a higher quality of the emotional annotation, with the ratings performed by five different annotators per each audio sample. Emozionalmente corpus is an open-source Italian speech emotional corpus that has been obtained through crowdsourcing methods, developed with the aim to assist the SER research in the Italian language (Catania et al., 2025).

Table 1. Emozionalmente Composition and Structure Attribute Description

Attribute	Description
Total Samples	6,902 audio recordings
Number of Speakers	431 amateur actors
Number of Emotions	7 categories (six basic emotions + neutral)
Emotions	Anger, Disgust, Fear, Joy, Sadness, Surprise, Neutral
Sentences per Speaker	18 different sentences
Sampling Rate	44.1 kHz

The recordings include emotional expressions of actors. Compared with spontaneous emotional datasets, there are several benefits in this procedure for having clear and well-defined emotion labels, which are suitable for supervised learning model training. The emotion labels are derived from the intended emotion expressed by speakers rather than perceptually validated labels from listeners. Previous research has shown that humans identify emotions in speech with an accuracy of 65% (Scherer, 2003). In the Emozionalmente dataset, the same result has been verified, with the annotated perceived emotion congruence being 66%. Training a model on weak labels, where annotations are influenced by human perception, inherently transfers the bias of human emotion recognition capabilities to the model itself. Based on this observation, we obtained a hard-core subset of the dataset by iteratively selecting files with the maximum annotation congruence. After selection, we balanced the datasets by equalling the number of samples per emotion, ending up with a final 508 samples per class and a total of 3556 audio files.

3.2 Data Augmentation

Data augmentation is a crucial strategy in SER allowing to overcome the challenges posed by limited and imbalanced datasets. By artificially increasing the diversity and volume of training data, augmentation techniques help improve model robustness, reduce overfitting, and enhance generalisation to real-world scenarios (Tris Atmaja & Sasou, 2022). After applying data augmentation, the kernel extracted from the Emozionalmente dataset was pushed to a total of 17,780 audio samples. The augmentation techniques that we used to mimic real-world distortions without destroying emotional cues included the following:

- **Time Stretching:** varying the playback speed within a range of 10%.
- **Gaussian Noise Injection:** adding small random noise to simulate background disturbances.
- **Vocal Tract Length Perturbation (VTLP):** altering the spectral characteristics to simulate variations in speaker's vocal tract.
- **Room Impulse Response (RIR):** applying real reverberation effects to improve model generalisation in noisy environments (Ko et al., 2017).

3.3 Preprocessing

To ensure speaker-independent training, we used *StratifiedGroupKFold*, which prevents speaker overlap across splits and preserves class balance, reducing imbalance bias. With the speaker-independent split, the model is made to generalise better to new speakers by not overfitting to speaker-dependent traits. Various random states are attempted to achieve the best class distribution, choosing the optimal split by reducing imbalance. After data splitting, the training set contains 14,670 samples, which account for 82.51% of the data, with a nearly balanced class distribution (imbalance ratio of 1.02). The validation set contains 1,410 samples (7.87%) and a slightly greater imbalance ratio of 1.17. The test set contains 1,709 samples (9.61%) with a similar imbalance ratio of 1.16. The process line-by-line is presented in the given pseudocode:

Table 2. Pseudocode of the algorithm for the split with StratifiedGroupKFold

Best Balanced Data Split Algorithm

Input: *Dataset df, maximum attempts max_attempts, seed seed*
Output: *Best balanced split (train_df, val_df, test_df)*
Initialise *best_balance* ←∞, *best_splits* ←None;
Set random seed using *seed*;
Generate *max_attempts* random states;
foreach *random state* in *random states* **do**
  Perform first stratified group split (5-fold);
  Select first split: *train_idx, temp_idx*;
  Extract validation-test subset *val_test_data* from *temp_idx*;
  Perform second stratified group split (2-fold) on *val_test_data*;
  Select first split: *val_idx, test_idx*;
  Create *train_df, val_df, test_df* using indices;
  Compute balance score as the sum of standard deviations:
    $balance\_score = \sigma(train\_df) + \sigma(val\_df) + \sigma(test\_df)$

  where $\sigma(X)$ is the standard deviation of class distribution in X;
  if *balance_score* < *best_balance* **then**
    Update *best_balance* and *best_splits*;

  end
end
return *best_splits*;

3.4 Model Selection

In order to replicate the experiment previously conducted by Catania (2025) with the Emozionalmente dataset and to perform a direct comparison to evaluate whether the techniques used lead to improved performance, we fine-tuned the *jonatasgrosman/wav2vec2-large-xlsr-53-italian* model by Grosman, J. (2021), a multilingual variant of Wav2Vec2 pre-trained on 53 languages as part of the XLSR (Cross-Lingual Speech Representations) initiative. This model was further fine-tuned on the Italian subset of Mozilla Common Voice 6.1 for Automatic Speech Recognition (ASR), making it particularly well-suited for speech tasks in Italian. The fine-tuning is done by training the model on our train dataset and freezing the lower convolutional layers of Wav2Vec2 to maintain the pre-learned speech representations. We add a fully connected classification head and fine-tune the whole model with a cross-entropy loss function with the AdamW optimiser. The training goes in the following steps:

1. **Audio Preprocessing**: samples are resampled to 16 kHz and trimmed or zero-padded to 4 seconds to have a fixed input length.
2. **Feature Extraction:** Wav2Vec2's processor converts raw waveforms into contextual embeddings.
3. **Classification Head:** a fully connected layer with 7 output neurons maps Transformer outputs to logits, which are converted to class probabilities using a softmax function during inference.
4. **Optimisation:** the model is trained using the AdamW optimiser and a learning rate of $5 * 10^{-5}$, batch size 32, and 10 epochs. Weight decay is 0.01 for regularisation of the model.
5. **Training Strategy:** the model is trained with an evaluation strategy per epoch, saving the best model based on eval loss. Mixed precision training is also enabled with FP16 for greater efficiency, and gradient checkpointing is used for memory conservation. Data loading is accelerated with 2 worker threads. The best model is loaded automatically at the conclusion of training.

4. RESULTS

In this section, we present a full evaluation of our fine-tuned model on a variety of datasets with various linguistic and acoustic properties. We try to verify the model's robustness and whether it can preserve high performance across various recording conditions, speaker populations, and emotional distributions. We begin by announcing the primary test set used for testing, and then move to further the assessment to popular SER datasets such as EMOVO (Italian) (Costantini et al., 2014), RAVDESS (English) (Livingstone & Russo, 2018), CREMA-D (English) (Cao et al., 2014), EmoDB (German) (Burkhardt et al., 2005), and CaFE (French) (Gournay et al., 2018). By incorporating a diverse set of corpora, we aim to provide a comprehensive study of the strengths and weaknesses of the model, particularly in cross-corpus scenarios where the performance would otherwise degrade. The outcomes of these tests will dictate the flexibility of the model and what can be further improved to increase its usefulness in practical reality. SER is crucial for human-computer interaction but hindered by inconsistent dataset splits and lack of benchmarks. To address this, emoBOX provides a multilingual toolkit and a large benchmark for evaluating SER models in intra and cross-corpus settings (Ma et al., 2024). EmoBOX also supports pre-computed dataset splits for intra-corpus experiments and new test set balancing for cross-corpus setups. It is provided with *emotion2vec* for balancing annotation disparities for equitable comparisons. EmoBOX, consisting of 32 emotion datasets across 14 languages, is the largest SER benchmark, enabling reproducibility and research. To validate model generalisation, the study attempted 10 Transformer-based models under cross-corpus validation settings.

4.1 Performance Across Datasets

The results in Table 3 show consistently high performance on the test set, with an impressive **accuracy of 93.56%** and strong macro-averaged metrics (precision, recall, F1-score all above 93%), reflecting the model's excellent ability to generalise within its own distribution. However, when evaluated across external datasets like EMOVO, RAVDESS, CaFE, CREMA-D, and EmoDB, the performance drops significantly, with accuracies ranging between **57% and 77%**. This highlights the persistent challenge of cross-corpus generalisation in SER models, where differences in recording conditions, speaker populations, and label definitions affect transferability.

Looking at specific datasets, **EmoDB** stands out with the highest cross-dataset performance (accuracy 76.71%, MCC 0.7285), suggesting better alignment with the model's learned features, whereas **CREMA-D** shows the lowest accuracy (57.20%) and MCC (0.4896), indicating more difficulty in generalisation.

Table 3. Performance Metrics

	Test set	EMOVO	RAVDESS	CaFE	CREMA-D	EmoDB
Metric	Value	Value	Value	Value	Value	Value
Accuracy	93.56%	61.73%	60.27%	64.96%	57.20%	76.71%
Precision (Macro Avg)	94.00%	66.13%	60.30%	69.12%	58.00%	81.00%
Recall (Macro Avg)	93.54%	61.73%	60.27%	64.38%	57.34%	77.99%
F1-Score (Macro Avg)	93.53%	61.12%	59.31%	64.80%	56.50%	77.22%
Weighted F1-Score	93.56%	61.02%	59.04%	64.12%	56.03%	77.00%
UAR	93.54%	61.73%	60.27%	64.38%	57.34%	77.99%
WAR	93.56%	61.73%	60.27%	64.96%	57.20%	76.71%
MCC	0.9251	0.5639	0.5396	0.5941	0.4896	0.7285
Prediction Correlation	0.9329	0.5677	0.5951	0.6939	0.5505	0.7933

Table 4 provides a breakdown by emotion, revealing interesting patterns:

- **Anger** and **surprise** are consistently well-predicted, with particularly high accuracy on the test set (94.8% and 99.6%, respectively) and reasonable transfer across datasets.
- **Disgust** and **joy** show much weaker generalisation, with some datasets (e.g., EMOVO, CaFE) scoring below 50%, pointing to either low representation in training or greater variability in how these emotions are expressed across corpora.
- **Fear** and **sadness** also show fluctuating performance, but notably **fear** reaches 95.4% on EmoDB, showing that for certain datasets, the model captures these emotions well.
- **Neutral** tends to hold stable across sets, likely benefiting from its more distinct acoustic features.

Overall, while the model demonstrates near-perfect in-distribution performance, the cross-corpus results highlight the **need for further improvements in robustness and generalisability**, particularly for emotions like disgust and joy. These findings reinforce the importance of designing SER models that can handle diverse and spontaneous speech conditions, critical for real-world applications such as telemedicine, where speaker diversity and recording variability are unavoidable.

Table 4. Emotions Accuracy

	Test set	EMOVO	RAVDESS	CaFE	CREMA-D	EmoDB
Emotion	Value	Value	Value	Value	Value	Value
Anger	94.80%	91.67%	80.21%	66.67%	51.70%	66.39%
Disgust	91.37%	44.05%	40.10%	62.50%	64.29%	70.45%
Fear	93.60%	63.10%	64.58%	81.94%	33.00%	95.38%
Joy	89.95%	46.43%	60.42%	31.25%	63.80%	90.00%
Neutral	96.41%	85.71%	38.54%	56.94%	76.50%	66.67%
Sadness	89.02%	48.81%	56.77%	64.58%	54.76%	79.03%
Surprise	99.58%	52.38%	81.25%	86.81%		

4.2 EmoBOX

In our analysis, we have chosen a sub-group of 10 datasets from the original 32-dataset emoBOX benchmark because of access limitations on a few corpora and the unavailability of obtaining certain datasets. The selection allows fair model performance comparison across diverse datasets, though sampling a subset introduces a margin of error from the full benchmark. Margin of error (e) for a finite population is derived by:

$$e = Z \times \sqrt{\frac{p(1-p)}{n}} \times \frac{N-n}{N-1}$$

where:
• **N** = 32 is the total number of datasets in the original benchmark,
• **n** = 10 is the number of datasets used in our evaluation,
• **Z** = 1.96 corresponds to a 95% confidence level,
• **p** = 0.5 represents the maximum variability assumption.

Using this formula, the margin of error in our study is 26.50%. So, although our results have important conclusions about model performance, the results must be taken with some caution since running the complete benchmark could result in somewhat different rankings. To mitigate this limitation, we are interested in relative model performance trends instead of absolute accuracy. Relative difference in performance attained within our chosen subset of dataset continues to be representative of overall trend and with given dataset diversity preserved. From the comparison, we observe that:

• Our model (Wav2Vec 2.0 large) achieves a significant improvement over its base counterpart (+10.89% accuracy) due to its increased depth (24 layers) and parameter count (317M).

• The large models (ours, Data2Vec 2.0 large, HuBERT large and WavLM large) benefit from deeper architectures, with the latter achieving the highest accuracy (73.57%) due to speaker-aware pretraining.

• Whisper large V3 surpasses all models in accuracy (78.76%), but its 1.55B parameters make it significantly more resource-intensive, limiting its practicality for SER.

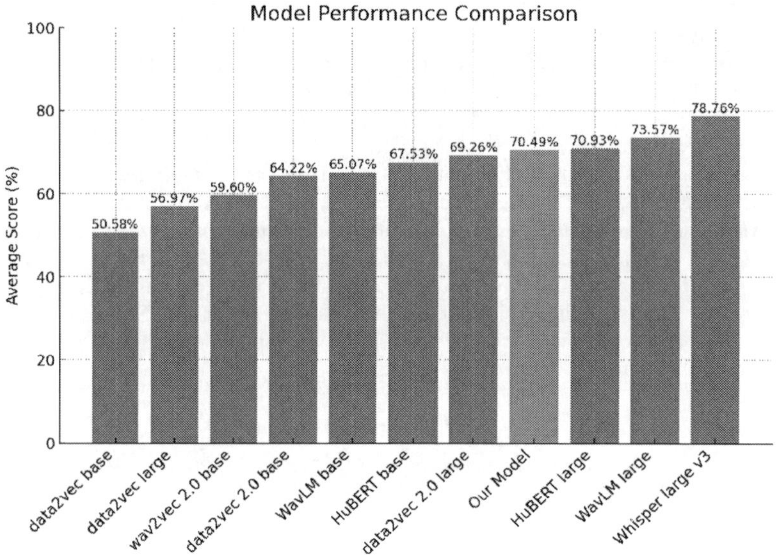

Figure 1. EmoBOX benchmark

5. DISCUSSION

In this paper, we demonstrated that high-quality Transformer-based SER models, when tuned on high-quality Italian emotional speech corpora, can achieve high in-domain performance and maintain cross-corpus comparability. While the gains (+13% accuracy) capture the benefits of data selection, augmentation, and balanced training, the observations also reveal the residual challenges of generalisability across datasets.

Compared to previous work (e.g., Catania et al., 2025; Wang et al., 2021), our model shows stronger resistance on spontaneous and multilingual test sets, while emotions like disgust and happiness still struggle with lower generalisation, possibly due to underrepresentation or cultural divergence in expression. This is in line with larger trends in SER research, where emotions with more subtle acoustic markers fall behind in classification performance.

One of the advantages of our approach is the optimisation of performance and efficiency: although larger models (e.g., Whisper large V3) are more accurate in raw terms, they come at significant resource cost, limiting practical use. Our model achieves a more practical compromise, which is particularly attractive to sensitive, real-time applications such as telemedicine and remote mental state monitoring.

Nonetheless, there are limitations. Use of acted affective data, although beneficial for clear-cut labels, may limit ecological validity when applied to spontaneous, in-the-wild speech. Furthermore, the subset of emoBOX datasets used is within a margin of error, and results should be interpreted carefully before full-benchmark validation is obtained. We advise interpreting the current comparative results primarily as indicative rather than definitive. A full-benchmark evaluation would be essential to confirm whether the observed model rankings and performance gains generalise across the entire benchmark. Incorporating such comprehensive validation represents an important step for future work.

Looking forward, combining domain-adaptive techniques and multimodal signals (e.g., facial affect) can potentially further enhance robustness. Furthermore, exploring semi-supervised or few-shot learning methods can help models better generalise to low-resource or novel domains, a critical milestone toward scalable, emotion-sensitive healthcare technologies.

6. CONCLUSION

This study examined the performance of Transformer-based SER models on Italian data, focusing specifically on dataset preparation, model fine-tuning, and cross-corpus evaluation. Through more strict selection criteria and targeted augmentation to balance emotional classes, an improvement in performance by up to 13%, compared to the former method using the same base architecture, was attained. On our in-domain test set, the model achieved 93.56% accuracy, showing the usefulness of painstakingly collected Italian speech data. When evaluated on external datasets, however, like EMOVO, accuracy fell to 61.73%, testifying to the notorious domain mismatch problem. Cross-corpus testing on alternative datasets revealed our model to still be competitive, beating its base model and holding the performance of larger models, but without the computational expense of models such as Whisper large V3. These results underscore the importance of careful data preparation in closing the gap between experimental control and real-world deployment. By illustrating how performance varies between datasets, we have defined a framework for improving generalisability and mitigating the degradation characteristic of cross-corpus environments. In particular, the combination of data filtering, class-balanced representation, and targeted augmentations was determined to be instrumental in improving both in-domain accuracy and robustness to novel corpora. Building on this foundation, future research can combine domain-adaptive methods and multimodal data to continue to enhance the robustness of SER in various clinical and nonclinical settings, continuing to move us toward an ultimately generalisable Speech Emotion Recognition algorithm.

REFERENCES

Atila, O., & Şengür, A. (2021). Attention guided 3D CNN-LSTM model for accurate speech-based emotion recognition. Applied Acoustics, 182, 108260.

Atmaja, B. T., & Sasou, A. (2022). Effects of data augmentations on speech emotion recognition. Sensors, 22(16), 5941.

Baevski, A., Zhou, Y., Mohamed, A., & Auli, M. (2020). wav2vec 2.0: A framework for self-supervised learning of speech representations. Advances in neural information processing systems, 33, 12449-12460.

Bakhshi, A., Wong, A. S., & Chalup, S. (2020). End-to-end speech emotion recognition based on time and frequency information using deep neural networks. In ECAI 2020 (pp. 969-975). IOS Press.

Burkhardt, F., Paeschke, A., Rolfes, M., Sendlmeier, W. F., & Weiss, B. (2005, September). A database of German emotional speech. In Interspeech (Vol. 5, pp. 1517-1520).

Cao, H., Cooper, D. G., Keutmann, M. K., Gur, R. C., Nenkova, A., & Verma, R. (2014). Crema-d: Crowd-sourced emotional multimodal actors dataset. IEEE transactions on affective computing, 5(4), 377-390.

Catania, F., Wilke, J. W., & Garzotto, F. (2025). Emozionalmente: A Crowdsourced Corpus of Simulated Emotional Speech in Italian. IEEE Transactions on Audio, Speech and Language Processing.

Chattopadhyay, S., Dey, A., & Basak, H. (2020). Optimizing speech emotion recognition using manta-ray based feature selection. arXiv preprint arXiv:2009.08909.

Costantini, G., Cesarini, V., & Casali, D. (2022). A Subset of Acoustic Features for Machine Learning-based and Statistical Approaches in Speech Emotion Recognition. In BIOSIGNALS (pp. 257-264).

Costantini, G., Iaderola, I., Paoloni, A., & Todisco, M. (2014). EMOVO corpus: an Italian emotional speech database. In Proceedings of the ninth international conference on language resources and evaluation (LREC'14) (pp. 3501-3504). European Language Resources Association (ELRA).

Dal Rí, F. A., Ciardi, F. C., & Conci, N. (2023). Speech emotion recognition and deep learning: an extensive validation using convolutional neural networks. IEEE Access, 11, 116638-116649.

Ekman, P. (1992). An argument for basic emotions. Cognition and Emotion, 6(3-4), 169–200.

Fanos, V., Dessì, A., Deledda, L., Lai, A., Ranzi, P., Avellino, I., ... & Colangelo, A. (2023). Postpartum depression screening through artificial intelligence: preliminary data through the Talking About algorithm. Journal of Pediatric and Neonatal Individualized Medicine, 12(2), 1-11.

Gournay, P., Lahaie, O., & Lefebvre, R. (2018, June). A canadian french emotional speech dataset. In Proceedings of the 9th ACM multimedia systems conference (pp. 399-402).

Grosman, J. (2021). Fine-tuned XLSR-53 large model for speech recognition in Italian. Hugging Face. https://huggingface.co/jonatasgrosman/wav2vec2-large-xlsr-53-italian.

Ko, T., Peddinti, V., Povey, D., Seltzer, M. L., & Khudanpur, S. (2017, March). A study on data augmentation of reverberant speech for robust speech recognition. In 2017 IEEE international conference on acoustics, speech and signal processing (ICASSP) (pp. 5220-5224). IEEE.

Livingstone, S. R., & Russo, F. A. (2018). The Ryerson Audio-Visual Database of Emotional Speech and Song (RAVDESS): A dynamic, multimodal set of facial and vocal expressions in North American English. PloS one, 13(5), e0196391.

Ma, Z., Chen, M., Zhang, H., Zheng, Z., Chen, W., Li, X., ... & Hain, T. (2024). Emobox: Multilingual multi-corpus speech emotion recognition toolkit and benchmark. arXiv preprint arXiv:2406.07162.

Scherer, K. R. (2003). Vocal communication of emotion: A review of research paradigms. Speech communication, 40(1-2), 227-256.

Wang, Y., Boumadane, A., & Heba, A. (2021). A fine-tuned wav2vec 2.0/hubert benchmark for speech emotion recognition, speaker verification and spoken language understanding. arXiv preprint arXiv:2111.02735.

COMPARATIVE STUDY OF ALGORITHMS FOR PREDICTION OF WEATHER CONDITIONS

Zirije Hasani, Alejna Zurnagji, Alejna Tashkeren, Bledi Buduri, Diellza Muqaj and Endrit Shemsedini
Faculty of Computer Science, University "Ukshin Hoti" Prizren
Prizren, Kosovo

ABSTRACT

Accurate weather forecasting is essential for informed decision-making in sectors such as agriculture, energy, disaster management, and environmental planning. With the advancement of data-driven methodologies, machine learning and statistical models have become increasingly important tools in weather prediction. This study presents a comparative analysis of three machine learning models—Prophet, SARIMA, and Linear Regression—for forecasting weather conditions using historical climate data from Kosovo. The models were evaluated across various weather parameters, including Temperature, Humidity, Wind Speed, and Air Pressure. Results indicate that Prophet consistently outperforms the other models in terms of minimizing prediction errors, as evidenced by its lowest Mean Squared Error (MSE) values. However, its performance is limited by poor R^2 values, which suggests that it struggles to explain the variance in the data, particularly for humidity and wind speed forecasts. SARIMA, while effective in some cases (notably wind speed), displays inconsistent results, with negative R^2 values, making it unsuitable for explaining the variance in the data. Linear Regression showed consistently poor performance across all parameters, with both high MSE and low R^2, indicating its ineffectiveness as a forecasting tool. Based on these findings, Prophet emerges as the most suitable model for weather forecasting in cases where minimizing prediction errors is the primary objective, particularly for temperature and air pressure. However, due to its limitations in explaining data variance, alternative models or approaches may be necessary for certain parameters. The study suggests that future research could investigate ensemble methods, neural networks, and real-time deployment strategies to improve weather forecasting, especially in resource-constrained regions.

KEYWORDS

Weather Forecasting, Machine Learning, SARIMA, Prophet, Linear Regression, Time Series Analysis, Climate Data

1. INTRODUCTION

Weather forecasting plays a vital role in modern society, enabling better planning, safety, and resource optimization. Traditional numerical weather prediction (NWP) methods rely heavily on physics-based models, which can be computationally intensive and sometimes limited in local-scale accuracy. In contrast, machine learning techniques offer data-driven alternatives capable of capturing complex patterns from historical data without the need for detailed physical models.

This study aims to compare the effectiveness of three widely used forecasting models—Linear Regression, SARIMA, and Prophet—for predicting weather conditions. The objective is to evaluate their accuracy, interpretability, and practical application in forecasting four essential meteorological variables: temperature, humidity, wind speed, and air pressure.

The data used includes four main weather variables: air pressure, humidity, temperature, and wind speed. For the implementation of the algorithms and analysis of the results, the Python language was used, integrating well-known libraries such as NumPy, Pandas, Scikit-learn, and PyCaret. Each algorithm was analyzed for its accuracy in forecasting the data for the year 2024. For example, Linear Regression was chosen for its simplicity and solid mathematical foundation, SARIMA for its ability to handle seasonality, and Prophet for its flexibility and ease of use with complex time series data.

The results of this paper not only help understand the relative performance of these algorithms but also contribute to scientific knowledge of their applications in specific climatic contexts, such as that of Kosovo. In the existing literature, the application of such methods for weather forecasting has been widely documented (e.g., (Hyndman, R. J., & Athanasopoulos, G., 2018) and (Taylor, S. J., & Letham, B., 2018)). However, few studies have addressed weather data from the Kosovo region, making this research an important contribution to the field of modeling and analyzing climatic data at the local level.

Next, the processes of data selection and preparation, algorithm implementation, and comparison of their results will be presented, emphasizing the practical and scientific implications of this research.

2. RELATED WORKS

Weather forecasting has evolved significantly with the adoption of machine learning (ML) techniques, which offer robust alternatives to traditional numerical and statistical models. Numerous studies have explored the application of different ML algorithms for predicting weather parameters such as temperature, humidity, rainfall, wind speed, and atmospheric pressure.

Several studies have compared traditional statistical models with modern machine learning algorithms for weather prediction. For example, Ahmed et al. (2019) conducted a comparative analysis between SARIMA and Artificial Neural Networks (ANN) for temperature forecasting and concluded that machine learning models often outperform statistical methods in capturing nonlinear patterns in data. Similarly, Shah et al. (2020) evaluated the performance of Support Vector Machines (SVM), Decision Trees, and Random Forests, finding ensemble models more robust for predicting humidity and wind speed.

The Prophet model, introduced by Facebook, has gained popularity due to its interpretability and effectiveness in handling seasonality and missing data. Taylor and Letham (2018) demonstrated its application across various domains, including weather prediction, showing comparable accuracy to SARIMA in seasonal data series.

In contrast, linear regression remains a foundational baseline model in forecasting. Although it cannot model complex dependencies, studies such as Kumar et al. (2018) have shown that it performs reasonably well in scenarios with stable climate trends and limited data variation.

Furthermore, hybrid approaches that combine statistical and machine learning models have also emerged. For instance, Zhang et al. (2021) proposed a hybrid LSTM-SARIMA model, achieving improved results for multivariate weather forecasting.

Despite these advancements, there is still a need for comparative studies focused on specific regional datasets, particularly in underrepresented areas like the Balkans. This study addresses this gap by evaluating the performance of Linear Regression, SARIMA, and Prophet on weather data from Kosovo, offering insights into their applicability in local forecasting systems.

Linear models like Multiple Linear Regression have been widely used for weather forecasting due to their simplicity and interpretability. For instance, Kumar and Ramesh (2019) applied linear regression models to predict daily temperature and humidity, achieving reasonable accuracy in short-term forecasts.

Time-series models such as ARIMA and its seasonal extension, SARIMA, have been popular for capturing temporal dependencies in meteorological data. Abhishek et al. (2012) demonstrated the use of SARIMA for predicting rainfall, emphasizing its suitability for seasonal data.

Recently, Facebook Prophet has emerged as a user-friendly and accurate forecasting tool. Its application in weather forecasting was explored by Nguyen et al. (2021), who showed its effectiveness in handling time-series data with seasonal patterns and missing values.

Machine learning models, especially tree-based and ensemble methods, have also gained traction. Zhang et al. (2017) compared Decision Trees, Random Forests, and Gradient Boosted Trees for temperature and humidity prediction, concluding that ensemble methods consistently outperformed simpler models.

Support Vector Machines (SVM) and Artificial Neural Networks (ANN) are commonly used for non-linear forecasting problems. For example, Shrestha and Solomatine (2006) used SVMs to forecast river flows as a proxy for rainfall estimation, while Kumar et al. (2013) implemented ANN models for monsoon rainfall prediction in India.

Deep learning has further enhanced predictive capabilities. Shi et al. (2015) introduced convolutional LSTM networks for precipitation nowcasting, achieving state-of-the-art results. Similarly, Chattopadhyay et al. (2020) leveraged deep neural networks to predict multiple meteorological variables with high accuracy.

Hybrid models combining statistical and ML methods have shown promising results. Ghanbari et al. (2020) proposed a hybrid SARIMA-ANN model to forecast air temperature, demonstrating that combining linear and non-linear models improves accuracy.

Comparative studies are particularly valuable in identifying suitable algorithms for specific use cases. Baboo and Shereef (2011) conducted a comparative analysis of ML models for rainfall prediction, highlighting trade-offs between accuracy and computational complexity. Similarly, Chattopadhyay and Chattopadhyay (2008) compared ANN and regression models for Indian monsoon forecasting.

Despite these advances, few studies focus on regional datasets, particularly in underrepresented areas like the Western Balkans. This study contributes to the literature by comparing Linear Regression, SARIMA, and Prophet using weather data from Kosovo, thereby filling a notable gap.

3. METHODOLOGY

3.1 Collection and Preprocessing

The dataset used in this study comprises hourly weather data collected from meteorological stations in the Republic of Kosovo, covering the period from January 2018 to December 2024. Key variables include temperature (°C), humidity (%), wind speed (m/s), and air pressure (hPa).

In the following table, we present part of our data used to evaluate algorithms. The data used in this research pertain to the Kosovo region, as the model is specifically being adapted for local conditions. Since weather patterns vary significantly between regions, models developed for other countries cannot be directly applied in this context.

Table 1. Sample of data used for experiments

Timestamp	*Air_Pressure*	*Humidity*	*Temperature*	*WindSpeed*
01.01.2018 00:00:00	974.446	100.2	-1.4	6.32
01.01.2018 01:00:00	974.248	100.2	-2.8	3.7
01.01.2018 02:00:00	974.182	100.2	-3	5.44
01.01.2018 03:00:00	973.652	100.2	-3.5	5.14
01.01.2018 04:00:00	973.189	100.2	-2.7	3.58
01.01.2018 05:00:00	972.726	100.2	-4.1	2.78
01.01.2018 06:00:00	972.064	100.2	-2	3.95

To ensure the dataset was suitable for accurate weather condition forecasting, several preprocessing steps were applied:

- Handling Missing Values through Interpolation: Missing or incomplete entries in the dataset were addressed using interpolation techniques. This method estimates and fills in missing values based on surrounding data points, allowing for a smoother and more consistent time series without introducing artificial trends or patterns.
- Normalizing Variables Where Appropriate: Selected weather variables, particularly those with wide-ranging values (e.g., temperature, pressure, wind speed), were normalized to a common scale. This step was crucial for ensuring that all features contributed equally during model training, especially for models sensitive to input magnitudes.
- Resampling to Daily Averages to Reduce Noise: The original data, which may have been recorded at irregular or high-frequency intervals, was resampled into daily averages. This helped reduce short-term fluctuations and noise, providing a more stable signal for forecasting algorithms to learn from.

- Splitting Data into Training (80%) and Testing (20%) Sets: To evaluate model performance, the dataset was divided into two subsets: 80% for training the models and 20% for testing. This split ensured that models were validated on unseen data, allowing for an unbiased assessment of forecasting accuracy.

3.2 Models Used

- **Linear Regression**: A baseline model that establishes a linear relationship between time (or lagged variables) and the target weather feature James et. al. (2023). **Linear Regression** is one of the simplest and most widely used methods for predicting numerical values. In this study, the model was applied to forecast key climate parameters—temperature, humidity, air pressure, and wind speed—using a linear function that relates past observations to future values. The model was utilized not only to predict future conditions for the year 2024 but also to analyze the relationships between different weather variables, helping to identify patterns and dependencies within the dataset.
- **SARIMA (Seasonal Autoregressive Integrated Moving Average)**: A time series model that accounts for both trend and seasonality in the data Noor et al (2022). Suitable for univariate forecasting tasks. **SARIMA (Seasonal Autoregressive Integrated Moving Average)** is a statistical model widely used for time series forecasting, particularly when the data exhibits strong seasonality. As an extension of the ARIMA model, SARIMA is well-suited for capturing both trend and seasonal patterns, making it an effective choice for weather-related time series. In this study, the SARIMA model was applied to forecast meteorological variables characterized by recurring seasonal cycles, leveraging its ability to model complex temporal structures in the data.
- **Facebook Prophet**: A forecasting tool developed by Meta, designed for business time series but adaptable to weather forecasting Toharudin et al. (2021). It captures trends, seasonality, and holiday effects using an additive model. **Prophet** is a model developed by Facebook, known for its ability to handle time series data with high variability and complexities such as unexpected seasonality, missing data, and changing trends. Prophet is capable of dealing with gaps and sudden shifts in data, providing accurate forecasts for both short-term and medium-term periods. Its robustness makes it an ideal tool for forecasting in contexts where data irregularities are common, such as weather prediction.

3.3 Evaluation Metrics

After implementing all three models, a detailed comparison of the performance of the models used for each weather parameter (temperature, humidity, air pressure, and wind speed) was carried out. The comparison was made using well-known metrics for evaluating the accuracy of forecasts:

R^2 Score (Coefficient of Determination) and Mean Squared Error (MSE) are two important metrics for assessing the accuracy of forecasting models, particularly in regression Chicco et al., 2020) Chain et al. (2014). Here's what each one indicates:

- **Mean Square Error (MSE)**

Mean Square Error (MSE) is a common metric used to evaluate the performance of regression models Chain et al. (2014). It measures the average of the squares of the differences between the predicted values and the actual values. A lower MSE indicates that the model's predictions are closer to the actual values, implying better performance.

The formula for MSE is:

$$MSE = \frac{1}{n}\sum(y_{true} - y_{pred})^2$$

Where:

- n is the number of observations
- y_{true} is the actual value
- y_{pred} is the predicted value

Because the errors are squared, larger errors have a disproportionately large effect on the MSE, making it sensitive to outliers.

- **R^2 Score (Coefficient of Determination)**

R^2 is a measure that indicates how well a model fits the actual data Chicco et al. (2020). It indicates the percentage of the variance in the dependent variable (output) that is explained by the model. R^2 can take values from 0 to 1, where:

- $R^2 = 1$ indicates that the model is perfect and fits the data completely.
- $R^2 = 0$ indicates that the model is no better than simply predicting the average of the data.
- $R^2 < 0$ this can occur when the model is very poor and performs worse than a simple model that uses the mean of the data.

The formula for R^2 is:

$$R^2 = 1 - \frac{\sum(y_{true} - y_{pred})^2}{\sum(y_{true} - \bar{y}_{true})^2}$$

Where:

- y_{true} these are the actual values.
- y_{pred} these are the values predicted by the model.
- $\bar{y}_{true}$ this is the average of the actual values.

A lower MSE value indicates that the prediction error is smaller and the model is more accurate. MSE is also more sensitive to large errors, as larger errors contribute to the total value quadratically.

After measuring and comparing the results, modeling the outcomes and analyzing the trends will be done to determine which model performed better in each aspect and to identify the most efficient method for forecasting weather conditions for the year 2024.

4. RESULT AND DISCUSSION

In this section, the results of the experiments conducted for forecasting weather conditions for the year 2024 will be presented, using three algorithms: Linear Regression, SARIMA, and Prophet. The results are based on the comparison of the performance of these models for four different climatic parameters: Temperature, Humidity, Air Pressure, and Wind Speed. The performance of each model has been evaluated through error and accuracy measurements such as RMSE (Root Mean Squared Error), MAE (Mean Absolute Error), and MAPE (Mean Absolute Percentage Error).

4.1 Evaluation of Model Performance

Each model was trained and tested on the four weather variables. Below is a summary of the key findings:

Table 2. Results from performance measurement of evaluation models

Model	Variable	MSE	R^2 Score
Linear Regression	Temperature	98.67	0.0061
SARIMA	Temperature	245.10	-2.522
Prophet	Temperature	5.99	0.547
Linear Regression	Humidity	419.08	8.68E-06
SARIMA	Humidity	217.547	-0.000366
Prophet	Humidity	30.445	-1.7500
Linear Regression	Wind speed	11290.72	0.0049
SARIMA	Wind speed	45.801	-20.556
Prophet	Wind speed	142.09	-7169.67
Linear Regression	Air pressure	57.423	0.089
SARIMA	Air pressure	151.65	-0.314
Prophet	Air pressure	13.7001	-0.98

TEMPERATURE

R^2:

- Prophet has the highest R^2 value (0.55), meaning this model explains about 55% of the variance in the data and is relatively accurate.
- SARIMA has a negative value (-2.52), indicating that the model performs worse than the average of the data and is not valid for predictions.
- Linear Regression has a very low R^2 value (0.0061), suggesting that it is unable to explain the variance in the data and is unsuitable for weather forecasting.

MSE:

- Prophet performs the best with the lowest MSE (5.99), indicating that this model has a smaller average error compared to the others.
- SARIMA has the highest MSE (245.10), showing very low accuracy in comparison to the other models.
- Linear Regression also has a high MSE (98.67), suggesting that this model is unsuitable for weather data.

Conclusion:

- Prophet is the best model in this case, as it has the lowest MSE and the highest R^2.
- SARIMA and Linear Regression are unsuitable for these data, as they have very high MSE and low or negative R^2.
- Prophet is a good choice for forecasting weather data for the upcoming year.

HUMIDITY

R^2:

- Prophet has a negative R^2 (-1.75), indicating that the model fails to explain the variance in the data and is not very useful for forecasting air humidity.
- SARIMA has an R^2 very close to zero (-0.00037), showing that it has very poor performance and is almost meaningless for this data.
- Linear Regression has a very small R^2 (8.68E-06), which is nearly zero, suggesting that this model is completely unsuitable for this case.

MSE:

- Prophet has the lowest MSE (30.45), indicating that it performs the best compared to the other models for this data.
- SARIMA has a much higher MSE (217.54), showing very low accuracy.
- Linear Regression has an even higher MSE (419.09), indicating large prediction errors and that it is not a suitable model for this data.

Conclusion:

- Prophet has the lowest MSE, but the negative R^2 shows that it is not an ideal model for this data.
- SARIMA and Linear Regression perform very poorly, with very high MSE and R^2 close to zero or negative, indicating that they are not suitable for forecasting air humidity.

WIND SPEED

R^2:

- SARIMA has a negative R^2 (-20.55), indicating that the model is not suitable for explaining the variance in the data.
- Prophet has an extremely low R^2 (-7169.67), which is a very poor value and shows that the model is almost meaningless for this data.
- Linear Regression has an R^2 very close to zero (0.0049), which is better than the other two models, but still indicates that the model does not fit the data well.

MSE:

- SARIMA has the lowest MSE (45.80), indicating that it is the most accurate model for forecasting wind speed.
- Prophet has the highest MSE (142.09), showing that it has a larger error compared to SARIMA, but it is still much better than Linear Regression.

- Linear Regression has an extremely high MSE (11290.72), indicating that its performance is very poor for this data and it is unsuitable for forecasting.

Conclusion:

- SARIMA has the lowest MSE, but the negative R^2 indicates that it does not explain the variance in the data well.
- Prophet has very poor performance, with a very low R^2 and high MSE.
- Linear Regression has a very high MSE, but its R^2 is closer to zero, showing that it is not a good model for this data.
- SARIMA is the best due to its lower MSE, but due to the negative R^2, none of the models are ideal.

AIR PREASUSRE

R^2:

- Linear Regression has the highest R^2 (0.089), which is very low, but at least positive, indicating that the model explains a small percentage of the variance in the data.
- SARIMA has a negative R^2 (-0.3149), showing that it does not explain the variance in the data well and performs poorly for this case.
- Prophet has the most negative R^2 (-0.98), indicating that it is the worst in terms of data fitting, despite having the lowest MSE.

MSE:

- Prophet has the lowest MSE (13.70), indicating that it performs the best compared to the other models for forecasting air pressure.
- Linear Regression has the highest MSE (57.42), showing that it has more error than Prophet but is much better than SARIMA.
- SARIMA has the highest MSE of all (151.65), indicating that it is not a suitable model for this data.

Conclusion:

- Prophet has the lowest MSE, making it better at reducing the absolute error, but the negative R^2 (-0.98) shows that it does not explain the data well.
- Linear Regression has an acceptable MSE and slightly better R^2, but still far from an accurate model.
- SARIMA performs the worst with the highest MSE and negative R^2, indicating that it is not suitable for this data.
- In this case, Prophet is better due to the lower MSE, but Linear Regression has a better R^2.

5. CONCLUSION

This study provides a comparative analysis of three machine learning models for weather condition forecasting using historical climate data from Kosovo. Each model demonstrated strengths depending on the weather variable and data characteristics.

Based on the comparison of the models (Prophet, SARIMA, and Linear Regression) across various weather parameters (Temperature, Humidity, Wind Speed, and Air Pressure), the following general conclusions can be drawn:

1. **Model Performance Across Parameters:**
    - **Prophet** consistently shows the lowest MSE across the different parameters, making it the best model for minimizing errors in predictions, especially in cases like temperature and air pressure. However, its R^2 values often indicate poor performance in terms of explaining the variance in the data, especially in humidity and wind speed forecasts.
    - **SARIMA** performs well in terms of MSE in some cases, such as wind speed, where it has the lowest MSE. However, its negative R^2 values suggest that it is not suitable for explaining the variance in the data, which reduces its effectiveness as a forecasting tool.
    - **Linear Regression** typically has low R^2 values, indicating that it is not a good model for explaining the variance in weather data. Its MSE values are generally high, indicating poor predictive accuracy, particularly in humidity and wind speed.

2. **Strengths and Weaknesses:**
    - **Prophet** is most effective in reducing prediction error (MSE) but struggles with explaining data variance (low or negative R^2), making it less reliable for understanding the underlying trends in the data.
    - **SARIMA** may be effective in some specific cases (e.g., wind speed), but its performance is inconsistent and often poor due to negative R^2 values, which means it fails to adequately model the variance of the data.
    - **Linear Regression** performs poorly overall, with both high MSE and low R^2 values across most weather parameters. It is not a reliable model for weather forecasting in this context.
3. **Recommendation:**
    - **Prophet** stands out as the most suitable model in cases where minimizing prediction error is the primary goal, particularly for temperature and air pressure. However, due to its poor R^2 performance in certain cases (humidity and wind speed), it may not always be the best model for all weather parameters.
    - **SARIMA** is not recommended due to its poor performance in explaining the data's variance, even though it performs well in terms of MSE in specific cases.
    - **Linear Regression** is unsuitable for this task overall, given its consistently high MSE and low R^2, and it should be avoided for forecasting weather conditions.

In conclusion, while no single model excels in all aspects, **Prophet** appears to be the best choice for forecasting weather data, especially in terms of minimizing errors, despite its limitations in explaining data variance.

Future research could explore ensemble methods, neural networks (e.g., LSTM), and real-time deployment scenarios for enhancing weather forecasting systems, particularly in regions with limited access to computational resources.

REFERENCES

Abdollahi, A., et al. (2020). Comparison of machine learning methods for weather forecasting. Journal of Environmental Management, 255.

Abhishek, K., Kumar, A., Ranjan, R., & Singh, K. (2012). A rainfall prediction model using artificial neural network. *2012 IEEE Control and System Graduate Research Colloquium*, 82–87.

Abiodun, O. I., Jantan, A., Omolara, A. E., Dada, K. V., Mohamed, N. A., & Arshad, H. (2018). State-of-the-art in artificial neural network applications: A survey. *Heliyon*, 4(11), e00938.

Ahmed, N. K., et al. (2010). Empirical comparison of machine learning models for time series forecasting. Econometrics and Statistics, 52-67.

Ahmed, N., Atiya, A. F., Gayar, N. E., & El-Shishiny, H. (2019). *An empirical comparison of machine learning models for time series forecasting*. Neural Computing and Applications, 31(5), 1305–1320. https://doi.org/10.1007/s00521-018-3523-3

Baboo, S. S., & Shereef, T. M. A. (2011). An efficient weather forecasting system using artificial neural network. *International Journal of Environmental Science and Development*, 2(4), 321–324.

Box, G. E., Jenkins, G. M., Reinsel, G. C., & Ljung, G. M. (2015). Time Series Analysis: Forecasting and Control. Wiley.

Chai, T., & Draxler, R. R. (2014). *Root mean square error (RMSE) or mean absolute error (MAE)? – Arguments against avoiding RMSE in the literature*. Geoscientific Model Development, 7(3), 1247–1250. https://doi.org/10.5194/gmd-7-1247-2014

Chattopadhyay, G., & Chattopadhyay, S. (2008). A comparative study of statistical and neural network techniques in predicting the average summer-monsoon rainfall over India. *International Journal of Climatology*, 28(4), 421–427.

Chattopadhyay, I., Ghosh, S., & Joshi, M. K. (2020). Machine learning and climate science: A review. *Environmental Research Letters*, 15(10), 103001.

Chicco, D., & Jurman, G. (2020). *The advantages of the Matthews correlation coefficient (MCC) over F1 score and accuracy in binary classification evaluation*. BMC Genomics, 21(6). https://doi.org/10.1186/s12864-019-6413-7

Di Persio, L., & Honchar, O. (2017). Artificial neural networks architectures for stock price prediction: Comparisons and applications. *International Journal of Circuits, Systems and Signal Processing*, 11, 403–413.

Fauzi, R., et al. (2021). Prophet model for rainfall prediction. Climate Research, 112-124.

Ghanbari, R. N., Vafakhah, M., & Shojaee, S. (2020). A hybrid SARIMA–ANN model for forecasting of temperature. *Theoretical and Applied Climatology*, 141(1), 303–313.

Hyndman, R. J., & Athanasopoulos, G. (2018). Forecasting: Principles and Practice. OTexts.

James, G., Witten, D., Hastie, T., Tibshirani, R., Taylor, J. (2023). Linear Regression. In: An Introduction to Statistical Learning. Springer Texts in Statistics. Springer, Cham. https://doi.org/10.1007/978-3-031-38747-0_3

Jung, J., et al. (2018). Comparative study of regression models for climate data prediction. Environmental Modeling & Assessment, 189-204.

Klein, A., et al. (2019). Application of SARIMA for short-term weather prediction. Meteorological Applications, 456-468.

Kumar, D., & Ramesh, C. (2019). Forecasting temperature and humidity using multiple linear regression. *International Journal of Innovative Technology and Exploring Engineering*, 8(6), 1392–1396.

Kumar, M., Jain, S. K., & Sharma, P. (2013). Long term rainfall prediction using artificial neural network. *International Journal of Computer Applications*, 69(9), 20–25.

Kumar, P., Rani, S., & Rani, P. (2018). *Linear regression analysis for weather forecasting.* International Journal of Computer Applications, 179(12), 28–32. https://doi.org/10.5120/ijca2018916520

LeCun, Y., Bengio, Y., & Hinton, G. (2015). Deep learning. *Nature*, 521(7553), 436–444.

Makridakis, S., et al. (2018). The M4 forecasting competition: Results and findings. International Journal of Forecasting, 802-838.

Nguyen, T. N., Nguyen, M. T., & Do, H. (2021). Application of Facebook Prophet model in weather forecasting. *International Journal of Advanced Computer Science and Applications*, 12(3), 192–199.

Noor, T. H., Almars, A. M., Alwateer, M., Almaliki, M., Gad, I., & Atlam, E.-S. (2022). SARIMA: A Seasonal Autoregressive Integrated Moving Average Model for Crime Analysis in Saudi Arabia. *Electronics*, *11*(23), 3986. https://doi.org/10.3390/electronics11233986

Rasp, S., Dueben, P. D., Scher, S., Weyn, J. A., Mouatadid, S., & Thuerey, N. (2020). WeatherBench: A benchmark dataset for data-driven weather forecasting. *Journal of Advances in Modeling Earth Systems*, 12(11), e2020MS002203.

Ravinesh, R., et al. (2020). Analysis of meteorological data using Prophet. Atmospheric Science Letters, 21(4), e987.

Shah, S. A., Shafiq, M., & Kim, K. (2020). *A comparative study of machine learning techniques for weather forecasting in smart cities.* Sustainability, 12(8), 3282. https://doi.org/10.3390/su12083282

Shi, X., Chen, Z., Wang, H., Yeung, D. Y., Wong, W. K., & Woo, W. C. (2015). Convolutional LSTM network: A machine learning approach for precipitation nowcasting. *Advances in Neural Information Processing Systems*, 28.

Shrestha, D. L., & Solomatine, D. P. (2006). Machine learning approaches for estimation of prediction interval for the model output. *Neural Networks*, 19(2), 225–235.

Taylor, S. J., & Letham, B. (2018). Forecasting at scale. he American Statistician, 37-45.

Taylor, S. J., & Letham, B. (2018). *Forecasting at scale*. The American Statistician, 72(1), 37–45. https://doi.org/10.1080/00031305.2017.1380080

Toharudin, T., Pontoh, R. S., Caraka, R. E., Zahroh, S., Lee, Y., & Chen, R. C. (2021). Employing long short-term memory and Facebook prophet model in air temperature forecasting. *Communications in Statistics - Simulation and Computation*, *52*(2), 279–290. https://doi.org/10.1080/03610918.2020.1854302

Zhang, G., Eddy Patuwo, B., & Hu, M. Y. (2017). Forecasting with artificial neural networks: The state of the art. *International Journal of Forecasting*, 14(1), 35–62.

Zhang, Y., Wu, L., & Wang, Z. (2021). *A hybrid model based on LSTM and SARIMA for weather forecasting*. Neural Processing Letters, 53(1), 561–577. https://doi.org/10.1007/s11063-020-10300-8

KPIS FOR THE GOVERNANCE AND STRATEGIC MANAGEMENT OF A PUBLIC APPLICATION PARK

Efrem Mbaki Luzayisu[1,2] and Stephan Tshitende Kalomba[1]
[1]*University of Kinshasa*
[2]*Catholic University of Congo, Democratic Republic of Congo*

ABSTRACT

This article proposes a structured approach for managing an application portfolio in the public financial sector, based on the definition of key performance indicators (KPIs). In a context where digital transformation is a major issue across all sectors, administrations, often lacking ERPs, are faced with managing a portfolio of heterogeneous applications. This complicates both the monitoring and evaluation of performance, as well as the management of technical incidents and business challenges. The article relies on recognized frameworks such as COBIT, ITIL, ANSSI, and the GDPR to structure these indicators around key dimensions such as availability, security, performance, and alignment with business needs. Calculation formulas, specific objectives, and classification thresholds (red, orange, green) are proposed for each KPI. A practical case, inspired by a Federal Public Service (FPS) in Belgium, illustrates the implementation of a BI dashboard that enables monitoring, alerting, and guiding application governance decisions, while facilitating alignment with the administration's strategic priorities.

KEYWORDS

Application Park, KPI, Public Sector, ITIL, ANSSI, GDPR, Dashboard, Supervision, Business Intelligence, IT Governance

1. INTRODUCTION

The digital transformation of public administrations has accelerated over the past two decades, driven by the combined effect of growing citizen demands, the evolution of digital technologies and e-government policies (OECD, 2023; United Nations, 2022). This dynamic has encouraged the proliferation of IT applications, deployed to meet specific and varied needs, ranging from the management of administrative procedures to the provision of personalized digital services. While this application expansion is an undeniable lever for modernization, it also leads to an increasing complexity of the information system (IS). Indeed, many applications have been developed or acquired independently, often without an overall vision or integration logic, resulting in a technological stack that is sometimes incoherent (Verdegem & Verleye, 2009). This heterogeneity makes it difficult to monitor, maintain, secure and optimize the application park.

In public administrations, this issue is all the more sensitive as the applications concerned support essential services to citizens (access to civil status, taxation, social security, employment, etc.). Any unavailability or malfunction of a service can compromise access to rights, cause delays in processing or damage user confidence (Berntzen & Johannessen, 2016). Given the organizational complexity that characterizes the public sector, it is difficult or easy to set up a single application map that supports its activities. Unlike private organizations that frequently use enterprise resource planning (ERP) systems, public entities operate mostly in a fragmented environment, without a single pivot system (Mazzei & Noble, 2017). This fragmentation makes it difficult to proactively detect anomalies, monitor application performance, and coordinate technical teams in the event of an incident.

In this context, the establishment of a system of technical and functional supervision becomes a necessity. Several solutions can be mobilized: audit logs generated at the database level, log files from operating systems or application logs. The latter, when centralized in tools such as the ELK suite — an acronym for Elasticsearch, Logstash and Kibana — or Graylog, provide consolidated visibility on critical IS events (Rouse, 2021). In addition, the use of monitoring solutions such as Prometheus, Grafana or Zabbix, combined with lightweight collection agents (e.g., Filebeat, Fluentd, Promtail) deployed on the servers, offers the possibility of triggering alerts in real time, correlating inter-application incidents and automating corrective actions. Such a distributed and centralized supervision architecture is a relevant response to the lack of an ERP in many public structures. It ensures compliance with regulatory requirements in terms of security and traceability, in particular those of the General Data Protection Regulation (GDPR) in Europe, as well as the recommendations of the National Agency for the Security of Information Systems (ANSSI) in France (CNIL, 2024). However, technology alone is not enough. To effectively manage the performance and availability of a complex application park, it is essential to set up relevant monitoring indicators, commonly known as KPIs (Key Performance Indicators). These indicators make it possible to objectively measure the state of health of the application system, to monitor its evolution, to anticipate degradation and to support the strategic decisions of the IT department (De Haes et Al, 2020; ISACA,2023). It is in this perspective that the present work is inscribed. We propose a structured approach aimed at defining and implementing performance indicators adapted to the management of an application park in a public context in Belgium. This approach is based on three axes: the identification of relevant performance dimensions, the formalization of simple and complex indicators with their calculation formulas, and the definition of alert thresholds allowing intuitive reading (color code: green, orange, red). Finally, we will illustrate the approach through a representative case study.

2. STATE OF THE ART

Historically, the management of an application estate has been based on technical approaches focused on infrastructure, incidents and updates. However, with the increasing complexity of information systems and the emergence of new paradigms (cloud computing, microservices, API-first, mobility), traditional approaches are gradually showing their limits (Böhmann et al., 2018). Effective governance of a set of applications today requires a more holistic vision, integrating technical, functional and organizational dimensions (Lankhorst, 2017). In the scientific literature, several studies have highlighted the problems related to the heterogeneity of applications, their lack of interoperability and the lack of consolidated visibility on their performance (European Commission., 2023). Research on IT monitoring and Application Performance Management (APM) emphasizes the importance of collecting real-time data on availability, latency, load and system errors (Gartner, 2023). However, these metrics, although useful, often remain too technical and difficult to interpret by functional or decision-making actors (Friedman et al., 2020).

Other studies have addressed Application Portfolio Management (APM), with an emphasis on streamlining applications, identifying redundancies, and prioritizing investments (Ross and Weill, 2002; Tamm et al., 2011). These approaches, although essential to the IS urbanization strategy, sometimes lack the granularity to allow daily operational management.In the public sector, the situation is even more complex. Administrations are facing strong citizen pressure, digital transparency obligations, and major budgetary constraints (OECD, 2023; United Nations, 2022). Few tools are specifically designed to meet their needs for large-scale application monitoring. Application referents, often responsible for several software programs at the same time, lack simple and relevant indicators to anticipate malfunctions, document incidents and guide their decisions (European Commission, 2021). It is in this context that the need for a KPI-based approach specific to application governance arises. According to the Figure 1 below, these indicators must allow a functional and operational reading of the performance of applications, taking into account the needs of end users, the business criticality of software solutions, and their possible interdependencies (van der Meulen & Rivera, 2014).

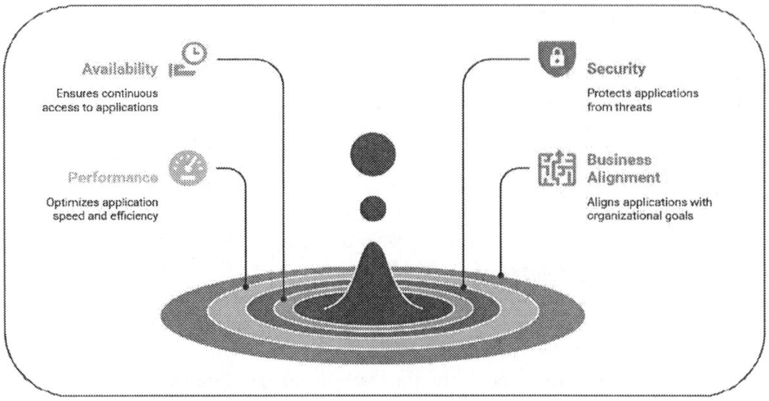

Figure 1. Towards an integrated solution for managing an application park through BI

3. CONTEXT AND CHALLENGES OF APPLICATION MANAGEMENT

In the era of the digital transformation of states, public administrations are called upon to modernize their infrastructures and optimize their processes to meet the growing expectations of citizens in terms of transparency, responsiveness and efficiency. In this context, the application park – all the software and systems used to support public service missions – becomes a strategic lever for performance and governance. However, the increasing complexity of these systems, the interdependencies between applications, as well as budgetary and regulatory constraints, make them particularly demanding to manage. Hence the importance of adopting a data-driven and results-based approach. This work focuses on the use of key performance indicators (KPIs) as decision-making tools in the strategic management and governance of the public application park. It highlights how these metrics not only measure the value and performance of applications, but also anticipate risks, align digital systems with the government's strategic objectives, and foster agile, sustainable, and transparent management (Almeida Marques, 2024; Niven et al., 2023). Taking a systemic and interdisciplinary approach, this study offers evaluation models, practical tools and recommendations for strengthened digital governance, aligned with the principles of good governance, innovation and accountability.

In public administrations, particularly in the financial sector (treasuries, tax administrations, budget agencies), the digital transformation has resulted in the proliferation of IT applications. The latter, sometimes designed in a heterogeneous way and without a coherent integration plan, make up a complex application park to administer. Faced with this challenge, Information Systems Departments (CIOs) must structure their governance approach and rely on reliable management instruments. Several reference frameworks can be used to frame this governance:

- ITIL (Information Technology Infrastructure Library) offers a repository of best practices in IT service management, focusing on incident, change and service catalog management (AXELOS, 2023).
- COBIT (Control Objectives for Information and Related Technologies) provides a global governance structure for information systems aligned with performance and compliance objectives.
- ANSSI (French National Agency for the Security of Information Systems) offers cybersecurity guides and requirements, which are very useful in contexts where availability and confidentiality are critical.
- The GDPR (General Data Protection Regulation) requires rigorous monitoring of traceability, access management and logs in systems processing personal data.

In Belgium, the federal services have joined forces with the company Smals with the aim of pooling their IT services while maintaining the possibility of respecting the local specificities of each ministry. In this context, performance indicators (KPIs) are put in place, among other things, to measure the availability of applications, their efficiency and the quality of their operation. Coupled with Business Intelligence (BI) solutions such as Power BI, Grafana or Kibana (as part of the ELK stack: Elasticsearch, Logstash, Kibana), they make it possible to generate dynamic dashboards, alerts, and make decisions based on consolidated data. With this in mind, we propose below a set of structuring KPIs to effectively manage a fleet of IT applications in a public financial environment. For each indicator, we specify the calculation formula, the recommended objective, and the critical thresholds to qualify the situation according to a color code (green, orange, red).

Table 1. Indicators for managing an application park in the public financial sector

N°	Indicator	Formula	Objective	Color thresholds
1	Application Availability Rate (%)	(Total Time - Downtime) / Total Time × 100	≥ 99.5%	Green≥ 99.5% Orange ≥ 98% Red < 98%
2	Mean Time to Incident Resolution (MTTR)	Sum of resolution times / number of incidents	≤ 4 hours	Green ≤ 4 a.m. Orange ≤ 8 a.m. Red > 8 a.m.
3	Critical Error Rate (Monthly)	Number of blocking errors / total number of × 100 processes	< 1%	Green< 1% Orange < 3% Red ≥ 3%
4	SLA compliance rate	Number of SLA-compliant services / total services × 100	≥ 95%	Green≥ 95% Orange ≥ 90% Red < 90%
5	Key Application Usage	Number of sessions / number of targeted users	≥ 80%	Green ≥ 80% Orange ≥ 60% Red < 60%
6	Average number of incidents per application	Number of incidents / number of applications	< 3	Green < 3 Orange < 5 Red ≥ 5
7	Unusable log rates	No. of unreadable or corrupted log files / total logs × 100	< 5%	Green < 5% Orange < 10% Red ≥ 10%
8	Technology obsolescence rate	Number of End-of-Support Apps / Total Apps× 100	< 10%	Green< 10% Orange < 20% Red ≥ 20%
9	Percentage of access traceability	(Number of apps with full logging / total apps) × 100	≥ 100%	Green = 100% Orange ≥ 90% Red < 90%
10	GDPR compliance rate	(Number of GDPR auditable apps / total apps processing personal data) × 100	≥ 95%	Green≥ 95% Orange ≥ 85% Red < 85%

This table can be easily integrated into an automated reporting platform using BI tools, connected to sources such as ITSM systems (ServiceNow, GLPI), monitoring tools (Zabbix, Prometheus) or centralized operating logs via ELK or Graylog. These indicators should be adapted according to the priorities of each organization. For example, a tax authority will place more emphasis on GDPR compliance and access traceability, while a public payment agency will focus on high availability and transactional reliability.

Finally, the indicator-based approach, combined with the use of governance frameworks and modern performance visualization tools, allows CIOs in the financial public sector to strengthen their control of the application park. It also promotes transparency with regard to internal (business lines, general management) and external (auditors, regulatory authorities) stakeholders, while being in line with the principles of quality, efficiency and safety.

Incident management within an application portfolio is a particularly relevant case study in the context of performance analysis and information system optimization insofar as there is a rich source of actionable data and measurable processes for any scientific approach aimed at evaluation and continuous improvement. Indeed, incident management involves a series of standardized activities that produce many organizational and technical indicators. Among these, we can mention: the total number of incidents, Mean Time to Acknowledge (MTTA), Mean Time to Resolve (MTTR), recurrence rate, application availability rate, and the percentage of incidents resolved at first contact. These metrics provide a solid foundation for analyzing service quality, team responsiveness, system resilience, and support process efficiency. Furthermore, this topic also allows for the integration of cross-disciplinary approaches, notably through economic analysis (average cost per incident), human factors (incidents caused by human error), and strategic impact (effects of critical incidents on business processes). The scientific interest thus lies in the multidimensional richness of this domain, which combines operational data, IT governance issues, and continuous improvement dynamics. Hence, the choice of incident management as a study example is not arbitrary. It is based on this domain's capacity to concretely illustrate key issues of performance, quality, and control of digital systems, while also enabling objective analysis through reliable, comparable, and interpretable indicators. In others terms, this choice is justified by the fact that such an environment offers a rich source of exploitable data and measurable processes—essential for any scientific approach aimed at evaluation and continuous improvement. Indeed, incident management involves a series of standardized activities that generate various technical and organizational indicators. Among these, we can mention: the total number of incidents, Mean Time to Acknowledge (MTTA), Mean Time to Resolve (MTTR), recurrence rate, application availability rate, and the percentage of incidents resolved at first contact. With these indicators, we provide a solid and reliable basis for evaluating the robustness of the systems, the speed of the teams, the efficiency of the technical support and the quality of the service. In addition, this topic can be addressed in a cross-sectional way, adding human (e.g., user errors), strategic (the impact of serious incidents on the organization's activities), and economic (such as the average cost of an incident) elements. In other words, the scientific interest of this subject comes from the fact that it is possible to combine concrete data, IT management issues and the desire for continuous improvement. To be even more precise, we recall below the formulas of the main indicators:

- Total Number of Incidents (TNI), Number of incidents recorded over a given period
- Mean Time to Acknowledge (MTTA), Average time taken to acknowledge an incident
  MTTA = (Sum of times between incident reporting and acknowledgment) / (Total number of incidents)
- Mean Time to Resolve (MTTR), Average time taken to resolve an incident
  MTTR = (Sum of times between incident reporting and resolution)/ (Total number of incidents)
- Incident Recurrence Rate (IRR)
  IRR = ((Number of recurring incidents)/(Total number of incidents)) X 100
- Application Availability Rate (AAR), Based on uptime without incidents
  AAR = (1-((Total downtime)/(Total observation time))) X 100
- First Contact Resolution Rate (FCRR)
  FCRR = (Number of incidents resolved at first contact)/(Total number of incidents) X 100
- Average Cost per Incident (ACI)
  ACI = (Total cost of incident management)/(Total number of incidents)

To continue with this example, we present below the implemented star schema used to structure the incident management data of the application portfolio. This schema, Figure 2, facilitates efficient querying and dynamic analysis of key performance indicators (KPIs), particularly for the calculation of indicators No. 2, No. 4, and No. 6 from Table 1. Furthermore, we showcase a sample dashboard page, Figure 3, designed with Power BI Desktop for 39578 incidents.

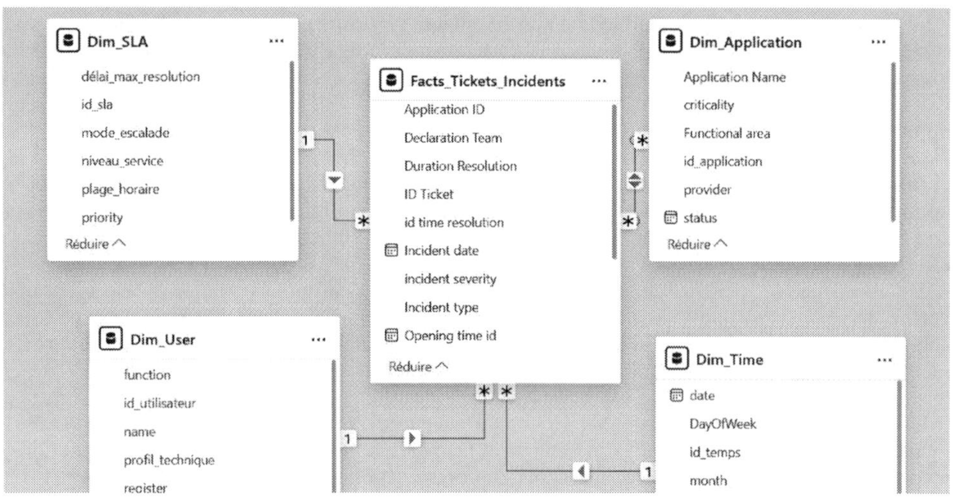

Figure 2. Star schema for KPIs 2, 4, and 6 with Power BI DeskTop

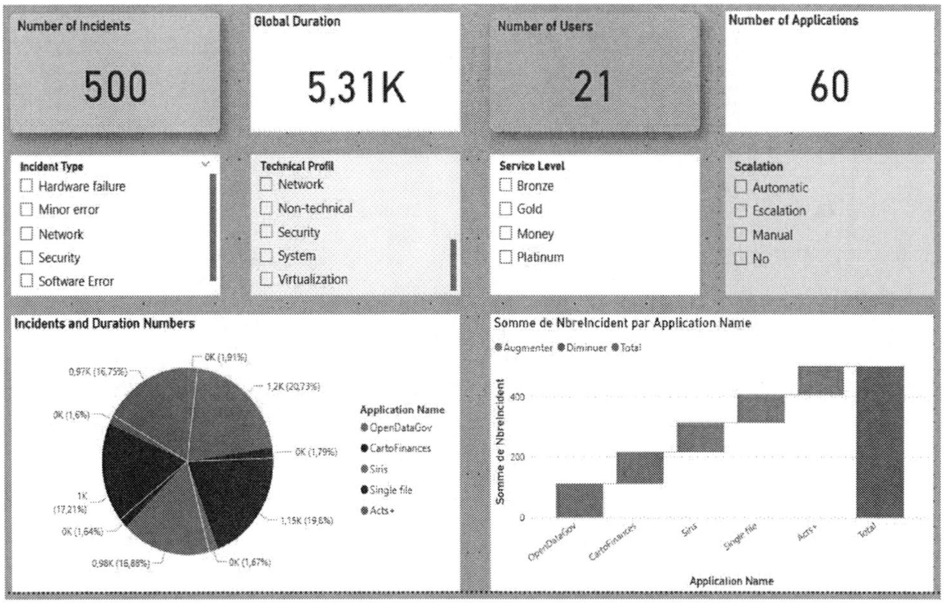

Figure 3. Sample Incident Management Dashboard with Power BI

4. USE CASE: DIGITAL PUBLIC ADMINISTRATION

To concretize the proposed approach and demonstrate the usefulness of the indicators in the operational and strategic management of an application park, we consider the case of an IT department within a Belgian Federal Public Service (FPS) in charge of taxation. The exposed data comes from the Cognos Audit database where the server records all user activities including user ID, administration, time of connection, BI instrument used and type, time of connection, etc. The data used also comes from the database centralizing the logs of the various applications of the FPS concerned. This administration manages about 120 interconnected business applications, used by both internal agents and citizens for tax returns, document consultation or online payment. The dashboard aims to provide a real-time, synthetic view of application performance, availability, obsolescence, and business value to help:

- Business managers to identify sticking points, underperformance or possible changes;
- The IT department to plan updates, rationalize the application portfolio, and prioritize interventions, with regard to the evolution of user needs;
- Management to orient IT budgets according to the real criticality of the applications.

The following dashboard, built using Power BI and fed by a combination of data from monitoring tools (Zabbix, Grafana), CMDB databases, business interviews and BI decision-making, presents some key indicators used in this case:

Table 2. Extract from the application dashboard – Case of an FPS in Belgium

App Name	Uptime (%)	Average time to resolution (h)	Number of incidents/month	Last updated (month)	Number of active users	Business Criticality Score (1-5)	Global Status
e-TaxFlow	99.6% ●	2h30 ●	● 3	Three months ●	23,400 ●	■ 5	What to Watch For
VAT return	97.8% ●	4h ●	● 6	18 months ■	5,600 ●	● 4	To be updated
PaymentTaxes	99.9% ●	1h30 ●	● 1	6 months ●	18,000 ●	■ 5	Criticism
Agent Directory	95.2% ■	6 hours ■	12 ■	30 months ■	500 ●	● 3	Obsolete
Tax archiving	98.4% ●	3 hours ●	● 4	24 months ■	700 ●	● 2	Secondary

It should be remembered that the terminals or thresholds used in this dashboard are not arbitrary. We refer to internationally recognized best practices, in particular those of the ANSSI, COBIT and ITIL standards. While integrating the realities of each business line, these thresholds define objective benchmarks for a rigorous and standardized evaluation of the operation and strategic importance of each application. In other words, these thresholds support proactive decision-making and help reduce risk and ensure strategic alignment between IT operations and organizational goals. Thus, they define a controlled balance between performance, risk and cost; a key factor for the efficient management of a modern application portfolio. In this context, the following observations should be made:

- The TaxPayments app is extremely critical. The data collected shows that it is performing well, but requires more monitoring compared to its transaction log in order to continuously ensure its integrity and security.
- It should be noted that the TaxOnWeb application generally works well despite a few security flaws, according to the recommendations of the ANSSI.
- Despite the efforts of the IAM team, the agent directory is showing signs of wear and tear. For example, a relatively long response time is mentioned during connections but also frequent outages due to a lack of regular updates.
- The dashboard is integrated with an automatic alert system (based on predefined thresholds) and generates weekly reports for IT committees and business units to support informed decision-making.

Ultimately, this practical case demonstrates that a structured, KPI-driven approach based on recognized standards enables:

- Objective prioritization of modernization initiatives,
- Clear, effective communication between IT and business teams,
- Proactive anticipation and prevention of service degradation,
- IT governance closely aligned with user needs and strategic objectives.

In other words, by combining internationally recognized frameworks (such as ITIL, ANSSI, and COBIT) with modern business intelligence tools, public administrations can rigorously manage their information systems while fulfilling their missions of transparency, performance, and service to citizens.

5. PERSPECTIVES AND RECOMMENDATIONS

This study provides sufficient evidence that a portfolio of heterogeneous applications of an IT system requires rigorous management, particularly in public administrations characterized by a high volume of interaction between civil servants and citizens. Under these conditions, it is imperative that application referents and IT service managers implement better strategies to overcome this challenge. Through this article, we show that the implementation of key performance indicators (KPIs) represents a structuring approach, allowing a rigorous monitoring and an informed management of decisions based on the data of the computer system. Indeed, starting from the application mapping and the collection of the needs of users and administrations, the method we propose is the determination of performance axes that should allow the definition of a set of relevant indicators. Thus, it becomes possible to anticipate failures more effectively, to react quickly to incidents, and to make the continuous improvement of the quality-of-service part of a sustainable dynamic. The practical case presented in this study illustrates in a concrete way the positive impact of this approach on operational efficiency as well as on user satisfaction and the strategic visibility of the information system. With a view to continuous optimization, several avenues of evolution can be considered:

- Automate KPI dashboards, using Business Intelligence tools (Power BI, Cognos, Tableau, SAS, Grafana, etc.) interconnected with application logs and ticketing systems;
- Set up intelligent alerts, capable of automatically detecting critical drifts or anomalies;
- Conduct a periodic evaluation of KPIs, regularly adapting them to changing business and technology needs;
- Integrate artificial intelligence to anticipate incidents, analyze user behavior or optimize the workload;
- Raise awareness among all stakeholders, by involving business agents in the monitoring of indicators, in a logic of shared governance.

Finally, it would be appropriate to extend this approach to other organizational contexts – such as the private sector, education or health care – with a view to enriching the basis of good practices and testing the robustness of the indicators in various environments.

considered:

- Automation of KPI dashboards: via BI tools (Power BI, Cognos, Tableau, SAS, Grafana, etc.) interconnected with application logs and ticketing systems.
- Implementation of intelligent alerts: in order to automatically detect critical drifts or anomalies.
- Periodic evaluation of KPIs: by regularly adapting the indicators to changes in business and technological needs.
- AI integration: to anticipate incidents, analyze user behavior, or optimize application load.
- Stakeholder awareness: involve business agents in the monitoring of indicators for shared governance.

Finally, it would be relevant to extend this approach to other organizational contexts (private sector, education, health, etc.) in order to enrich the basis of good practices and validate the robustness of KPIs in various environments.

ACKNOWLEDGEMENT

We express our deep gratitude to God, Creator of Heaven and Earth, for His grace and faithfulness throughout our academic and professional journeys.

REFERENCES

Almeida Marques, P. C. de., & Oliveira, P. (2024). Integrating artificial intelligence and the balanced scorecard: Strengthening organisational resilience in times of crisis. SSRN Electronic Journal.

AXELOS, (2023). ITIL 4: Digital and IT strategy. TSO (The Stationery Office).

Berntzen, L. & Johannessen, M.R., (2016). The Role of Citizen Service Centres in Public Sector Digital Transformation.

Böhmann, T., Leimeister, J.M. & Möslein, K., (2018). The New Frontiers of Service Science. Business & Information Systems Engineering.

Bon, J. van, (2012). ITIL Lifecycle Suite. Van Haren Publishing.

Brynjolfsson, E. & McAfee, A., (2014). The Second Machine Age: Work, Progress, and Prosperity in a Time of Brilliant Technologies. W.W. Norton & Company.

CIO Council, (2017). Application Rationalization Playbook: Getting the Most from Your Application Portfolio.

CNIL, (2024). Guide de la sécurité des données personnelles. Commission Nationale de l'Informatique et des Libertés, https://www.cnil.fr

De Haes, S., & Van Grembergen, W., (2020). Enterprise governance of IT: Achieving strategic alignment and value (2nd ed.). Springer. [https://doi.org/10.1007/978-3-030-45685-5] (https://doi.org/10.1007/978-3-030-45685-5)

Devaux, S.A., (2014). Earned Value Project Management (4th Ed.). Management Concept.

Dumas, M., La Rosa, M., Mendling, J. & Reijers, H.A., (2018). Fundamentals of Business Process Management (2nd Ed.). Springer.

European Commission, (2021). Interoperable Europe Act – Proposal and Impact Assessment.

European Commission, (2023). Interoperable Europe Act – Final text and implementation guidelines, Publications Office of the European Union.

Friedman, R. et al., (2020). Application Performance Management for Cloud-Native Systems. IEEE Cloud Computing. Gartner, 2022. Best Practices for Application Portfolio Management.

Gartner, (2023). Magic quadrant for application performance monitoring and observability, Gartner Research

Hammer, M. & Champy, J., (2006). Reengineering the Corporation: Manifesto for Business Revolution. HarperBusiness. IBM Institute for Business Value, 2020. Accelerating Digital Government Transformation.

ISACA, (2023). COBIT 2019 design guide: Designing an information and technology governance solution (2nd ed.). ISACA Publishing.

ISO/IEC, (2011). ISO/IEC 25010: Systems and software engineering — Systems and software Quality Requirements and Evaluation (SQuaRE) — System and software quality models.

ISO/IEC, (2015). ISO/IEC 38500: Information technology — Governance of IT for the organization. ITIL Foundation, 2019. ITIL Foundation 4th Edition. AXELOS Global Best Practice.

Lankhorst, M., (2017). Enterprise Architecture at Work: Modelling, Communication and Analysis. Springer. Luftman, J., 2003. Competing in the Information Age: Align in the Sand. Oxford University Press.

Mazzei, M.J. & Noble, D., (2017). Big Data Dreams: A Framework for Corporate Strategy. Business Horizons, 60(3), pp. 405–414.

McKinsey & Company, (2020). Unlocking Success in Digital Transformations.

Niven, P. R., & Lamorte, B. (2023). Thirty years with the balanced scorecard: What we have learned. Business Horizons, 66(1), 123–132

OECD, (2023). Digital government index 2023: Enhancing the resilience and responsiveness of governments. OECD Publishing. [https://doi.org/10.1787/3f21dff7-en] (https://doi.org/10.1787/3f21dff7-en)

Office of Government Commerce, (2009). ITIL Service Operation. TSO (The Stationery Office). OGC, 2007. Service Design - ITIL V3. The Stationery Office.

Proceedings of the 9th International Conference on Theory and Practice of Electronic Governance, ACM, pp. 251–254. Boar, B.H., (2001). The Art of Strategic Planning for Information Technology. Wiley.

Ross, J.W. & Weill, P., (2002). Six IT Decisions Your IT People Shouldn't Make. Harvard Business Review.

Ross, J.W., Weill, P. & Robertson, D.C., (2006). Enterprise Architecture as Strategy. Harvard Business Review Press.

Rouse, M., (2021). What is the ELK Stack? SearchITOperations. TechTarget.

Tamm, T., Seddon, P. B. et al., (2011). How Does Enterprise Architecture Add Value to Organisations? Communications of the Association for Information Systems.

United Nations, (2022). E-Government survey 2022: The future of digital government. United Nations. [https://publicadministration.un.org/egovkb] (https://publicadministration.un.org/egovkb)

van der Meulen, R. & Rivera, J., (2014). Key Metrics for Application Strategy. Gartner Research.

Van Grembergen, W. & De Haes, S., (2009). Enterprise Governance of Information Technology: Achieving Strategic Alignment and Value. Springer.

Verdegem, P. & Verleye, G., (2009). User-Centered E-Government in Practice: A Comprehensive Model for Measuring User Satisfaction. Government Information Quarterly, 26(3), pp. 487–497.

Weill, P. & Ross, J.W., (2004). IT Governance: How Top Performers Manage IT Decision Rights for Superior Results. Harvard Business School Press.

INVESTIGATING VISUAL VIDEO COMPONENTS IN THE CONTEXT OF HYPERSTIMULATION IN CHILDREN'S VIDEOS

Binh Minh Tran and Emanuel Santos
Concordia International School Hanoi
Van Tri Golf Estate, Kim Nỗ, Đông Anh, Hà Nội, Vietnam

ABSTRACT

In recent years, with the widespread use of the internet in the lives of children, which often provides useful material for parents and teachers to teach children, there is mounting concern from parents over the addictive nature of some children's media content on YouTube, calling them "hyperstimulating", particularly Cocomelon, the 2nd most viewed YouTube channel in the world with more than 196 billion views. However, research specifically in this area is scarce. This paper tries to analyse the visual aspects of these concerns and compare those from Cocomelon's videos, which received the main brunt of criticism of addiction, to those of other children's shows and YouTube content similar to Cocomelon's format. While this paper does not conclusively establish a link between the visual aspects to hyperstimulation, it does confirm that Cocomelon, compared to the 5 other sources of children's content, has a higher average value of scene changes and faster camera-pace.

KEYWORDS

Cocomelon, Addiction, Hyper-stimulation, Scene changes, Camera pace, Data Analysis

1. INTRODUCTION

In recent years, parents have become more concerned with the content their children consume on the internet. The YouTube channel Cocomelon, the second most viewed YouTube channel in the world and the most popular creator of children's content, has drawn much criticism; several parents have reported noticeable changes in their child's behaviour after watching its videos (Lloyd, 2023). News articles, as well as TikTok, followed to report on this issue, even going as far as to advise parents against showing their kids Cocomelon videos, calling these videos "over-stimulating" or "hyper-stimulating" (Leah Goulis, 2023). News articles bought several experts who claim Cocomelon's addictive nature is due to its fast-paced nature: the quick scene sequences, quick camera changes, bright colour usages, repetition, and upbeat music tunes (Leah Goulis, 2023) (Hopkins, 2023). Some even go further to suggest the addictive nature of Cocomelon could have psychological effects on children's brain development (Travers, 2024).

However, one problem that almost all of these above claims have is that they are subjective. The evidence of how the reported specific components of fast-pacedness and vibrant colours in Cocomelon's videos cause children's cognitive development, or even how prominent the specific components are in Cocomelon's videos is almost little to none. In a recent YouTube video that tries to investigate why Cocomelon is harmful to children, a majority of its evidence stems from blogs of concerned parents or the aforementioned articles that still lack backed-up research (SunnyV2, 2024). It is also important to note how all news articles that report on Cocomelon's addictiveness to children reference a source—the supposed start of concern for Cocomelon's addictiveness to children—that is now unavailable.

Furthermore, research on Cocomelon and children's media content on YouTube is mixed. Some experts disagree that the fast-pacedness of Cocomelon causes addiction (Hopkins, 2023) (Mayer & 2021, 2022). Research suggests that children's content is inherently addictive. For example, (Soni & Srivastava, 2022) found that children tend to look away less frequently when watching Cocomelon videos, indicating a strong engagement with the content. The primary challenge is determining what triggers the addiction or behavioural changes associated with watching Cocomelon videos.

This paper explores quantitative evidence that differentiates Cocomelon from other channels by examining metrics typically associated with addictive behaviour. This paper investigates the visual elements in children's media that are purported to contribute to hyper-stimulation, particularly in videos like Cocomelon's. Specifically, it examines two factors highlighted as key contributors: rapid scene transitions and swift camera movements (Capanna-Hodge, 2023). These metrics are compared between Cocomelon and other YouTube Kids content creators. The paper also compares these to the Disney Show Bluey, which is regarded as a preferable alternative to Cocomelon (Abonita, 2024).

2. DATA COLLECTION & TRANSFORMATION

In this study, YouTube videos from 5 children's content YouTube channels Cocomelon (Cocomelon, n.d.), Pinkfong (Pinkfong, n.d.), BabyChaCha (LiaChaCha, n.d.), SuperSimpleSongs (SuperSimpleSongs, n.d.), Bebefinn (Bebefinn, n.d.), and YouTube clips from Bluey (Bluey, n.d.) are selected. These channels were chosen due to their focus on musical and narrative content, as well as their substantial global viewership, collectively amassing approximately one billion views (Social Blade, n.d.).

2.1 Video Collection and Extraction

The top 20 videos of each channel in the popular category on YouTube were collected. The selection criteria required that only original videos, rather than compilations or remixes of other content, be included in the study. The Pytube library (Ghose et al., 2023) was used to download videos based on provided link URLs. For the visual data, the OpenCV library (Bradski, 2000) was extensively used to separate frames and analyse the colour values of each frame. OpenCV library also provides video data, specifically the FPS of the video. This data helps with converting the total number of frames a video has into the video's time length.

2.2 Scene Changes

This study defines a scene change as an abrupt change in the camera view or perspective of the video. Using the OpenCV library, downloaded videos are first separated into frames. Each frame is then divided into a 100 by 100 grid. An average value for each of the RGB colour values is extracted from each box within the grid. After collecting all RGB values of every frame from every video, the RGB difference between frames is used to find if a scene change exists between two compared frames. The difference between two RGB values is found by calculating their Euclidean distance. The differences between frames in each box from the 100 by 100 grid are averaged to find their average difference. A scene change is detected when the average difference between frames exceeds 70, a threshold determined through extensive video testing and finding the optimal RGB difference value that splits scene changes and non-scene changes while leaving additional room for variation without or limited detection of false positives, approximately only 0-2.

Additionally, the time between each scene is also recorded every time a scene change is detected. This is then averaged to find the average time interval of a scene in a video.

2.3 Camera Speed Data

Comparing frames over a time interval shows how still a frame is between scene changes or how still objects are on the screen. This idea is illustrated in Figures 1 and 2.

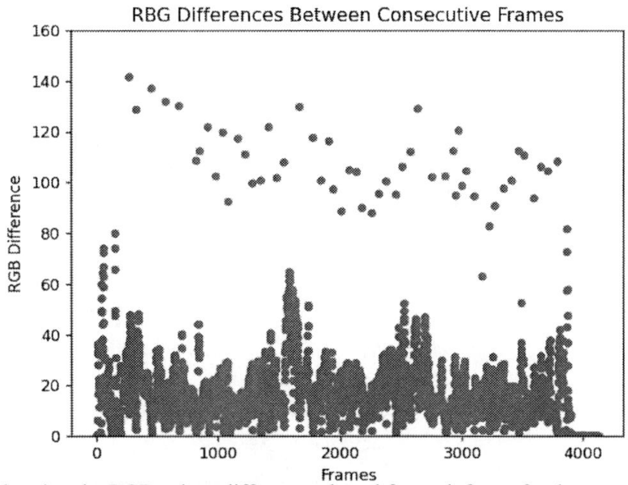

Figure 1. Graph showing the RGB values difference plotted for each frame for the sample Cocomelon video

Figure 1 shows the average RGB colour differences, represented on the left vertical axis "RGB Differences", between each frame of Cocomelon's "Bath Song" video (Cocomelon, 2018), represented on the bottom horizontal axis "Frames" with each frame representing a chronological numerical value. Using our previously determined threshold of 70 to determine if the transition between two frames is a scene change, we can see a difference between scene changes, the scattered data points over the top half of Figure 1, and continuous frames, the clustered data points at the bottom half of Figure 1.

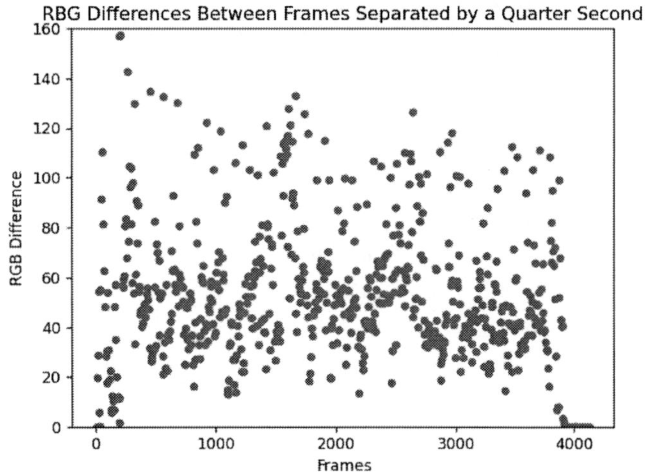

Figure 2. Graph showing the RGB value difference between frames separated by a quarter of a second interval for the sample Cocomelon video

Figure 2 shows the average RGB colour differences like Figure 1, but now between frames separated by a quarter of a second interval. While this results in fewer frame-data points, the data points are more scattered, and the difference between scene changes and continuous scenes is harder to indicate. Separating frames by time intervals shows two potential characteristics of the video—a fast-moving camera perspective or many fast-moving objects—both of which indicate fast-pacedness. Importantly, using our threshold of 70 for the detection of scene change would result in a higher value of scene change false positives. To quantify these, both the standard deviation and scene changes per minute are used to show the variance across different time intervals—of no time interval, a quarter of a second, half a second, a second, and two seconds which will be later represented as all_frames, half_half_second, half_second, per_second, and two_seconds—again capturing how fast-paced the videos are.

3. DATA ANALYSIS

The following visualisations were done in Tableau. Some of the graphs are also generated using the Matplotlib library (Hunter, 2007).

The data analysis is divided into 3 main areas: the first section analyses the frequency of scene changes detected every minute, the second section discusses the time length of a scene, and the third section focuses on the camera pace and object's stillness of a scene.

3.1 Scene Changes Frequency

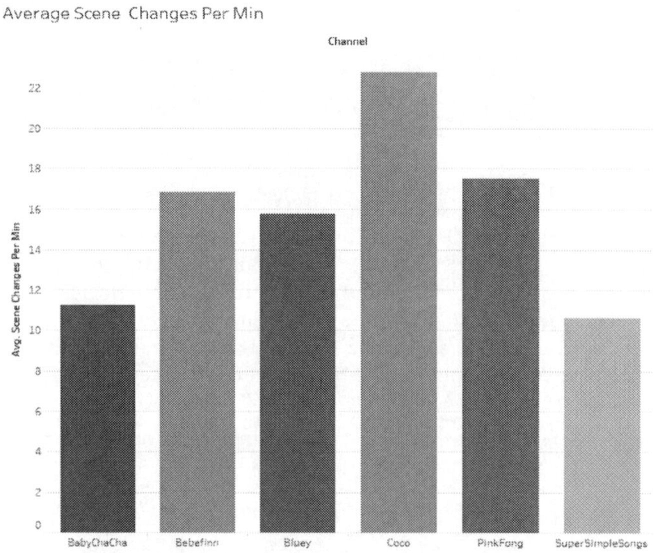

Figure 3. Graph showing the average scene changes per minute of the six selected sources of children's content

Figure 3 shows the average scene changes per minute of videos from each creator. Figure 3 shows a clear sign that Cocomelon has the highest amount of scene changes per minute. This confirms previous statements about Cocomelon's higher frequency of scene changes: compared to 5 other sources of children's media content creators, Cocomelon is the highest in the number of scene changes, around 23 scenes per minute. In a study on children viewing fast-paced television content, researchers observed that frequent scene-switching generates a substantial cognitive load for the developing brain, potentially influencing its functioning (Lillard & Peterson, 2011).

Although Bluey does not exhibit the highest frequency of scene changes, it aligns with Bebefinn and Pinkfong, the creator of the "Baby Shark" music video, in its average rate. Furthermore, Bluey demonstrates a similar scene-change frequency to Bebefinn, possibly owing to Bebefinn's inclusion of narrative-driven content rather than exclusively musical segments. Consequently, Bebefinn provides a more direct basis for comparison with Bluey than other children's content producers that primarily focus on music videos.

SuperSimpleSongs has a special case of a video with no scene changes detected. This also matches the channel's videos, as the channel is more focused on creating music than on animations and scene transitions. Many videos are still frames with only the characters or letters moving.

3.2 Time Length of a Scene

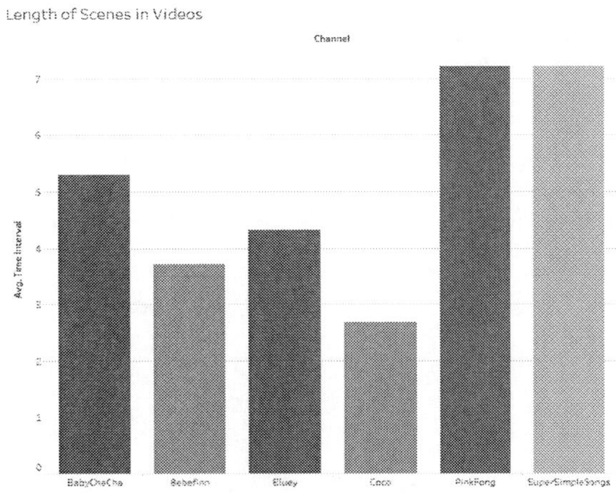

Figure 4. Graph showing the average time length of a scene in videos of the six selected sources of children's content

Figure 4 shows an inverse relationship between the length of scenes, represented on the vertical axis of Figure 4, and the number of scenes per minute in Figure 3. Cocomelon, which exhibits the highest number of scene changes per minute from Figure 3, maintains an average scene length of 2.6 seconds. SuperSimpleSongs demonstrates the longest time interval of a scene of around 7 seconds, which goes along with SuperSimpleSongs' low frequency of scene changes per minute from Figure 3. However, these observations may be influenced by variations in total video duration. For instance, Pinkfong, ranking second in scene changes, features one of the longest average scene lengths. Nonetheless, for a majority of the channels' collected video data, an inverse relationship is shown between the average time of scenes and the number of scene changes per minute. The shortness of time length shown in Cocomelon provides evidence to claims of its scenes being shorter compared to its contemporaries.

3.3 Camera Pace and Object's Stillness in a Scene

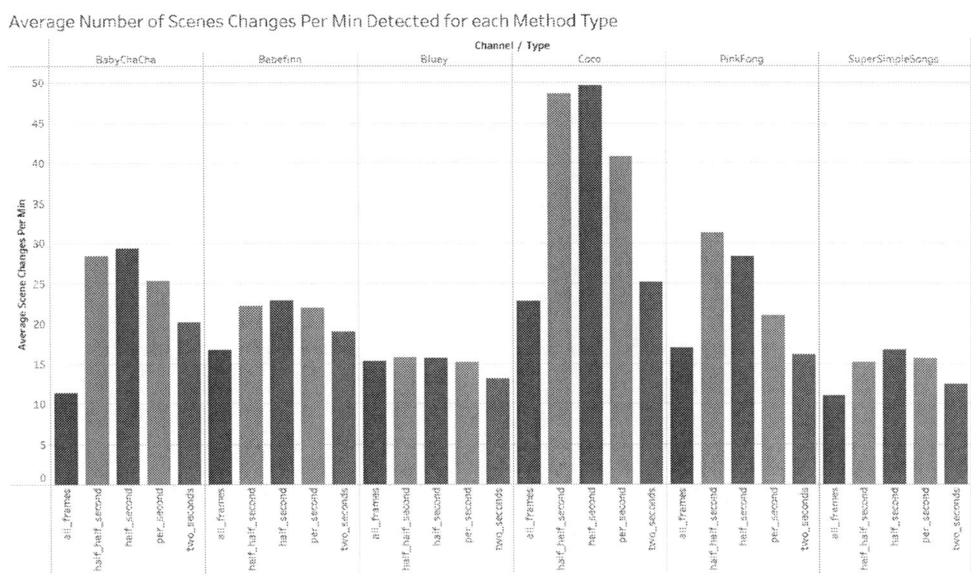

Figure 5. Graph showing the average number of scene changes per minute on the y-axis based on time intervals

Figure 5 shows a clear increase in the number of scenes detected for Cocomelon when it is switched from all_frames to half_half_second. This represents the previous two example graphs of Figures 1 and 2, which show that after separating the frames over a time interval, RGB difference data points become more scattered, increasing false positives of scene change detection. In the case of Cocomelon's videos, this suggests that the camera perspective or the objects within the scene are moving rapidly, leading to an increase in the number of scene changes per minute from 22 to 48 when comparing the all_frames analysis with the half_half_second method. This phenomenon can also be seen in channels like BabyChaCha and Pinkfong where a substantial jump of at least twice the original figure in all_frames. This is in contrast to Bebefinn, Bluey, and SuperSimpleSongs where there is only a small increase. Bluey saw the least amount of change, maintaining the figure at around 16. This means in one consecutive scene of a Bluey video, camera perspectives and objects move slowly and remain relatively still.

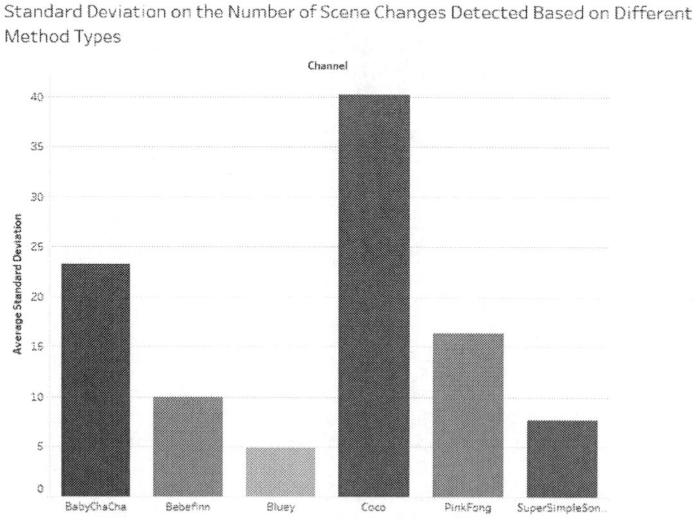

Figure 6. Graph showing the standard deviation on the number of scene changes detected based on the different time intervals

Figure 6 shows the average standard deviation of the number of scene changes per minute based on the different time intervals of each video from each creator of children's content. In Figure 6, Cocomelon has the highest standard deviation out of all other creators of children's content, a standard deviation value that is almost twice as large as BabyChaCha's and three times as large as Pinkfong's. While Figure 5's figures may be influenced by the base frequency of scene changes from each video, Figure 6 shows a clearer picture that the spread in the number of scene changes detected based on different time intervals was much bigger for Cocomelon compared to BabyChaCha and Pinkfong, which the author analysed in Figure 5 to show almost similar fast-pacedness to Cocomelon. This shows that, compared to other creators of children's media, Cocomelon's videos are faster-paced through camera and object movement within a scene.

4. DISCUSSION ON VISUAL CAUSES OF HYPERSTIMULATION IN CHILDREN'S MEDIA

The analysis shows a general pattern of Cocomelon having a higher number of scene changes in a video and faster camera movements or moving objects in a frame than the selected children's content on average. This aligns with Soni & Srivastava's (2022) paper, which found Cocomelon's scenes to last by 2 seconds, which matches close to this paper's figure of 2.6 seconds.

While this paper does not prove a link between these video components causing behavioural changes in children, previous studies found that fast-paced content had affected children's cognitive behaviours. In Lillard and Peterson's (2011) experiment, by watching fast-paced television, the children's executive function became temporarily impaired. In Lillard et al. 's (2015) experiment, they also found that scenes that are too short can overload children's brains, lowering their executive function as they are continually introduced to new content for each scene. In a recent literature review, the paper states that fast-paced content could disrupt the development of children's brains and cautions parents over their child's screen time and usage of online children's content (Swider-Cios et al., 2023). The findings of previous research (Soni & Srivastava, 2022)(Lillard & Peterson, 2011)(Swider-Cios et al., 2023) along with this paper could explain the various concerns raised by parents from their children's changed behaviours after watching Cocomelon that were documented from the YouTube investigation (SunnyV2, 2024).

However, formal research into this particular topic is still limited, as experts pointed out the need for a more conclusive link between fast-paced content and behavioural changes in children (Hopkins, 2023) (Mayer & 2021, 2022). Not all experts assert that scene changes and video pace are the reason for the hyperstimulation of children's videos. Other experts claim it to be the fantasy-like setting that Cocomelon creates that causes hyperstimulation (Geddes, 2019). However, this pushes the question of how fantasy-like a video has to be to become addictive. All of the children's shows that the authors selected for this study are fantasy-like cartoons: if Bluey is a world in which animals act and talk like human beings, what difference does Bluey have to Cocomelon?

Similarly, the definition of "fast-paced" and visual video components should also be posed in similar regards: how many scene changes are too much, how fast should the camera and objects move within a scene before it is too addictive, and how short can scenes be until it overloads a child's executive function? In previous studies, there is no formal definition or criteria for "fast-paced" content. It seems the label of "fast-paced" is based on the author's perception at that time. In the Lillard and Peterson (2011) experiment, they argued that the 1977 experiment, which found no correlation between fast content and executive functions, could be different as the 1977 experiment used the episodes from Sesame Street, in contrast to the 2011 experiment, which used SpongeBob SquarePants.

More research is necessary into the criteria of what makes a video too fast-paced, too fantasy-like, or of any component for it to be a cause of addiction in children's media.

5. CONCLUSION AND FUTURE WORK

The findings from this paper prove claims made from the media reporting on Cocomelon that comparatively it has more scene changes, its scenes are shorter, and it is more fast-paced than other shows. While the question of what criteria would qualify a video as being "fast-paced" is yet to be understood, there is a consensus of criticisms, both among experts and the media, against Cocomelon's content, labelling it as "fast-paced". This consensus and the findings from this paper show that Cocomelon videos are different from their contemporaries, such as in scene changes and pace, which this paper studies to be true. Nonetheless, the link between addictiveness and the studied visual components, which had been claimed to cause addiction or behavioural changes, or any other reason, needs to be studied further before making any official claim over any content being "too fast-paced" and causing harm to children.

This paper does not have an adequate metric to measure the viewer's attention or addiction to a video. Therefore, future steps would include making a clear definition for "fast-paced" content and finding if there is a correlation between "fast-paced" related video components to addiction and by how much.

Additionally, while this paper had only focused on the visual aspects, research should also be conducted on the audio aspects of children's content, if there are noticeable qualities between the music used in each children's video and how they relate to the children's videos' addictiveness.

REFERENCES

Abonita, L. (2024, June 14). *Is Bluey better or Cocomelon?* Sandilake Clothing. https://www.sandilakeclothing.com/blogs/sandilake/is-bluey-better-or-cocmelon

Bebefinn. (n.d.). *Bebefinn - Nursery Rhymes & Kids Songs - YouTube.* Www.youtube.com. https://www.youtube.com/channel/UC-cY4X2sLECUhishGNiyofQ

Bluey. (n.d.). *Bluey - Official Channel - YouTube.* www.youtube.com. https://www.youtube.com/@BlueyOfficialChannel

Bradski, G. (2000). *Bradski, G. (2000) The OpenCV Library. Dr. Dobb's Journal of Software Tools, 120; 122-125. - References - Scientific Research Publishing.* www.scirp.org. https://www.scirp.org/reference/ReferencesPapers?ReferenceID=1692176

Capanna-Hodge, D. R. (2023, March 21). Screen Time, Autism and CocoMelon. Medium. https://medium.com/@drroseanncapannahodge/screen-time-autism-and-cocomelon-566c7740d822

Cocomelon. (n.d.). *Cocomelon - Nursery Rhymes - YouTube.* www.youtube.com. https://www.youtube.com/channel/UCbCmjCuTUZos6Inko4u57UQ?feature=gws_kp_artist&feature=gws_kp_artist

Cocomelon. (2018). Yes Yes Playground Song | CoComelon Nursery Rhymes & Kids Songs. In www.youtube.com. https://www.youtube.com/watch?v=KTOwPz-zMWY

Geddes, L. (2019, December 9). *What's so fascinating about weird children's TV shows?* Www.bbc.com. https://www.bbc.com/future/article/20191206-why-children-find-weird-television-so-mesmerising

Ghose, R., Dahlin, T. F., & Ficano, N. (2023, May 8). *pytube 15.0.0: Python 3 Library for Downloading YouTube Videos.* PyPI. https://pypi.org/project/pytube/

Goulis, L. (2023, March). TikToker reveals the worrying reason your kids are addicted to CoComelon. New York Post. https://nypost.com/2023/03/01/worrying-reason-your-kid-cant-stop-watching-cocomelon

Hopkins, A. (2023, March 13). *Is "CoComelon" harmful to children? DC pediatric psychologist weighs in.* WJLA. https://wjla.com/news/local/cocomelon-controversy-speech-delays-behavioral-issues-harmless-noise-emotions-facial-expressions-parents-netflix-youtube-tv-show-cakids-children-sesame-street-pediatric-mental-health-kids-screen-time

Hunter, J. D. (2007). Matplotlib: A 2D Graphics Environment. *Computing in Science & Engineering*, *9*(3), 90–95.

LiaChaCha. (n.d.). *LiaChaCha - Nursery Rhymes & Baby Songs.* YouTube. https://www.youtube.com/channel/UC1DY-gFnouXe065NMlLdvbA

Lillard, A. S., Li, H., & Boguszewski, K. (2015). Television and Children's Executive Function. Advances in Child Development and Behavior, 48, 219–248. https://doi.org/10.1016/bs.acdb.2014.11.006

Lillard, A. S., & Peterson, J. (2011). The Immediate Impact of Different Types of Television on Young Children's Executive Function. *PEDIATRICS*, *128*(4), 644–649. https://doi.org/10.1542/peds.2010-1919

Lloyd, S. (2023, April 29). *"CoComelon Made My Kid a Zombie."* Newsweek. https://www.newsweek.com/cocomelon-kids-zombie-addiction-adhd-speech-delay-autism-1792511

Mayer, B. A., & 2021. (2022, December 12). Some Think "CoComelon" is Too Stimulating for Their Kids. We Asked an Expert To Weigh In (M. Felder, Ed.). Parents. https://www.parents.com/news/some-think-cocomelon-is-too-stimulating-for-their-kids-we-asked-an-expert-to-weigh-insome-think-cocomelon-is-too-stimulating-for-their-kids-we-asked-an-expert-to-weigh-in/

Pinkfong. (n.d.). *Pinkfong Baby Shark - Kids' Songs & Stories - YouTube.* Www.youtube.com. https://www.youtube.com/channel/UCcdwLMPsaU2ezNSJU1nFoBQ

Social Blade. (n.d.). Track YouTube analytics, future predictions, & live subscriber counts - Social Blade. Socialblade.com. https://socialblade.com/youtube/

Soni, A., & Srivastava, S. (2022). Growing Cartoon Trends Pose as Hyper-stimulant to Adolescents that Cultivates Addiction, Impair Creativity, and Inhibits Arithmetic Efficiency. *PsyArXiv*, 5–7. https://doi.org/10.31234/osf.io/zpfs5

SunnyV2. (2024, September 19). *Cocomelon - The Most Evil Channel On YouTube.* YouTube. https://www.youtube.com/watch?v=YEFptHp0AmM

SuperSimpleSongs. (n.d.). *Super Simple Songs - Kids Songs - YouTube.* Www.youtube.com. https://www.youtube.com/@SuperSimpleSongs

Swider-Cios, E., Vermeij, A., & Sitskoorn, M. M. (2023). Young children and screen-based media: The impact on cognitive and socioemotional development and the importance of parental mediation. *Cognitive Development*, *66*(66), 101319. https://doi.org/10.1016/j.cogdev.2023.101319

Travers, M. (2024, May 17). *Why Kids' Shows Like "Cocomelon" Hamper Critical Brain Development.* Forbes. https://www.forbes.com/sites/traversmark/2024/05/17/why-kids-shows-like-cocomelon-hamper-critical-brain-development/?sh=21bcc70e536b

INTELLIGENT SYSTEMS PLATFORM ENHANCED BY DIGITAL TWINS AND GENERATIVE AI TOWARDS TO ON-INDUSTRY EXPERIMENTATION

João Pedro Fusco
University of Marilia
Dolores Guerreiro de Oliveira Street, 361, Marilia, São Paulo, Brazil

ABSTRACT

The article addresses the synergistic relationship between generative artificial intelligent and digital twins, exploring how their combined capabilities and applied in intelligent systems can revolutionize industrial processes by enabling On-industry Experimentation. Digital twins, virtual representations of real-world assets, have already proven invaluable for simulation and optimization. By integrating with generative AI, these applications become even more autonomous and user-friendly. The article highlights how generative AI can improve digital twins in several ways. It can generate large amounts of synthetic data to train machine learning models, improving the accuracy of predictions and simulations. Additionally, generative AI can be used to create new design variations or optimize existing ones in the digital twin environment. This enables rapid prototyping and experimentation, speeding up product development cycles. By combining the power of generative AI with the fidelity of digital twins, industries can achieve unprecedented levels of automation, efficiency and innovation. This article also introduces the concept of On-industry Experimentation as new approaches to industry research and innovation that are embedded in manufacturing management and reflects new demands for decentralized and inclusive research that bridges sources of knowledge and fosters open innovation. The article ends by presenting and discussing a conceptual computing platform for intelligent systems in On-industry Experimentation. This computational architecture outlines the key characteristics and definitions of the proposed model, based on methodologies and computational technologies that integrate generative artificial intelligence and digital twins potential applications and challenges in this emerging field, providing valuable information for researchers and practitioners.

KEYWORDS

Intelligent Systems, Digital Twin, Generative Artificial Intelligence, On-industry Experimentation, Industry 4.0

1. INTRODUCTION

The reconfiguration of interactions between industry professionals in manufacturing and researchers, coupled with the need to address complexity and uncertainty through collaborative investigation of manufacturing challenges, represents one of the main hurdles in implementing open innovation.

In this context, intelligent systems play a central role, as, through cyber-physical technologies, they enable a technological infrastructure capable of supporting on-industry experimentation (OIE). This approach allows for direct experimentation in production environments, facilitating dynamic adaptation to complex industrial scenarios and promoting a continuous cycle of innovation and process improvement.

OIE describes new approaches to research and innovation in industry, which are incorporated into real-world manufacturing management and reflects new demands for decentralized and collaborative research that leads to the conduct of practical tests and research directly in the operational environment, promoting open innovation in real-time environments.

This scenario, which addresses complexity and uncertainty in industrial manufacturing, is made possible by intelligent systems that explore the integration of digital twins (DT) and generative artificial intelligence (GAI) to optimize processes and support agile development cycles. The exploration of the potential of integrating computational technologies into implementing intelligent systems based on DT with advanced GAI techniques for experimentation and open innovation in industrial manufacturing still lacks investigation and studies that demonstrate its technical and sustainable feasibility.

By presenting the OIE concept as a manifestation of collaborative experimental research applied directly in the production environment, the aim is to explore the integration of GAI in scenarios of virtual replicas of assets and processes; analyze potential application scenarios and methodologies; and propose a computational architecture with virtualized representations of the industrial environment for real-time sensing, monitoring, and simulation.

The literature shows that digital twins have established themselves as an essential technology in the context of Industry 4.0, providing virtual replicas of physical systems capable of simulating, monitoring, and predicting the behavior of assets, processes, and systems in real-time. Since its inception, this technology has evolved from a tool focused on product lifecycle management to integrated platforms that incorporate machine learning and advanced simulations. These capabilities broaden its application potential, enabling process optimization, data-driven decision-making, and the prediction of future scenarios, making it a strategic resource for the digital transformation of manufacturing (Huang et al., 2024).

The synergy between GAI and DT has driven operational efficiency and innovation in sectors such as manufacturing, healthcare, and smart cities (Huang et al., 2024; Korada, 2024). This integration allows DTs not only to reflect the current state of physical systems but also to anticipate future scenarios through generative algorithms capable of processing large volumes of data in real-time. For example, deep learning techniques based on GAI have been used for the creation of synthetic data and modeling of critical scenarios in manufacturing environments (Huang et al., 2024; Ding, 2024).

Faced with this scenario, the need for a new paradigm of collaborative and applied experimentation emerges, which transcends traditional models of simulation and technological validation. As demonstrated in Lacoste et al. (2022), which explore this concept in agribusiness, on-industry experimentation is presented as an innovative approach for conducting experiments directly in the industrial environment.

In this article, a conceptual computing platform of intelligent systems for experimentation in industry is presented. This computational architecture details the main characteristics and definitions of the proposed model, based on methodologies and computational technologies that integrate generative artificial intelligence and digital twins. The proposal aims to establish a solid foundation for the development of intelligent systems capable of supporting collaborative experimentation in industry, providing an innovative and dynamic environment for process optimization and real-time decision-making.

2. GENERATIVE ARTIFICIAL INTELLIGENCE AND DIGITAL TWINS APPLIED IN MANUFACTURING ENVIRONMENTS

Generative artificial intelligence models (GAIM) can generate text, images, and video responses similar to human responses in response to requests based on information generated from large datasets (Makatura et al., 2023).

From the perspective of building intelligent systems, GAIM share similar functionalities with knowledge management systems, in relation to the processes of acquisition, processing, interpretation, consultation, and generation of knowledge (Li et al., 2011). GAI models can be used in the design and manufacturing process, due to their skills in creating and contributing to an iterative process in process development.

However, limitations are observed in quantitative processing, accuracy, and verification of the generated knowledge. (Wang et al., 2023) demonstrates that the performance of intelligent systems based on GAIM, applied in the manufacturing domain, shows superior results in providing information, generating coherent and structured content, and proposing initial solutions, but when delving into questions in a scenario of critical analysis and intrinsic details within this manufacturing domain, the answers tend to be unreliable with a lack of traceability and verifiability, representing an opportunity for investigation and improvement of this process.

In this context, DTs, defined as an integrated simulation strategy of an object, system, or process that mirrors the existence of physical models, historical information, and real-time data (Liu et al., 2023), enabled an effort to improve the process of generating solutions in the manufacturing scenario through the integration of GAIM with DT technology to evaluate and verify the solutions generated, thus increasing the reliability and explainability of intelligent systems (Xu et al., 2023).

The application of GAIM in digital twin scenarios aimed at intelligent systems can mitigate this challenge, as it allows the generation of solutions based on machine learning algorithms trained with large volumes of data. In the context of DTs, this approach enables the construction of more accurate and adaptable representations of physical systems, enhancing simulation and prediction capabilities.

Thus, the integration between GAI and DT is presented as a promising solution in the convergence between the physical and digital domains, characterizing cyber-physical systems. This approach enables the creation of highly faithful representations of the production process, allowing detailed analysis of the behavior of industrial assets, the generation of more accurate data, and the optimization of process configurations, resulting in greater efficiency and adaptability in manufacturing.

3. INTELLIGENT SYSTEMS SUPPORTING ON-INDUSTRY EXPERIMENTATION

In the context of the current industrial revolution, manufacturing companies face significant challenges, such as the increasing complexity of production processes, the need for greater flexibility in human work, and the urgency for sustainable production practices. These demands impose obstacles on the industrial sector, since traditional processes have limitations in dealing with uncertainties, managing operational complexity, and efficiently utilizing the large amount of data generated throughout the production chain (Liang et al., 2015).

Applications of intelligent systems based on DT and GAI in the context of cyber-physical environments are reviewed and discussed as an effort to follow in supporting the current revolution in the manufacturing industry (Mad Yusoh et al., 2021). Intelligent systems, in this context, include the applications of DT and GAI techniques in production fault recognition (Yijun et al., 2024), simulating complex scenarios and anticipating future conditions, proposing predictive solutions, simulating events, and providing actionable insights with agents based on large language models (LLM) for decision-making (Stergiou & Psannis, 2022).

The incorporation of artificial intelligence techniques enables the development of advanced models for decision-making, prediction, and optimization, resulting in more efficient processes and more consistent decisions. In addition, the use of hybrid approaches, combining fuzzy logic, case-based reasoning, and neural networks, allows for improving the prediction of scenarios, increasing the quality and reliability of products, as well as the operational efficiency of production systems (Mad Yusoh et al., 2021).

In the practice of continuous experimentation, which is based on the idea of multiple tests that result in the deployment of a system or product with the possibility of always running one or more different versions of the solution to evaluate their performance (Gaimo et al., 2020), with the aim of improving the cyber-physical system in the long term through a series of incremental improvements validated in the field of use (Mattos et al., 2018), intelligent systems present the necessary conditions to enable autonomous agents based on large sets of historical and real-time data, generated from sensing environments of physical objects replicated in digital twins.

As an evolution of continuous experimentation, innovation processes are urgently needed in manufacturing to face the technological challenges of Industry 4.0. The need to broaden the concept of open innovation is considered using intelligent systems that allow internal and external stakeholders to collaboratively participate in research, development, and innovation (RD&I) processes through virtualized environments of manufacturing systems with access to real-time data and with the possibility of proposing customized solutions that can be tested in real environments.

3.1 On-Industry Experimentation

The concept of on-industry experimentation is presented as a new manifestation of collaborative experimental research. At its core is a potential global community that makes it possible to build productive relationships between industry professionals, R&D, scientists, and startups to develop the innovation paths needed to solve the challenges of the manufacturing industry.

OIE is the result of accumulated improvements in several domains that collaboratively realize a transformational change in recognizing the power of collective intelligence, that is, industries mainly depend on their internal R&D departments to conceive and develop new products and bring them to market, however, open innovation emphasizes collaboration and partnerships with researchers, science and technology institutions, and startups, leveraging external knowledge and contributions.

In the context of OIE, open innovation adopts a more inclusive and integrated approach to the investigation and experimentation environment, actively seeking collaborative approaches to the use of real and real-time data and the application of solutions in the manufacturing production environment through virtual representations of the real scenario. Frequently, this change is catalyzed by the analytical, learning, and autonomous decision support opportunities presented by digital technologies such as digital twins and artificial intelligence applied in cyber-physical systems.

OIE allows RD&I to occur directly in production processes and at significant scales, rather than in minimized representations of the real environment that are designed externally. The interests of industry and external OIE actors are explicitly recognized as a prerequisite for negotiating their alignment and building productive collaborative relationships.

Researchers and other interested parties add value to the experimental process by providing specialized skills and external perspectives to help local RD&I evaluate ideas in their real scenarios, where the empirical knowledge and experiential learning of industry professionals are complemented by the suggestion of metrics and experimental designs, performing analyses and documenting experiences, interpreting results and expanding horizons, proposing opportunities and next steps in the open innovation process.

3.2 Digital Twins

OIE is enabled by intelligent systems that are based on the idea of digital twin scenarios through which we can operate a system in cloud computing as a virtual representation for management and analysis of industrial databases based on IoT (Internet of Things) produced through sensing and industrial procedures (Stergiou & Psannis, 2022; Zhang et al., 2023).

The application of DT for the development of intelligent systems in the manufacturing process in OIE can be applied in i) status monitoring in which digital replicas of physical assets are used in which machines are continuously monitored using IoT and sensing (Yijun et al., 2024); ii) simulation of the virtualization of machines, processes, and products that are created to represent real configurations. Simulation allows for the joint proposition of design, development, and testing of new products and processes using their digital representation before applying to real physical assets (Wang & Chen, 2021); and iii) visualization of physical assets, from dashboards and data analytics with real-time data and alert systems to monitor and debug an operational problem.

In the context of intelligent systems that use cyber-physical scenarios to support the deployment of OIE, DTs offer access and communication, analysis, and intelligence capabilities (Yijun et al., 2024). The access and communication layer are responsible for interacting with production processes and gaining access to data through sensors and open interfaces on the status of a physical asset to update its representation in the DT.

In intelligent systems, DTs operate as cyber-physical agents capable of accessing, analyzing, and interpreting the current state of their physical counterpart, enhanced using GAI. Faced with irregularities or opportunities for continuous improvement and innovation, the DT can interact directly with the operational environment, performing virtual simulations and executing processes adaptively. This capability allows for the identification of optimized solutions before their implementation in the real environment, reducing risks, improving production efficiency, and driving innovation in industrial systems.

3.3 Generative Artificial Intelligence Integrated with Digital Twins

Generative artificial intelligence as a computational technology that can generate content, solutions, and insights from a large dataset, which goes beyond simply analyzing or acting on existing data in intelligent systems, with the aim of assisting or replacing humans in creating diverse, personalized, and scalable content more quickly and at a lower cost (Brown et al., 2021), has the potential to optimize and maximize the results of intelligent systems based on digital twins in the context of OIE.

Similarly, based on datasets generated by DT systems from production environments, real-time perception and acquisition of static data, produced by a network of sensors in IoT environments, advanced GAI technologies enable the generation of diverse content and solutions in predicting failures, mitigating operational problems, and proposing improvements in the production process, while intensifying the validation of collaborative experimentation from dialogues in open innovation processes, the basis for the adoption of the OIE concept.

The application of GAI in DT in the industrial sector in OIE scenarios represents a fundamental evolution for the creation of intelligent systems that extend reflections of their physical representations with predictive and adaptive functionalities (Li et al., 2022). Advanced connectivity and pre-training techniques for personalized models highlight a coordinated evolution towards a deeper and more meaningful integration. This integration ensures that digital twins can leverage the full spectrum of generative AI potential, from improving real-time decision-making to facilitating long-term strategic planning (Zhou & Wang, 2023).

The integration of GAI in DT marks a fundamental advance in the construction of intelligent systems that enable the implementation of on-industry experimentation processes. In an open manufacturing innovation environment, intelligent systems can be improved by allowing users to interact in natural language, generating high-quality knowledge in the context of planning or operating manufacturing processes (Chen et al., 2024).

4. CONCEPTUAL PLATFORM OF INTELLIGENT SYSTEMS

This section presents the characteristics and definition of the conceptual platform based on methodologies and computational technologies of generative artificial intelligence integrated with digital twins, a platform for building intelligent systems is proposed to support an on-industry experimentation process.

The conceptual platform is a cyber-physical system that, through an intelligent system that integrates DT technologies and GAI models, creates a collaborative experimentation environment in industrial production processes.

From access to the status of the production environment in real-time and the generation of content and insights based on large datasets, together with virtualization, simulation, and real-time verification, the intelligent system allows for integration between actors in the industrial innovation process. This integration enables investigation and experimentation enabled by data management, human-computer interaction, process knowledge from a large volume of historical and real-time data, verification and optimization of feedback, and generation of insights and contributions to innovation. Its main objective is to provide an open innovation model in a real environment and optimize the productivity and quality improvement of production processes.

From this definition, the architecture of the platform is presented as shown in Figure 1, comprising functional modules, namely i) physical layer; ii) control layer; iii) digital twin layer; iv) generative artificial intelligence layer; and v) user layer.

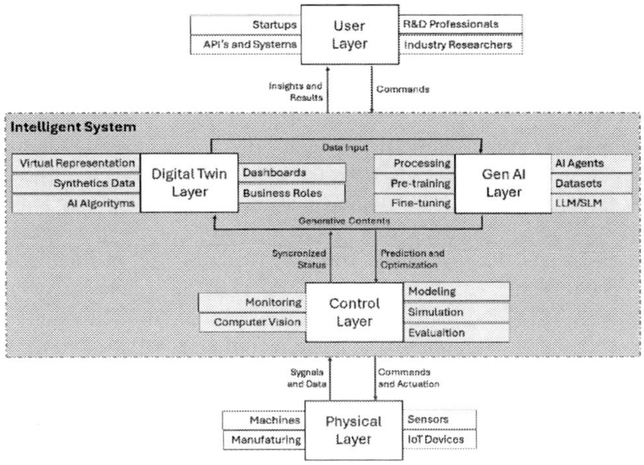

Figure 1. On-Industry Experimentation-enabling Intelligent System Conceptual Platform

4.1 Functional Modules of the Platform

The physical layer is an essential module in a cyber-physical manufacturing system, representing the physical assets of the production environment as tangible elements that can be monitored, controlled, or influenced through information and actions in the digital world. These assets may include machines, equipment, production processes, products, among others.

In the context of an intelligent system enabled by DT, the physical layer is equipped with IoT devices that sense and continuously monitor the state and performance of these assets. These sensors collect real-time data and signals, such as temperature, vibration, pressure, and other variables critical to the production process.

The data generated by IoT devices is transmitted to an intelligent system, where it will be digitally represented for i) DT construction; and ii) composing a large dataset for GAI technologies.

The control layer acts as a logical module in the intelligent system, performing the monitoring, modeling, and processing of data and signals obtained in real-time from the physical layer, complements the data acquisition of physical assets through computer vision systems that use cameras, sensors, and image processing algorithms to visualize and interpret the real world, allowing to expand the synchronized status of the physical asset through the informational representation of the asset based on visual information.

The control layer also acts in the intervention of the physical asset from instructions and commands generated by the integration of DT and GAI that allow for prediction, simulation, optimization, and evaluation in the execution of commands in the physical layer.

The DT layer is defined as a complex and interconnected module that emerges as a crucial element in the architecture of the proposed intelligent system. Integrated with the generative artificial intelligence layer, it articulates between the physical environment and the digital environment, creating the interface between system expert users and the physical production environment.

The synthetic representation of the physical world can be made by three-dimensional models enabling immersive visualization that expands the understanding of the real environment. Using virtual reality and augmented reality technologies, users visualize and interact with physical assets intuitively and immersively, as if they were physically present on site.

In addition to the sensed data of the physical asset, this layer generates synthetic data to simulate scenarios, test hypotheses, and predict the behavior of the physical asset in different operating conditions from generative data of the GAI layer to automatically generate improvement recommendations, optimize processes, diagnose problems, and even design new products or processes, presented through intelligent dashboards that provide insights into asset performance, identify bottlenecks, alert to possible failures and anomalies, assist in decision-making, support predictive maintenance and production optimization, and allow for simulating the impact of these insights in the virtual environment before being applied to the physical layer.

The GAI layer acts as an engine of creation and optimization, driving strategic decision-making and continuous experimentation. This process happens because GAI is capable of processing large volumes of data, both historical and real-time, coming from DT and the physical layer. This capability allows for identifying patterns, trends, and anomalies that would be difficult to detect by other means. The generative AI layer uses pre-training and fine-tuning techniques of machine learning models, such as LLM and SLM (Small Language Models), to continuously adapt to changes in the industrial environment and improve its ability to generate relevant and accurate content.

The GAI layer can be integrated with intelligent agents that act as interfaces between the digital and physical worlds. These agents can receive commands from the GAI, execute tasks in the physical layer, and provide feedback to the DT, creating a cycle of learning and continuous improvement. This layer can generate creative and relevant content for various purposes, such as 3D models of products, process simulations, analysis reports, improvement recommendations, and even new solutions to industrial problems. It can also generate virtual models of industrial processes based on historical and real-time data, allowing for simulating different scenarios and optimizing process performance even before they are implemented in the physical world.

The GAI layer, when applied to an intelligent system composed of a DT in the context of a cyber-physical environment in the manufacturing industry, represents a significant advance in the search for more efficient, innovative, and adaptable processes. By enabling modeling, simulation, evaluation, and innovation of industrial processes, GAI empowers companies to make more strategic decisions and reach new levels of manufacturing excellence.

In an advanced industrial ecosystem with the characteristics of collaborative experimentation for open innovation, the user layer serves as the crucial interface that allows human interaction with a complex intelligent system, composed of DT, GAI in a cyber-physical environment. This layer, composed of startups, APIs, RD&I professionals, and industry researchers, plays a fundamental role in the validation, homologation, and collaborative implementation of strategic decisions.

The user layer offers intuitive and accessible interfaces, allowing various stakeholders to interact with the intelligent system. They can simulate complex scenarios in the virtualized environment of the DT, testing different hypotheses and validating the effectiveness of new strategies. The user layer allows for the homologation of decision-making in a safe and controlled environment, before being applied in the physical layer.

Based on the insights and results obtained, users can generate commands for the GAI layer and for the DT layer. These commands may include requests to generate new models, simulate processes, optimize parameters, or even create new products.

The user layer acts as an innovation laboratory, where startups, researchers, and industry professionals can collaborate to develop new solutions and improve existing processes. User feedback is fundamental for the continuous improvement of the intelligent system, ensuring that it meets the needs of the industry.

5. CONCLUSION

The intersection between generative artificial intelligence and digital twins represents a significant milestone in the evolution of Industry 4.0, potentiating new forms of experimentation and innovation within production environments. This article explores how the integration of these technologies can enable the emergence of intelligent systems capable of supporting decision-making, optimizing processes, and creating highly faithful virtual environments for simulation, prediction, and validation of industrial solutions.

The on-industry experimentation approach introduces a new paradigm in industrial research and development, promoting collaborative and applied research experience directly in production processes. Using DT and GAI, OIE enables the creation of digital environments that replicate in real-time the operations of the physical world, enabling a two-way interaction where solutions can be simulated and refined before their effective implementation. This approach not only reduces costs and minimizes risks but also accelerates innovation and industrial improvement cycles.

The challenges found in the application of GAI and DT in industrial environments highlight the need to improve aspects such as scalability, reliability, and transparency of generative models. The massive volume of data generated by sensors and IoT devices requires robust computational infrastructure and advanced machine learning algorithms to ensure the accuracy of simulations and the validity of predictions. Furthermore, the incorporation of GAI into DT systems should consider strategies to mitigate algorithmic biases, ensuring that the decisions taken are explainable, traceable, and reliable.

Another fundamental aspect for the advancement of OIE lies in the interaction between different stakeholders, including researchers, industry professionals, and startups. Collective intelligence emerges as an essential resource for the co-creation of innovative solutions, encouraging collaboration between experts from different domains and allowing for the continuous exchange of knowledge between academia and industry. The adoption of intelligent systems based on DT and GAI, within an open and participatory model of innovation, has the potential to significantly transform production processes, making them more flexible, responsive, and efficient.

The proposition of the conceptual platform, presented in this article, structures a cyber-physical model to enable OIE, integrating functional layers that interconnect physical assets, digital representations, predictive modeling, and human-computer interaction. This structure promotes a dynamic flow of data and insights, allowing for continuous experimentation and optimization of industrial operations. With the application of personalized language models and predictive AI, intelligent systems can identify patterns, predict anomalies, and generate actionable recommendations that improve operational efficiency and strategic management of production processes.

The transformative potential of the convergence between DT and GAI lies in the ability to create a highly interconnected and responsive industrial ecosystem, where collaborative experimentation enables significant advances in productivity, safety, and innovation. The development of robust frameworks for the implementation of OIE in modern manufacturing inaugurates a promising path for the evolution of Industry 4.0, leveraging the power of digital technologies to redefine the paradigms of industrial production. However, challenges such as the standardization of protocols, accuracy of generative artificial intelligence models, culture of industry RD&I areas, data governance, and the adoption of appropriate regulatory models remain critical aspects that must be addressed to ensure the full exploitation of these technologies.

REFERENCES

Ali MI, Patel P, Breslin JG, Harik R, Sheth A. (2021). Cognitive digital twins for smart manufacturing. IEEE Intell Syst.; 36(2):96–100. DOI: 10.1109/MIS.2021.3062437.

Cheng Y, Zhang C, Zhang Z, Meng X, Hong S, Li W, Wang Z, Wang Z, Yin F, Zhao J, He X. (2024). Exploring large language model-based intelligent agents: Definitions, methods, and prospects. 10.48550/arXiv.2401.03428.

Ding W. (2024). Generative AI for critical digital twins [dissertation]. Pittsburgh: Carnegie Mellon University; DOI: 10.1184/R1/25892518.v1.

Gaimo F, Andrade H, Berger C. (2020). Continuous experimentation and the cyber-physical systems challenge: An overview of the literature and the industrial perspective. J Syst Softw. 170:110781. DOI: 10.1016/j.jss.2020.110781.

Huang X, Liu Y, Wang Z, Chen J. (2024). When digital twin meets generative AI: Intelligent closed-loop network management. IEEE Network. Available from: https://ieeexplore.ieee.org/document/10818633.

Korada L. (2024). Role of generative AI in the digital twin landscape and how it accelerates adoption. J Artif Intell Mach Learn Data Sci. 2(1):902–6. DOI: 10.51219/JAIMLD/laxminarayana-korada/217.

Lacoste M, Cook S, McNee M. (2022). On-farm experimentation to transform global agriculture. Nat Food. 3:11–8. DOI: 10.1038/s43016-021-00424-4.

Li BM, Xie SQ, Xu X. (2011). Recent development of knowledge-based systems, methods and tools for one-of-a-kind production. Knowl Based Syst. 24:1108–19. DOI: 10.1016/j.knosys.2011.05.005.

Liang SY, Rajora M, Zou P. (2015). Intelligent manufacturing system for next generation factories. Adv Intell Syst Res.

Liu S, Bao J, Zheng P. A (2023). review of digital twin-driven machining: From digitization to intellectualization. J Manuf Syst. 67:361–78. DOI: 10.1016/j.jmsy.2023.02.010.

Lv Z, Wu J, Li Y, Song H. (2022). Cross-layer optimization for industrial Internet of Things in real scene digital twins. IEEE Internet Things J. DOI: 10.1109/jiot.2022.3152634

Mad Yusoh SS, Abd Wahab D, Adil Habeeb H, Azman AH. (2021). Intelligent systems for additive manufacturing-based repair in remanufacturing: a systematic review of its potential. PeerJ Comput Sci. 7:e808. DOI: 10.7717/peerj-cs.808.

Makatura L, Foshey M, Wang B, et al. (2023). How can large language models help humans in design and manufacturing? arXiv [Preprint]. DOI: 10.48550/arXiv.2307.14377.

Mattos DI, Bosch J, Olsson HH. (2018). Challenges and strategies for undertaking continuous experimentation to embedded systems: Industry and research perspectives. In: International Conference on Agile Software Development. Springer. p. 277–92.

Stergiou CL, Psannis KE. (2022). Digital twin intelligent system for industrial internet of things-based big data management and analysis in cloud environments. Virtual Reality Intell Hardw. 4(4):279–91. DOI: 10.1016/j.vrih.2022.05.003.

Wang X, Anwer N, Dai Y, Liu A. (2023). ChatGPT for design, manufacturing, and education. Procedia CIRP. 2023; 119:7–14. DOI: 10.1016/j.procir.2023.04.001.

Wang Y, Pan Y, Yan M, et al. (2023). A survey on ChatGPT: AI-generated contents, challenges, and solutions. arXiv [Preprint]. DOI: 10.48550/arXiv.2305.18339.

Xia K, Sacco C, Kirkpatrick M, Saidy C, Nguyen L, Kircaliali A, Harik R. (2021). A digital twin to train deep reinforcement learning agent for smart manufacturing plants: Environment, interfaces and intelligence. J Manuf Syst. 58 (Part B):210–30. DOI: 10.1016/j.jmsy.2020.06.012.

Xu Q, Zhou G, Zhang C. (2023). Generative AI and digital twin integrated intelligent process planning: A conceptual framework. Research Square [Preprint]. Nov 29. DOI: 10.21203/rs.3.rs-3652246/v1.

Yijun H, Zhang J, Chen X, Lam A, Chen B. (2024). From simulation to prediction: Enhancing digital twins with advanced generative AI technologies. Proceedings of the IEEE International Conference on Control and Automation. (2024):490–5. DOI: 10.1109/ICCA62789.2024.10591881.

SMART HERITAGE: ADVANCED TECHNOLOGIES FOR SUSTAINABILITY IN HISTORIC CENTERS

Stefania De Medici and Giuseppe Cataldi
Department of Civil Engineering and Architecture (DICAR), University of Catania, Italy

ABSTRACT

Climate change and increasing urbanisation are major challenges for the conservation of the architectural heritage. The need to improve the energy efficiency of existing buildings clashes with the protection of their historic value, requiring innovative solutions that balance conservation and sustainability. This study analyses international best practices for the energy retrofitting of buildings, with a focus on smart cities and the integration of new technologies in historic centres. Through a multidisciplinary approach, the research explores the role of contemporary art and design as tools for ecological transition, promoting a model of sustainable and interactive cities. The analysis of case studies shows how the use of innovative materials, energy monitoring systems and participatory strategies can transform historic buildings into smart buildings, ensuring energy efficiency and cultural enhancement. The results highlight the potential of smart technologies in the management of historic urban spaces, fostering dialogue between tradition and innovation. In this context, the synergy between architectural design, cultural production and sustainability emerges as a key element in addressing contemporary climate challenges and outlining a new resilient and ecological urban paradigm.

KEYWORDS

Architectural Heritage, Energy Efficiency, Sustainable Renovation, Digital Innovation, Urban Regeneration

1. INTRODUCTION

The abnormal global warming of the last century is defined by the United Nations as the direct or indirect result of human activity (United Nations 1992). The European Green Deal aims to achieve climate neutrality, promote a zero-emissions economy, protect natural capital and improve the well-being of citizens (European Commission 2019). Nevertheless, in the professional practice of the building sector, the technological progress related to the energy efficiency of buildings does not consider the needs of environmental protection and cultural heritage preservation (Lucchi, Polo López, and Franco 2020). The need to adapt the performance of the built environment to contemporary requirements highlights the vulnerability of historic centres and can lead to significant changes in the built environment.

Recent decades have shown the importance of reconnecting with the local area, suggesting that the digital revolution can help overcome some of the problems associated with urbanisation (Snis, Olsson, and Bernhard, 2021). Contemporary architecture should respond to the need for efficiency and optimisation, developing smart cities that are able to combine sustainability and wellbeing, designed with a logic similar to that of a microchip (Ratti and Claudel, 2016). Designed in relation to its geographical and socio-economic context, each historic building is the result of local building knowledge. Their renovation is a challenge to find a balance between preserving their identity and the need to adapt to new requirements. Nowadays, the need for energy efficiency has become a priority in the context of environmental protection.

At European level, initiatives have been developed to improve the efficiency of existing buildings. Nevertheless, generalised strategies could neglect the identity of each building, which is essential for restoration and renovation. Moreover, in Italy there are more than 3 million historic buildings built before 1945, which have high energy consumption and represent about 25% of the national building stock, making their renovation an urgent priority (Trovò, 2023).

From a regulatory point of view, Europe began to regulate energy efficiency in construction in 1993 with Directive 1993/76/EC (SAVE) and then developed a series of EU laws (EPBD I-II-III, EED) to promote energy certification, renewable sources and energy diversification. Unfortunately, current regulations do not provide specific indications for heritage buildings, leading to problems when renovating them and encouraging the abandonment of the building heritage.

One of the major innovations in the construction industry today is the use of sustainable materials and renewable resources. The growing demand for comfort and energy savings makes the integration of building efficiency and monitoring strategies essential. Individuals, communities, governments and corporations need to work in synergy to address the climate crisis, balancing conservation and transformation.

2. RESEARCH AIM

In the process of regenerating historic centres, the reuse of buildings and the redevelopment of urban spaces serve as key tools for restoring the identity of places. The aim is not only to preserve the existing heritage, but also to redefine its role within the contemporary urban fabric by integrating new design cultures and emerging technologies. This approach aims to develop innovative models for the management and use of historic spaces, addressing sustainability, accessibility and adaptability to new social and economic paradigms.

This research aims to analyse international best practices for the ecological transition of the built environment, and to identify design criteria and strategies capable of balancing functionality, aesthetic value and educational potential. Particular attention will be paid to the role of contemporary art and design as catalysts for accelerating the green and smart transformation of historic cities. The multidisciplinary approach adopted combines contemporary design and technological innovation to promote a modern use of urban spaces, while ensuring a balance between heritage conservation and the implementation of project solutions that enhance sustainability and resilience.

3. MATERIALS AND METHODS

To achieve this goal, some best practices have been identified and analysed as study objects in order to define the criteria that guided design choices. The three categories selected represent key dimensions of sustainable urban transformation: a) *Smart Hubs for Public Space* emphasize the integration of technology, community participation, and flexible design to create inclusive and adaptive urban environments; b) *Smart Playgrounds for Educational Purposes* highlight the role of interactive design in fostering environmental awareness and engagement through play; and c) *Smart Buildings for Energy Efficiency* demonstrate how technological innovation can enhance sustainability while preserving architectural heritage. Together, they provide a comprehensive framework for rethinking the built environment through a balance of sustainability, innovation, and social impact. Examining case studies offers concrete examples of effective strategies, allowing us to learn from solutions in similar contexts. The study of case studies provides inspiration and stimulates creativity to improve the quality of the design and implementation of new devices. The best practices selected are based on the search for models that do not hide the innovative elements. They are oriented towards the search for a new aesthetic in the balance between the old and the new (De Medici 2021).

4. BEST PRACTICES ANALYSIS

Designing and transforming the space around us means reflecting on the effects brought about by society and the technological progress in our era. Nowadays, as we witness the evolutionary transition of communities, we need to design and build urban spaces bearing in mind which features buildings and cities must have in order to be able to accommodate new functions. A multidisciplinary approach enabled the research to address the issue of sustainability in historic cities and its implementation through the analysis of best practices. In accordance with the methodology proposed by Bretschneider et al., the best practices have been collected, analysed and interpreted so as to ensure completeness and comparability (Bretschneider, Marc-Aurele, and Wu 2005). This methodology showed the richness of the projects, their thematic and design creativity, as well as the ability to innovate by responding to the current needs of the environment with a view to the future and openness to possible opportunities.

4.1 Smart Hub for Public Space

Table 1. Pure Tension Pavilion, worldwide (Orhon 2016)

Object: freestanding marquee **Designer:** Synthesis Design + Architecture **Date:** 2014	 Figure 1. Pavilion front view	 Figure 2. Pavilion detail
Sustainable materials Community involvement Renewable energy	The pavilion (figure 1) consists of an aluminium frame and a double-layer polyester mesh membrane. Weighing 70 kg, it is foldable and can be assembled in less than an hour.	Analysing the solar incidence allowed to map the most exposed areas, allowing to optimise the photovoltaic module layout for a total power of about 450 Wp (figure 2).

Table 2. AU Peace and Security Building, Addis Adaba (Orhon 2016)

Object: tensile structure for shading **Designer:** GIZ **Date:** 2015	 Figure 3. Courtyard view	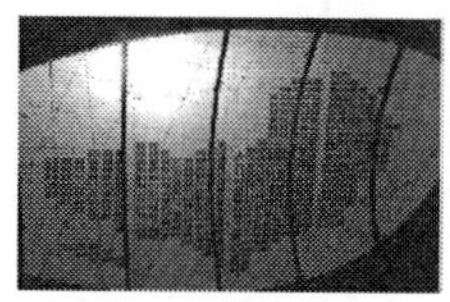 Figure 4. Modules detail
Sustainable materials Community involvement Renewable energy	The OPV shading membrane (Figure 3) above the courtyard is one of the building's main distinctive features. The active sunshade, which is shaped like the African continent, is also the logo of the African Union.	The building's photovoltaic system (Figure 4) is made up of transparent modules with flexible organic solar cells. It generates enough energy to power the building's LED lighting.

Table 3. German Pavilion EXPO 2015, Milan (Orhon 2016)

Object: tensile structures **Designers:** Schmidhuber + Milla & Partner + Nüssli **Date:** 2015	 Figure 5. Overall view	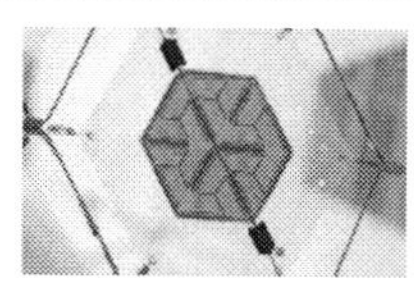 Figure 6. Module detail
Educational purpose Sustainable materials	These structures, which resemble trees growing out of the ground, act as energy containers made of a membrane that evokes the concept of plants in bud: 'Idea Seedlings' (Figure 5).	Using semi-transparent flexible solar cells, these structures can produce a total power of 5 kWp. The total area of the cells is over 300 m^2 (Figure 6).

4.2 Smart Playground for Educational Purpose

Table 4. The Pool, worldwide (Jen Lewin Studio 2008)

<table>
<tr><td colspan="2">Object: work of art
Designer: Jen Lewin studio
Date: 2008</td><td>
Figure 7. Overall view</td><td>
Figure 8. Pad detail</td></tr>
<tr><td></td><td>Interactivity</td><td rowspan="3">In this interactive installation, the movement of users creates swirls of light and colour (Figure 7). By moving on the pads, users activate light effects, generating dynamic and engaging effects.</td><td rowspan="3">Each pad (Figure 8) is equipped with sensors and RGB LEDs that react to pressure and speed of movement. Powered by solar panels or small generators.</td></tr>
<tr><td></td><td>Community involvement</td></tr>
<tr><td></td><td>Energy self-sufficiency</td></tr>
</table>

Table 5. Energy Carousel, Dordrecht (Ecosistemaurbano 2012)

<table>
<tr><td colspan="2">Object: unconventional carousel
Designer: Ecosistemaurbano
Date: 2012</td><td>
Figure 9. Structure detail</td><td>
Figure 10. Overall view</td></tr>
<tr><td></td><td>Educational purpose</td><td rowspan="3">Interactive play structure designed to transform children's kinetic energy into electricity. Consisting of a structure of ropes and fabrics (Figure 9), it stores the energy generated by movement to provide light at night.</td><td rowspan="3">Encourage children to explore alternative methods of producing sustainable energy (Figure 10). The intensity and colour of the lights change according to the energy accumulated during the day, resulting in a dynamic and engaging visual experience.</td></tr>
<tr><td></td><td>Community involvement</td></tr>
<tr><td></td><td>Energy self-sufficiency</td></tr>
</table>

Table 6. Feeling the Energy, Milan (Ratti 2022)

<table>
<tr><td colspan="2">Object: multi-sensory installation
Designer: CRA – Carlo Ratti Associati e Italo Rota
Date: 2022</td><td>
Figure 11. Overall view</td><td>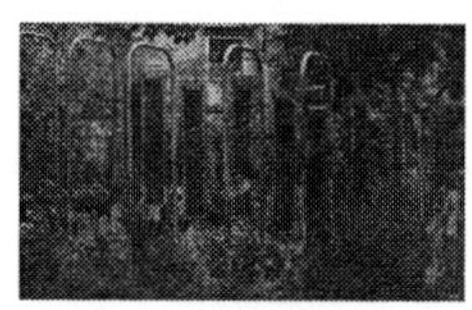
Figure 12. Overall view</td></tr>
<tr><td></td><td>Educational purpose</td><td rowspan="2">A multi-sensory experience in which sound, light and wind allow visitors to perceive energy in different ways (Figure 11). It celebrates the collaboration between man and nature in the production of vital energy.</td><td rowspan="2">The path, made of 500 metres of antibacterial copper, encourages experimentation with renewable energy. Through wind, sun and movement, it produces energy that is reused for lighting (Figure 12).</td></tr>
<tr><td></td><td>Interactivity</td></tr>
</table>

4.3 Smart Building for Energy Efficiency

Table 7. Apple Store, Paris (Foster + Partners 2018)

Object: heritage building renovation **Designer:** Foster + Partners **Date:** 2018	 Figure 13. Kaléidoscope detail	 Figure 14. View of the courtyard
Energy management Renewable energy Rainwater management	The *Kaléidoscope* solar skylight (Figure 13) is made up of mirrored pyramids to reflect natural light onto the inner courtyard walls. The integrated photovoltaic panels generate clean energy.	The skylight optimises incoming sunlight and energy production (Figure 14). A rainwater harvesting system is included in the building to irrigate trees, watering green walls and flushing toilets.

Table 8. Apple Store, Rome (Foster + Partners 2021)

Object: heritage building renovation **Designer:** Foster + Partners **Date:** 2021	 Figure 15. Building view	 Figure 16. Internal view
Energy management Lighting control	Built in 1873, the building (Figure 15) became the headquarters of the renowned Caffè Aragno, a meeting place for Rome's intellectuals and artists. Today it has been refurbished as an Apple Store.	The interior design is a combination of old and new: the ancient oculus, once a skylight (Figure 16), has been reactivated with an LED system reproducing the natural variations of light.

Table 9. Agnelli Foundation HQ, Turin (Ratti 2017)

Object: heritage building renovation **Designer:** CRA – Carlo Ratti Associati **Date:** 2017	 Figure 17. Building view	 Figure 18. Internal view
System Integration Indoor air recirculation HVAC system	The building is the headquarters of the Agnelli Foundation (Figure 17) where education and training spaces are available. A new approach to architecture, where IoT technologies are integrated to create an interactive and adaptive environment.	Through several IoT sensors, the building (Figure 18) is able to adapt to the needs of its users in real time, by monitoring temperature and CO2 levels. A geolocation system allows for customised control of light and air conditioning.

5. DISCUSSION AND RESULTS

Technological innovation can play a key role in the energy retrofitting of architectural heritage, helping to address the challenge of climate change. Using advanced technologies such as energy efficiency systems, renewable energy sources and sustainable materials can reduce environmental impact without compromising the preservation of historic urban contexts. Cultural production, through community involvement, education and awareness-raising, can foster an understanding of the importance of sustainability and renewable energy sources, promoting an integrated approach to the conservation of the architectural heritage and the built environment.

As a result of the analysis of international best practice, the research identified recurring design criteria for applying the Green Deal to the built heritage. Key principles include the use of sustainable materials, renewable energy and energy self-sufficiency to ensure long-term environmental benefits. Strategies such as rainwater management, HVAC optimisation, indoor air recirculation and lighting control enhance resource efficiency and occupant comfort. In addition, interactivity, community involvement and an educational purpose promote engagement and awareness, while systems integration improves overall performance. These criteria provide a comprehensive framework for balancing heritage conservation with sustainability and innovation.

Table 10. Design criteria for performance improvement of heritage buildings

	Sustainable materials		Interactivity		Energy management
	Community involvement		Energy self-sufficiency		Rainwater management
	Renewable energy		System Integration		Indoor air recirculation
	Educational purpose		Lighting control		HVAC system

5.1 Smart Hub for Public Space

An integrated approach combining smart technologies, sustainability and community involvement can create a progressive and sustainable urban space. This concept is reflected in several examples, starting with Volvo's Pure Tension Pavilion (Tab. 1), which uses a lightweight structure and photovoltaic fabrics for autonomous energy production, demonstrating how sustainable materials and contemporary design innovation can promote energy-efficient solutions. Similarly, the AU Peace and Security Building (Tab. 2) uses organic and translucent photovoltaic modules, highlighting an integrated approach between architecture and sustainability. A smart space, capable of integrating sustainable energy production systems to improve the energy efficiency and sustainability of the spaces surrounding the buildings, is a common feature of these projects. However, while the Volvo Pure Tension Pavilion explores modular and temporary solutions, the AU Peace and Security Building and its focus on sustainability are more rooted in permanent architecture, demonstrating two different ways of addressing the same need. At the same time, the structures created inside the German Pavilion at EXPO 2015 (Tab. 3) introduce an advanced photovoltaic system integrated into the structural membrane. A design element capable of generating energy through translucent surfaces optimised to capture sunlight. This solution demonstrates how design can become an active element in energy production, contributing to the creation of self-sufficient spaces and optimising the use of renewable resources.

The projects analysed show that flexibility in urban design is essential to respond to technological developments and the needs of the community, making cities more inclusive, sustainable and resilient. The integration of innovative design, renewable energy and smart technologies allows the transformation of built spaces into dynamic and adaptive elements capable of reducing environmental impact and promoting collaboration between different disciplines such as art, architecture and engineering.

5.2 Smart Playground for Educational Purpose

The development of a smart playground for educational purposes represents an innovative approach that seamlessly integrates learning, play, and energy sustainability. This concept is exemplified by various artistic and design installations that engage users through physical interaction, making them active participants in both the creative and energy-generating processes. A notable example is The Pool by Jen Lewin Studio (Table 4), an interactive installation composed of illuminated, motion-sensitive platforms that transform human movement into an immersive experience. Similarly, Feeling the Energy by Carlo Ratti Associati (Table 6) utilizes advanced technology to visualize real-time energy production and consumption, fostering environmental awareness through a playful and accessible language.

Another striking example is the Energy Carousel by Ecosistema Urbano (Table 5), an interactive ride that converts children's motion into electricity, demonstrating how play can serve as both a tool for engagement and a means of energy generation. By focusing on physical interaction and child participation, this approach underscores the role of creativity in activating urban spaces, leveraging interactive technologies and gamification strategies to educate and involve the community. Ultimately, the fusion of art, technology, and sustainability enables the creation of educational, interactive environments that cultivate environmental awareness while transforming play into an intuitive and engaging energy experience.

5.3 Smart Building for Energy Efficiency

The transition toward smart buildings focused on energy efficiency marks a crucial intersection between sustainability and heritage conservation. Retrofitting historic structures with intelligent technologies not only safeguards their cultural and architectural value but also significantly reduces environmental impact. A striking example is the renovation of the Apple Stores in Paris (Table 7) and Rome (Table 8) by Foster + Partners, which illustrates how the redevelopment of historic buildings can seamlessly integrate advanced technological solutions without compromising their architectural integrity. These projects successfully balance innovation and conservation by enhancing energy performance through the incorporation of energy-generating design elements and smart systems that minimize resource consumption.

A key aspect of these interventions is the optimization of energy use through intelligent lighting, efficient thermal regulation, and real-time energy monitoring. The Agnelli Foundation (Table 9), designed by Carlo Ratti Associati, exemplifies intelligent renovation by employing sensor-based lighting and climate control systems that dynamically adjust to user needs, reducing energy consumption while enhancing indoor comfort. The restoration of existing buildings involves a multi-layered approach, from improving energy efficiency and integrating renewable energy sources to implementing automation solutions and designing outdoor spaces that complement the architectural context.

While Foster + Partners' Apple Store renovations showcase how restoration and technological innovation can coexist without diminishing historical character, the Agnelli Foundation represents a cutting-edge model of smart technology applied to dynamic space management. These examples highlight how smart buildings can act as a bridge between past and future, demonstrating that energy-efficient redevelopment relies on the intelligent fusion of heritage and technology. Ultimately, these interventions aim to enhance both the environmental and architectural quality of buildings, ensuring well-being, safety, and long-term sustainability through integrated, advanced solutions.

6. CONCLUSIONS

Climate change stands as one of the most urgent and complex challenges of the 21st century, driving a relentless pursuit of technological innovation. However, it is crucial to direct these advancements toward lasting and pragmatic goals, avoiding ephemeral or utopian visions. The demand for balance between technological innovation and the conservation of historical contexts has opened a compelling and intricate field of studies, where tradition and innovation converge. At the heart of this research is the idea that the intersection of traditional culture and contemporary creative production can serve as a foundation for a new design paradigm, to align with the principles of the Green Deal and embody the vision of the European New Bauhaus.

The combination of seemingly contradictory cultural dimensions fosters a continuous dialogue, which is reflected in cultural and technological innovations.

The analysis of case studies underscores the pivotal role of design and contemporary art in the ecological transition of the built environment. These disciplines emerge as powerful drivers of sustainability, not only in terms of energy efficiency but also in shaping the future of smart cities. The examined experiences demonstrate how the integration of aesthetic, technological, and participatory solutions can successfully merge energy retrofitting with heritage enhancement, creating more interactive and resilient urban spaces. The examples of smart playgrounds illustrate how design can serve as an educational tool, transforming complex concepts of energy production and consumption into playful, interactive experiences. Likewise, interventions in historic buildings highlight the potential of contemporary design and architecture in redefining the relationship between past and future, blending smart technologies and innovative materials to ensure both seamless integration and thoughtful transformation of the built environment.

Looking ahead, the imperative to explore and adapt emerging technologies to historical urban areas becomes increasingly evident. When applied with sensitivity and respect, technological innovation can serve as a powerful tool for preserving and enriching cultural heritage, transforming it into a dynamic and sustainable entity. In the broader vision of the smart city, the integration of intelligent technologies within historic contexts enables the preservation of cultural identity while enhancing urban efficiency and sustainability.

This research has shown how contemporary design can act as a universal language, encouraging collective participation, facilitating the spread of sustainable practices and cultivating ecological awareness. Within this framework, contemporary art and design emerge not only as instruments of energy retrofitting, but also as catalysts for aesthetic transformation, capable of embedding sustainability in everyday and functional experiences. The city of the future will thus be defined by adaptive, intelligent and participatory spaces, where the dialogue between tradition and technological innovation, culture and nature will serve as a driving force for a more ecological and inclusive model of urban transformation.

REFERENCES

Bretschneider, Stuart, Frederick J. Marc-Aurele, and Jiannan Wu. (2005). "'Best Practices' Research: A Methodological Guide for the Perplexed." *Journal of Public Administration Research and Theory*. https://doi.org/10.1093/jopart/mui017.

De Medici, Stefania. (2021). "Italian Architectural Heritage and Photovoltaic Systems. Matching Style with Sustainability." *Sustainability (Switzerland)* 13 (4). https://doi.org/10.3390/su13042108.

Ecosistemaurbano. (2012). "Energy Carousel." 2012. www.ecosistemaurbano.com/energy-carousel.

European Commission. (2019). "The European Green Deal. Communication from the Commission to the European Parliament, the European Council, the Council, the European Economic and Social Committee and the Committee of the Regions." Brussels, Belgium: European Commission.

Foster + Partners. (2018). "Apple Champs Elysees." 2018. www.fosterandpartners.com/projects/apple-champs-elysees.

———. (2021). "Apple via Del Corso." 2021. www.fosterandpartners.com/projects/apple-via-del-corso.

Jen Lewin Studio. (2008). "The Pool." 2008. www.jenlewinstudio.com/portfolio/the-pool.

Lucchi, Elena, Cristina S. Polo López, and Giovanna Franco. (2020). "A Conceptual Framework on the Integration of Solar Energy Systems in Heritage Sites and Buildings." In *International Conference Florence Heri-Tech: The Future of Heritage Science and Technologies*. Florence: IOP Publishing. https://doi.org/10.1088/1757-899X/949/1/012113.

Orhon, Ahmet. (2016). *Integration of Photovoltaics into Tensile and Inflatable Structures*.

Ratti, Carlo. (2017). "Fondazione Agnelli." 2017. www.carloratti.com/project/fondazione-agnelli.

———. (2022). "Feeling the Energy." 2022. www.carloratti.com/project/feeling-the-energy.

Ratti, Carlo, and Matthew Claudel. (2016). *The City of Tomorrow: Sensors, Networks, Hackers, and the Future of Urban Life. The City of Tomorrow: Sensors, Networks, Hackers, and the Future of Urban Life*.

Snis, Ulrika Lundh, Anna Karin Olsson, and Iréne Bernhard. (2021). "Becoming a Smart Old Town – How to Manage Stakeholder Collaboration and Cultural Heritage." *Journal of Cultural Heritage Management and Sustainable Development* 11 (4). https://doi.org/10.1108/JCHMSD-10-2020-0148.

Trovò, Francesco. (2023). "Historic Building and Green Energy. Strategies to Make Supply from Renewable Sources Compatible with Conservation." *Technical Annals* 1 (2). https://doi.org/10.12681/ta.33918.

United Nations. (1992). *United Nations Framework Convention On Climate Change*. https://unfccc.int/resource/docs/convkp/conveng.pdf.

AGENTIC SMART CITY SIMULATOR FROM USER LOCALE DATA

Akash Sahoo and Rabi Mahapatra
Texas A&M University, College Station, USA

ABSTRACT

Smartphone data is becoming increasingly valuable as smartphones are now widely owned and frequently used. This data can offer deeper insights into user behavior by capturing information related to their daily activities. It can be leveraged for behavior or intent prediction, location forecasting, and personalized recommendations for places and activities. In this work, we developed an Android application to collect data from the smartphone's wireless sensors and GPS. By applying deep learning techniques to this complex and diverse behavioral data, we can extract meaningful patterns and insights. This paper presents two major contributions: first, a method to predict a user's location and behavior based on historical location and sensor data; and second, the development of a dynamic city simulator. The simulator models various institutions, including residential, commercial, and educational buildings, as well as transportation and public services. Within the simulator, different roles are assigned to virtual individuals, and environmental factors such as traffic congestion, crime, and other dynamic events are randomly introduced to realistically simulate user behavior. We use Pydantic AI to create agents, roles and use Qwen model as the underlying LLM to simulate them.

KEYWORDS

Internet of Things, Android, Location Modeling, LLM Fine-Tuning, City Simulator, Multi Agents Interaction

1. INTRODUCTION

Smartphones have become integral to modern life, with advanced sensors enabling seamless data collection to enhance user experience. These include precise location tracking, activity monitoring, and deep learning-enabled camera functions. With widespread GPS integration, smartphones have catalyzed research in human activity recognition (HAR), supporting diverse applications like navigation, emergency alerts, behavior monitoring, and disease tracking.

HAR has found applications in pervasive healthcare, elderly activity monitoring, physical activity recognition, travel mode identification, and indoor localization. Smartphones and wearables now allow for continuous monitoring of health metrics such as heart rate and sleep cycles, offering insights into user behavior and wellness.

Beyond health, smartphones enable personalized recommendations across domains like shopping, entertainment, and recreation. Sensor data—including GPS, gyroscope, and accelerometer—is used to classify user activities and infer both location and intent, opening avenues for intelligent services and automation.

Modeling human behavior in urban environments has gained traction, thanks to advances in agent-based modeling, mobile sensing, and large language models (LLMs). Our work proposes a modular city simulator that integrates agentic frameworks and large-scale behavioral modeling to emulate daily life activities for residents, workers, law enforcement, and visitors.

At the core of the simulator are generative agents with roles and preferences, embedded in structured environments such as residential or commercial zones. These agents interact with the environment and exhibit emergent behaviors that mirror real-world urban dynamics. LLMs enhance this by interpreting sparse sensor data into semantically rich behaviors, enabling complex reasoning like route planning and emergency response. This paper has the following contributions:

i) we designed a framework that predicts a user's location/intents by combining GPS data from their smartphone with Wi-Fi and Bluetooth wireless sensor information;

ii) we developed an android application and integrated it with Google Cloud Platform (GCP) for data collection and storage;

iii) we proposed a deep learning architecture using Residual Network (ResNet) and Long Short-Term Memory (LSTM), and benchmarked the approach;

iv) We use deep learning models and fine-tune OpenAI GPT for prediction of the location & intents

v) Build City Simulator and used LLMs as various roles for Agents.

2. RELATED WORK

Location data from mobile phones has been widely used to analyze behavioral patterns and predict future activities. Xie used clustering and map integration to identify regions of interest and predict user locations. Xu and Han proposed a semantic-behavior framework using BiLSTM and a similarity-based Markov model. Similarly, Alvarez-Lozano et al. utilized Hidden Markov Models and POI histories to forecast the next user location, confirming the Markovian nature of behavior when categorized by time and day.

Liu et al. applied spatiotemporal dynamic time warping to classify user behavior using Call Detail Records (CDRs), identifying dining, working, entertainment, and resting patterns, as well as transport mode. Deep learning methods have also been used for indoor activity prediction—Li et al. employed CNN, ANN, and LSTM on spatial location data to recognize six basic human activities.

The increasing popularity of location-based social networks (LBSNs) has enabled the modeling of spatial-temporal activity preferences (STAP). Yang et al. proposed a STAP model that captures user intent based on historical time and location data. Likewise, deep neural networks have been effective for recognizing indoor activities and localizing users through sensor fusion.

Past research has used machine learning and deep learning to predict activities or locations from smartphone sensor data. Given the strong performance of CNNs and residual networks in such tasks, our work builds on these by employing ResNet-based architectures.

The rise of large language models (LLMs) has transformed mobile computing and virtual assistant capabilities. Our framework draws from behavior simulation studies that use LLMs and agent-based design. Guozhen et al. proposed a simulation taxonomy addressing limitations of current behavior modeling. Park et al. introduced generative agents capable of planning and self-reflection, later extended to 1,000-person simulations, demonstrating scalability.

SocioVerse showed the potential of LLMs in modeling millions of real-world agents, using structured prompts and large-scale behavioral data to simulate societal patterns. Our approach extends this vision with a focus on real-time mobile sensor integration to enhance realism in urban simulations.

In summary, we unify advances in location prediction, activity recognition, and LLM-driven agent modeling into a framework that simulates intelligent, dynamic behavior grounded in real-world sensor data.

3. METHODS

This paper discusses how smartphone-based sensor data can predict location activity, including wireless sensors like Bluetooth, Wi-Fi, and location services. The research includes data collection, storage and transformation, data processing involving neural networks and machine learning, and future location prediction. The steps are defined and explained in detail in the *Experiments* section. This section mainly focuses on the various methods utilized in the research and their architecture. The study employs four deep learning architectures and evaluates their performance from experiments conducted on data collected through an Android application. We have also used machine learning algorithms like Random Forest Classifier, Support Vector Machines (SVMs), and XGBoost Classifier to compare their performance with our deep learning models.

This study employs multiple machine learning techniques, including a Random Forest classifier that reduces overfitting by averaging results from multiple decision trees, and XGBoost, valued for its speed and robustness in combining weak learners. Support Vector Machines (SVMs) are also used for their versatility in classification and regression tasks. Additionally, the study introduces deep learning algorithms, including a proposed model, discussed in subsequent sections.

3.1 Deep Learning Methods for Modeling User Location

LSTM (Long Short-Term Memory): A three-layer LSTM model is used to capture temporal dependencies in sequential activity data. Each layer has 100 units with dropout rates of 40% and 50%, followed by dense layers with ReLU and softmax activations. The model uses the Adam optimizer and categorical cross-entropy loss function.

CNN + LSTM: This model combines 1D convolutional layers (with filter sizes 64, 128, and 64) for feature extraction with two stacked LSTM layers for temporal modeling. Max pooling and dropout are used for regularization. The architecture concludes with dense layers for final classification.

Convolutional LSTM: Unlike the CNN+LSTM stack, Conv-LSTM performs convolution operations within the LSTM cells themselves. The model includes three Conv-LSTM layers with filter sizes 128, 64, and 128, each followed by dropout. The flattened output is fed into dense layers for classification.

ResNet + LSTM: This architecture incorporates residual blocks to address the degradation problem in deep convolutional networks. The ResNet output is flattened and passed through three LSTM layers, each followed by dropout, enhancing both feature extraction and temporal learning capabilities.

Large Language Models (LLMs): An OpenAI GPT-4o-mini model was fine-tuned using 100k training samples with system/user/assistant prompts to evaluate its effectiveness for activity prediction tasks. Although LLMs are not naturally optimized for time-series data, the model's performance was assessed using a separate 25k test set. Fine-tuning cost, time, and training loss trends were documented.

4. CITY SIMULATOR

We propose to build a city simulator to simulate various users, roles, and activities so that we can improve intent prediction and better understand the user. Our City Simulator is designed to model complex urban environments using autonomous AI agents. It integrates real-time simulations of human behavior, transportation, services, and environmental factors, providing an interactive platform for visualization and analysis. We use the Pydantic AI framework to model classes and their properties.

4.1 Simulator Features

Table 1. Simulator components and features

Feature/ Category	Type	Functionality
Interactive City Map	Feature	A dynamic city visualization containing various categorized locations, enabling users to observe agent movements and interactions.
AI Agents	Feature	Each agent operates autonomously, driven by individual personalities, goals, and structured daily routines.
Dynamic Systems	Feature	Emergent behaviors arise from systems such as crime incidents, traffic congestion, and weather fluctuations, affecting both agents and the environment.
Real-time Visualization	Feature	A graphical user interface (GUI) supports real-time simulation with interactive tools for filtering agents, zooming, and time controls for accelerated or historical replay.
Residential	Location Type	Apartments, houses, and senior living facilities where residents and workers reside.
Commercial	Location Type	Public venues such as gyms, grocery stores, restaurants, coffee shops, and bars that serve as workplaces or recreational spots.
Educational	Location Type	Institutions including elementary schools, high schools, and colleges attended by student agents.

Transportation	Location Type	Key nodes such as airports, bus stops, and structured bus routes facilitating agent travel across the city.
Public Services	Location Type	Essential infrastructure such as police stations, hospitals, and parks which support societal needs and agent welfare.
Residents	Agent Type	Individuals who live in residential locations and follow daily routines that may include work, shopping, and leisure activities.
Workers	Agent Type	Agents employed within commercial or public service sectors who commute to their workplaces and engage in city life.
Students	Agent Type	Young agents enrolled in educational institutions, following academic schedules and participating in extracurricular activities.
Law Enforcement	Agent Type	Agents tasked with maintaining order, patrolling areas, responding to crimes, and managing public safety.
Criminals	Agent Type	Agents exhibiting adversarial behavior, engaging in unlawful activities, and interacting dynamically with law enforcement.
Visitors	Agent Type	Temporary agents entering the city through transportation hubs, engaging primarily in tourism, shopping, or business.

4.2 Code Components

Table 2. Code structure

Folder	Functionality
/src/agents	Contains agent-related classes including person, personality, goals, needs, and agent management. Responsible for autonomous agent behavior, decision-making, and interactions.
/src/engine	Core simulation engine components including the main simulation coordinator, time system, weather system, crime system, traffic system, and event system.
/src/gui	Graphical user interface components for rendering the simulation, handling user input, and displaying information.
/src/map	Map-related components including locations, road networks, and pathfinding functionality.
/src/utils	Utility functions and helper classes used throughout the project, including configuration loading.
/assets	Static assets like images, icons, and other resources used by the simulation.
/docs	Documentation for the project.
/config	Configuration files that define simulation parameters and settings.
/android_app	Codebase for the Android app used to collect sensor and location data from users.
/geodata	Storage of geolocation data from users for intent prediction and activity analysis.
/models	Pre-trained models built using geodata and sensor data for intent and activity prediction.
/config_llms	LLM APIs and config, number of agents, roles, simulator environment variables.

5. EXPERIMENTS

We evaluated the wireless sensors and location data originating from smartphones based on the accuracy of the test data evaluation metric. We used the different models described in the Methods section to conduct the experiments necessary for the evaluation. The focus of this section will be on data and how it is handled.

5.1 Data Collection using Android Application

An Android application was developed to collect wireless sensor and location data from smartphones running Android 12 or higher. The application requests user permissions for location access, activity recognition, and foreground services. It utilizes core Android services such as BluetoothAdapter, BluetoothManager, and BluetoothDevice for Bluetooth sensors, WifiManager for Wi-Fi status, and Location for GPS data. Google Firebase services, including FirebaseAuth, Firebase User, and Firebase Firestore, are employed for user authentication and cloud-based data storage.

The collected parameters include timestamp, latitude, longitude, charging type, charging status, Wi-Fi status, Bluetooth status, and Bluetooth device type. The application also detects the status of Wi-Fi and Bluetooth and classifies Bluetooth devices into categories such as audio/video, computer, health, networking, peripheral, phone, and wearable.

Users authenticate through their Google accounts, with all authentication and authorization managed via Firebase. Data is transmitted to Cloud Firestore every 60 seconds or when a user moves more than 10 meters. A total of 147,322 data points were collected from 23 consenting users over a period of three weeks.

Figure 1 shows three different views of the Android app interface, including the activity log (Figure 1b) and location log (Figure 1c) available for user review. Figure 1 visualizes the locations visited by a user over a 24-hour period.

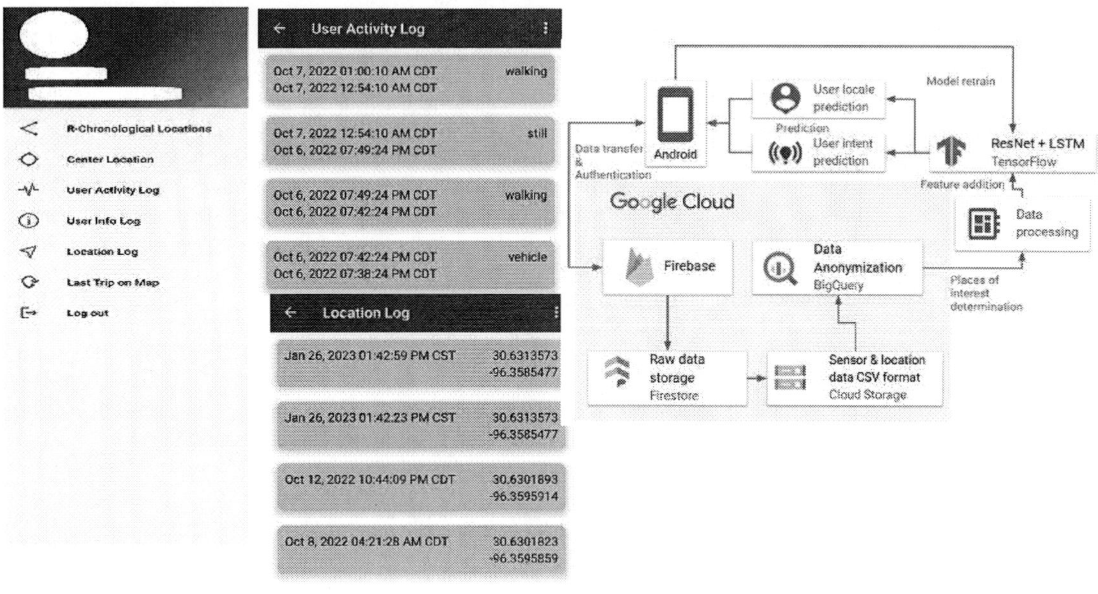

Figure 1. Data Collection from Android App and Backend for storage

5.2 Data Storage, Movement, Anonymization, and Processing

5.2.1 Data Processing

Data is stored in SQL database and private personal information is removed and anonymous id is given. For model training, the collected dataset undergoes preprocessing to ensure quality and consistency. Users with only a single unique location in their dataset are removed as they provide insufficient movement information.

Data for each user is sorted chronologically based on timestamps, from earliest to latest. This prepares the dataset for training temporal models.

Figure 2. One Day movement of a user.

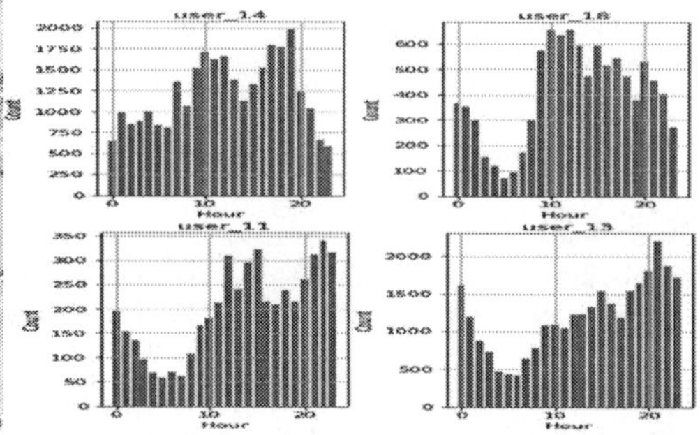

Figure 3. Hourly categorization of data per user

5.2.2 Place of Interest Determination and Data Attribute Processing

Smartphone GPS data was clustered using DBSCAN with a 200m radius and Haversine distance, generating 170 valid user location clusters. Convex hulls were computed via GeoPandas, and reverse geocoding (Geopy) mapped centroids to seven place types: Home, Restaurant, University, Office, Recreational Center, Airport, and Supermarket. Invalid clusters (43) were discarded. Data attributes were processed into machine-readable categories: charging types (1–3), Bluetooth states (1–5), and binary values for Wi-Fi and charging status. Temporal features such as hour (0–23), weekday (0–6), and weekend flags were also encoded to support location and intent prediction models.

5.2.3 Simulation of the Environment

The simulator environment includes a dynamic city map populated with diverse agent types such as residents, workers, students, law enforcement, criminals, and visitors. Each agent is instantiated as a Pydantic model, ensuring strict validation of properties such as location, mood, goals, and schedules.

The Qwen model is hosted via the Ollama framework, enabling local and efficient inference. Agents interact with each other by generating context-aware dialogue and behavior choices through structured prompts sent to Qwen. The prompts are dynamically constructed based on each agent's current state and environment. After implementing all data processing methods, the final dataset contains 90 641 rows of data. It has all data points arranged in ascending order according to each user's timestamp to preserve the chronology of events. The final dataset is divided into training and testing halves at a 7: 3 ratio to train 'the models. We used batch sizes of 128 for the ResNet + LSTM model and 64 for the rest of the three deep learning models for 80 epochs. We trained each of the several models for various hyper-parameter values and identified the model with the highest accuracy as the best. Agents are modeled using Pydantic's 'Base- Model' to ensure type safety and data validation. We then use LLM and ask Decide: 1. What action should the agent take next? 2. Which location should they go to? 3. Brief reasoning for this decision.

The steps for agent is acted as Action—Location—Reasoning

6. RESULTS

We evaluated the models described using wireless sensor and location data. Table 4 summarizes accuracy results across deep learning and machine learning models for both location and intent prediction.

Among deep learning models, the **ResNet + LSTM** architecture achieved the highest accuracy: **90.06%** for location and **90.26%** for intent prediction. It outperformed CNN + LSTM, LSTM, and Conv-LSTM. The **CNN + LSTM** also performed well, followed by LSTM and Conv-LSTM, which had similar performance.

Among machine learning models, **Random Forest Classifier** performed best, surpassing **SVM** and **XGBoost.** The fine-tuned **GPT-4o-mini** model achieved 81.98% for location and 82.70% for intent prediction, demonstrating LLM potential despite being less optimized for time-series data.

Figures illustrate the loss vs. epoch trends during training. The simulator further superimposed this predictive data on city agents, using LLMs to infer user intent and behavior in dynamic urban contexts.

Table 3. Models and their accuracy for Location predictions

Model	Location Prediction Accuracy (%)	Intent Prediction Accuracy (%)
LSTM	89.574	89.699
CNN + LSTM	90.044	90.179
Convolutional LSTM	89.563	89.653
ResNet + LSTM	90.058	90.261
OpenAI GPT-4o-mini Finetuned	81.980	82.700
SVM	89.431	89.391
XGBoost Classifier	81.745	81.583
Random Forest Classifier	89.913	90.012

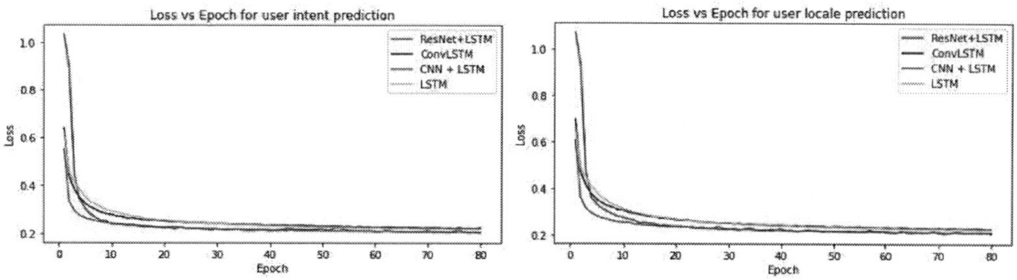

Figure 4. Loss versus Epoch training for Intent and Locale prediction

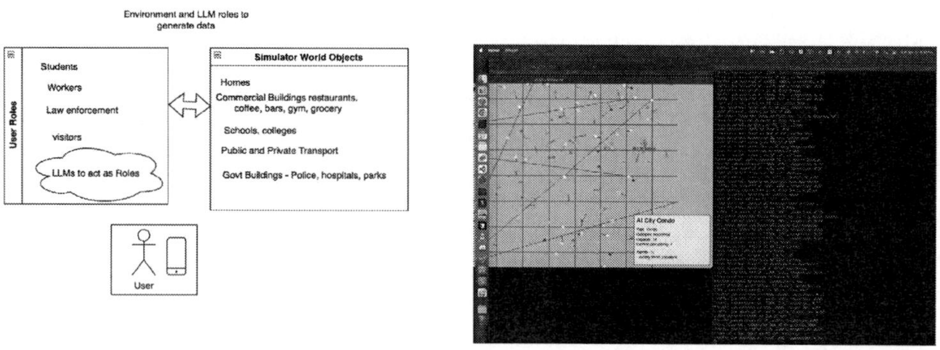

Figure 5. Simulator architecture

7. CONCLUSION

This paper presents a framework for user location and intent prediction using smartphone sensor data. We developed an Android app integrated with Google Cloud to collect and process wireless sensor data, ensuring user privacy through anonymization. Our proposed deep learning model—ResNet combined with LSTM—outperformed other models with over 90% accuracy in predicting both user locale and intent.

We also introduced a city simulator powered by LLM-driven agents to model realistic urban behavior based on user data. Future work will focus on enabling custom user intents, distilling LLMs for on-device inference, and modeling social relationships and preferences for deeper simulation insights.

REFERENCES

Alvarez-Lozano, J., García-Macías, J.A. and Chávez, E., (2012). User location forecasting at points of interest. *Proceedings of the RecSys Workshop on Personalizing the Local Mobile Experience (LocalPeMA '12).* New York, NY, USA, pp. 7–12.

Guozhen, Z., Zihan, Y., Nian, L., Fudan, Y., Qingyue, L., Depeng, J. and Yong, L., (2024). Human Behavior Simulation: Objectives, Methodologies, and Open Problems. *Proceedings of ArXiv preprint.* pp. 2412.07788.

Li, J., Guo, Y. and Qi, Y., (2019). Using Neural Networks for Indoor Human Activity Recognition with Spatial Location Information. *Proceedings of the 11th International Conference on Intelligent Human-Machine Systems and Cybernetics (IHMSC).* Hangzhou, China, pp. 146–149.

Liu, Y., Zhao, N., Han, W. and Zhu, Q., (2021). User Behavior Patterns and Activity Recognition Based on Mobile Big Data. *Proceedings of the International Conference on Electronic Information Technology and Smart Agriculture (ICEITSA).* Huaihua, China, pp. 203–209.

Najafi, B., Aminian, K., Paraschiv-Ionescu, A., Loew, F., Bula, C.J. and Robert, P., (2003). Ambulatory system for human motion analysis using a kinematic sensor: Monitoring of daily physical activity in the elderly. *Proceedings of IEEE Transactions on Biomedical Engineering.* pp. 711–723.

Osmani, V., Balasubramaniam, S. and Botvich, D., (2008). Human activity recognition in pervasive health-care: Supporting efficient remote collaboration. *Proceedings of Journal of Network & Computer Applications.* pp. 628–655.

Park, J.S., O'Brien, J.C., Cai, C.J., Morris, M.R., Liang, P. and Bernstein, M.S., (2023). Generative Agents: Interactive Simulacra of Human Behavior. *Proceedings of ArXiv preprint.* pp. 2304.03442.

Park, J.S., Zou, C.Q., Shaw, A., Hill, B.M., Cai, C., Morris, M.R., Willer, R., Liang, P. and Bernstein, M.S., (2024). Generative Agent Simulations of 1,000 People. *Proceedings of ArXiv preprint.* pp. 2411.10109.

Wen, H., Li, Y., Liu, G., Zhao, S., Yu, T., Li, T.J.-J., Jiang, S., Liu, Y., Zhang, Y. and Liu, Y., (2024). AutoDroid: LLM-powered Task Automation in Android. *Proceedings of ArXiv preprint.* pp. 2308.15272.

Xie, R., (2019). Exploring Mobile Phone Data for Urban Activity Analysis: Applications from Individual Activity Pattern to Group Activity Regularity. *Proceedings of IEEE SmartWorld, SCALCOM, UIC, ATC, CBDCom, IOP, SCI.* Leicester, UK, pp. 1213–1218.

Xu, M. and Han, J., (2020). Next Location Recommendation Based on Semantic-Behavior Prediction. *Proceedings of the 5th International Conference on Big Data and Computing (ICBDC '20).* New York, NY, USA, pp. 65–73.

Yang, D., Zhang, D., Zheng, V.W. and Yu, Z., (2015). Modeling User Activity Preference by Leveraging User Spatial Temporal Characteristics in LBSNs. *Proceedings of IEEE Transactions on Systems, Man, and Cybernetics: Systems.* pp. 129–142.

Yang, J., Cheng, K., Chen, J., Zhou, B. and Li, Q., (2018). Smartphones based Online Activity Recognition for Indoor Localization using Deep Convolutional Neural Network. *Proceedings of Ubiquitous Positioning, Indoor Navigation and Location-Based Services (UPINLBS).* Wuhan, China, pp. 1–7.

Zhang, X., Lin, J., Mou, X., Yang, S., Liu, X., Sun, L., Lyu, H., Yang, Y., Qi, W., Chen, Y., Li, G., Yan, L., Hu, Y., Chen, S., Wang, Y., Huang, X., Luo, J., Tang, S., Wu, L., Zhou, B. and Wei, Z., (2025). SocioVerse: A World Model for Social Simulation Powered by LLM Agents and A Pool of 10 Million Real-World Users. *Proceedings of ArXiv preprint.* pp. 2504.10157.

MONITORING BEHAVIOURAL CHANGE IN TO-BE-CLIMATE-NEUTRAL SMART CITIES ADAPTING THE TRANSTHEORETICAL MODEL OF BEHAVIOUR CHANGE (TTM)

Rohit Mondal, Sabrina Bresciani and Francesca Rizzo
Department of Design, Politecnico di Milano, Italy
Via Durando 10, 20158, Milano, Italy

ABSTRACT

In response to the challenges posed by climate change, cities are increasingly adopting systemic perspectives that incorporate bottom-up strategies, such as social innovation, moving away from traditional, technocratic and siloed, top-down approaches to climate neutrality. However, achieving large-scale socio-technical transformations necessitates addressing behavioural change among citizens. While behavioural change has traditionally been studied at the individual level through the widespread adaptation of models such as the Theory of Planned Behaviour (TPB) or the Transtheoretical Model of Behaviour Change (TTM) across disciplines, existing literature offers limited insights into their role and adaption in systemic transformations, particularly within the context of climate transitions. Nonetheless, there is a growing consensus—both in academic discourse and within the public sector, including policymakers and climate transition teams—on the critical need to integrate understanding and assessment of behavioural change into systemic climate transition action plans and strategies. Cities committed to the EU Mission of achieving 100 climate-neutral and smart cities by 2030 are actively selecting greenhouse gas (GHG) and non-GHG indicators to monitor, evaluate, and learn the systemic impacts of their climate actions. This study examines the indicator selections of 39 cities within the EU-funded NetZeroCities project using an adaptation of the TTM to categorise indicators of behavioural change into the stages of change of the TTM. Specifically, it aims to (1) provide emerging evidence on cities' interest in monitoring behavioural change in relation to climate actions and (2) identify the stage of behavioural change that cities across Europe intend to assess. The study's novel findings contribute not only to the transdisciplinary literature but also offer valuable insights for policymaking, supporting more effective and holistic climate transition strategies.

KEYWORDS

Monitoring, Behavioural Change, Systemic Transformation, Climate Neutrality, Transtheoretical Model of Behaviour Change (TTM)

1. INTRODUCTION

The 2030 target of the Paris Agreement, which seeks to limit global warming to 1.5 degrees Celsius (IPCC, 2022; United Nations, 2023), remains an enormous and ongoing challenge. Conventional climate strategies, predominantly reliant on technological, infrastructural, and political frameworks, have thus far proven inadequate in delivering substantial progress (Repo and Matschoss, 2019; Newell et al., 2021). As Newell et al. (2021) highlight, achieving climate goals necessitates profound shifts in lifestyles, policy frameworks, service provision, and technological innovation.

Household-related emissions, which constitute between 50% and 75% of global greenhouse gas emissions, underscore the critical role of behavioural change in mitigating climate impacts (Hertwich and Peters, 2009; Hirano et al., 2016; Akenji et al., 2019; Ivanova et al., 2020) In recognition of this, international organisations and climate initiatives are increasingly advocating for bottom-up approaches that foster systemic behavioural transformations at the bottom-up level (Bernauer, 2013; Sabel and Victor, 2017). However, the extent to which such interventions can be implemented with sufficient speed and efficiency to achieve significant global progress remains uncertain.

A critical lesson from previous shortcomings is the need to distinguish between "shallow" and "deep" scaling of behavioural interventions, as outlined by Newell et al. (2021). Shallow scaling fails to account for the contextual determinants of behaviour, relying on surface-level mechanisms such as political incentives, market forces, or nudges (Schwartz, 1977; Cialdini et al., 1990). In contrast, deep scaling aims to bring about societal transformation by embedding sustainable norms and values within cultural and institutional practices. Such an approach necessitates a non-linear, multidimensional approach that accounts for the complexities of social and psychological dynamics.

Additionally, individual-level barriers, including financial constraints, inadequate infrastructure, and ingrained social norms, limit the capacity for behavioural change (Cialdini et al., 1990; Gifford & Nilsson, 2014). Cognitive biases and psychological factors further complicate this process, reinforcing the need for robust theoretical frameworks to guide effective, systemic interventions (Kahneman, 2012; Haselton et al., 2015).

Despite increasing recognition of the importance of behavioural change within systemic transformations, the integration of behavioural insights into climate policy remains underexplored (European Commission, Joint Research Centre, 2025). However, there is growing consensus for the early incorporation of BI into the policy cycle to enhance the effectiveness of interventions. The application of BI, especially when combined with a systemic approach to climate neutrality, has demonstrated efficacy in climate policy areas that are fundamentally dependent on human behaviour for success. This challenges the prevailing misconception that behavioural interventions primarily focus on adapting individual behaviour to existing systems, rather than transforming the systems themselves (Rutter et al., 2017).

Cities, particularly in Europe, are in the early stages of adopting a systemic approach to climate neutrality as part of their commitment to the EU Mission, "100 Climate-Neutral and Smart Cities by 2030" (European Commission, 2021). This study examines the role of behavioural change within systemic transformations for climate neutrality, drawing on early evidence from the EU-funded NetZeroCities (NZC) project (https://netzerocities.eu/). Specifically, it explores how cities are planning to implement and monitor behavioural change interventions.

The NetZeroCities project, funded by the European Commission, supports EU cities in achieving climate neutrality by offering tools, expertise, and a shared knowledge platform. Beyond reducing GHG emissions, it aims to drive systemic change through social innovation and citizen participation for a just transition. This study addresses the research question: *What is the emerging evidence of cities' plans to monitor and evaluate the impact of their climate actions on citizens' behavioural change?*

Cities have submitted their planned climate actions through pilot projects and selected indicators to track progress toward climate neutrality. By analysing the plans submitted by 39 European cities, specifically their selection of indicators based on the NetZeroCities standardised indicator set (Neumann et al., 2022) and customised indicators proposed by the cities, this study identifies emerging trends in how behavioural change is being addressed by municipalities. Specifically, it reveals: (1) patterns in indicator selection related to citizen behavioural change, and (2) the behavioural change stages cities aim to assess, based on an adapted Transtheoretical Model (Prochaska et al., 1994). These insights contribute to the broader discourse on behavioural change in systemic transformations and support more holistic, evidence-based policymaking.

2. APPLYING THE TRANSTHEORETICAL MODEL TO STUDY BEHAVIOURAL CHANGE

Prochaska et al.'s Transtheoretical Model of Change (1994), originally developed to assess an individual's readiness to adopt healthier behaviour within the context of psychotherapy, proposes five distinct stages of behavioural change. Additionally, the model incorporates a stage signifying "Termination", which represents the failure to maintain the implemented behavioural change, as well as the concept of "Relapse", denoting a regression from the "Action" or "Maintenance" stage to an earlier phase. TTM has been widely adopted across disciplines such as consumer studies (Mitrega et al., 2022), climate education (Inman and Jeffrey, 2006) and transportation (Rizki et al., 2024) owing to its generalisability across domains. In this study, the model is used to categorise behavioural change indicators within the context of climate neutrality to assess which stage of behavioural change they correspond to (see Figure 1).

2.1 Stages of Change

The Transtheoretical Model (TTM) outlines six stages of behavioural change: (1) Precontemplation, where individuals are unaware of the negative impacts of their behaviour and show no intent to change; (2) Contemplation, where they begin to recognise the need for change but remain uncertain; (3) Preparation, marked by planning and initial steps towards action; (4) Action, where concrete behavioural changes are made; (5) Maintenance, where these changes are sustained and protected from relapse; and (6) Termination, in which the change fails to hold, leading to regression. This study has adapted the model by grouping the first five stages into three consolidated categories (Figure 1), partly aligning with the I-change model (Vries et al., 2005).

2.2 Adaptation of the Model

a. Awareness: This stage encompasses interventions corresponding to the first two stages of the TTM, namely Precontemplation and Contemplation. It aligns with soft climate adaptation strategies (Sovacool, 2011), which primarily involve initiatives that nudge or raise awareness among citizens. Such interventions aim to inform and encourage behavioural change without necessarily prompting immediate action. For instance, increased nudging towards sustainable commuting within organisations exemplifies this stage. The intervention seeks to nudge citizens towards Pro-Environmental Behaviours or PEBs (Gifford & Nilsson, 2014), yet it does not track whether individuals are actively engaging in or practising the behaviour.

b. Intention: This category includes interventions that span the Contemplation and Action stages of the TTM. Interventions at this level are more structured than those in the Awareness stage, as they focus on increasing the active involvement of citizens in adopting PEBs. However, such interventions remain voluntary and, while they may motivate, inspire, and prepare citizens for PEB, they do not ensure sustained systemic adoption. An example of this is the Citizens' Capacitation Programme aimed at carbon neutrality. Success is measured by the number of citizens actively participating in the programme. Although the intervention fosters engagement towards climate neutrality, it does not enforce sustained participation.

c. Action: This final stage encompasses the Action to Maintenance phases of the TTM and includes interventions designed to ensure a higher degree of citizen participation, either through hard and soft policy measures or normative influence. For example, an intervention that encourages an increased number of employees to shift to sustainable commuting would be classified here. This stage typically involves enforceable measures that seek to ensure consistent and sustained behavioural change within a specific citizen group.

The above 3 categories were used not only to mark indicators of behavioural change but also to specify at what behavioural stage the interventions are likely to make an impact based on an adaptation of TTM.

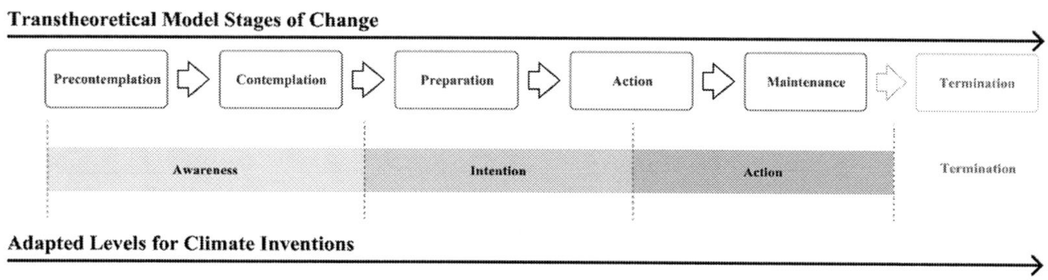

Figure 1. Transtheoretical Model Stages of Change adapted to study climate interventions towards systemic behavioural change. (Source: Adapted from the TTM stages of change by Prochaska et al., 1994)

3. EMPIRICAL DATA SOURCES

The 39 cities of the NZC project included in this study are the first cohort of Pilot cities which submitted their initial proposal refinement documents to strategise their climate actions and select monitoring, evaluation and learning indicators (Chaudhary et al., 2022).

The pilot projects (described at: https://netzerocities.eu/pilot-cities-cohort-1-2022/) in this study include: *Net Zero Investment Co-Innovation Lab* (Bristol, UK), *Budapest CARES* (Budapest, Hungary), *Blueprint for Net-Zero Apartment-block Neighbourhoods* (Cluj-Napoca, Romania), *Green and Carbon Neutral Building Transition Guide – Istanbul Model* (Istanbul, Türkiye), *NEUTRON* (Kozani, Greece), *SCALE UP* (Uppsala, Sweden), *Systemic change towards sustainable commuting in Lahti* (Lahti, Finland), *Let'sGOv* (Bologna; Bergamo; Florence; Milan; Padova; Parma; Prato; Rome; Turin, Italy), *Leuven's Pilot Project* (Leuven, Belgium), *The Initiation of Sustainable Energy Community for the City of Liberec* (Liberec, Czech Republic), *Limassol City Cooling Challenge* (Limassol, Cyprus), *Net Zero Malmö Pilot* (Malmö, Sweden), *Together Toward Climate Neutrality* (Nantes, France), *NEEST* (Krakow; Łódź; Rzeszów; Warsaw; Wrocław, Poland), *Creating NetZero Vision for Rivne* (Rivne, Ukraine), *URBANEW* (Barcelona; Madrid; Seville; Valencia; Valladolid; Vitoria-Gasteiz; Zaragoza, Spain), *1.5-Degree City* (Turku, Finland), *The North Star* (Umeå, Sweden), and *UP-SCALE* (Kranj; Ljubljana; Velenje, Slovenia).

Pilot cities were provided with templates to complete based on their current or planned pilot activities. This study focuses specifically on analysing the submitted non-GHG and Co-benefit indicators (Table 1).

Table 1. Content categories of the entire document template for cities to report indicator selections (Neumann et al., 2022)

1. Direct Impacts & Indicators for Reporting	*2. Indirect Impacts and Indicators for Reporting*	*3. Outcomes for Sensemaking & Qualitative Insights Reporting*
1.1 Long-term GHG Impacts (Standardised)	**2.1 Co-Benefits (Standardised)**	3.1 Early Outcomes (Customised according to city/project)
1.2 Long-term GHG Impacts (Customised according to city/project)	**2.2 Co-Benefits (Customised according to city/project)**	3.2 Later Outcomes (Customised according to city/project)
1.3 Long-term non-GHG Impacts (Customised according to city/project)		

4. DATA ANALYSIS METHODS

The analysis followed a three-step process: (1) categorisation of the indicators related to behavioral change from (sections, 1.3, 2.1, and 2.2 of) the pilot projects submitted by all 39 cities based on a compilation of all the indicators submitted (Mondal et al., 2024); (2) coding (based on the adapted Transtheoretical Model) of the dataset to assess the indicator of behavioural change (Carter et al., 2014); and (3) synthesis to extract key insights. Inputs were compiled in a spreadsheet, with each intervention reviewed and its indicator selections recorded. Indicators related to behavioural change were identified and independently coded by two researchers into the adapted TTM categories—Awareness, Intention, and Action (Table 2). Discrepancies in coding were resolved through an organised discussion. The final coded dataset (through unanimous agreement) was then analysed to identify recurring trends in indicator selection.

5. KEY FINDINGS AND DISCUSSIONS

A previous work (Mondal et al. 2024) mapped climate actions and their associated indicators selected by the cities (analysing them in relation to the UNSDGs), and this analysis focuses specifically on the indicators which are related to behavioural change based on 198 interventions submitted by 39 European cities in the initial phase of their pilot projects. Among all the indicators selected by cities, 35% (18 out of 52) of long-term non-GHG interventions and 32% (18 out of 56) of customised co-benefit interventions were linked to direct or indirect systemic behavioural changes. In contrast, only 4.4% (4 out of 90) of standardised co-benefit interventions addressed behavioural change. Of all indicators identified as behaviour-related, 27.78% aligned with *Awareness*, 50% with *Intention*, and 22.23% with *Action*, reflecting the varying depths of engagement across the interventions (Figure 2).

Table 2. TTM Stages of Behavioural Change Coding (Awareness, Intention or Action) of Cities' Indicator Selections (Mondal et al., 2024) related to Behavioural Change

Indicator Selections by Cities	Custom metric/unit of measurement	TTM Stages
Customised according to city/project		
Co-Design Events	Number	*Awareness*
Share of citizens with eco-friendly behaviors	%	*Awareness, Intention*
Resident engagement in energy and climate conscious actions	Surveys (Likert Scale 1-5: 4)	*Action*
Degree of satisfaction and acceptance of residents affected by the actions in the project	Surveys (Likert Scale 1-5 : 4)	
participation percentage - No. of stakeholders	Number	*Awareness*
No. of users of public transport system	Total # of users	*Action*
No. of interventions in case organisation during the project	No. of interventions	*Awareness*
Incentives/Work benefits offered by the employers (value/employee)	euros/employee	*Intention*
Share of employees in participating in interventions	% of all employees in case organisations	*Action*
Increase in grassroots initiatives	n° of new grassroots initiatives in energy	*Awareness*
Level of sense of agency among stakeholders and residents	Likert scale; 5 scales to be determined in survey	*Action*
Social spectrograph	NA	*Intention*
Number of neighbourhoods with partnerships	NA	*Awareness*
Improved citizen participation per city / district (estimations)	# of citizens engaged through the Pilot activities	*Behaviour*
UTU research team will collect data regarding behaviour change, using both objective (e.g. electronic traffic counting) and subjective (e.g. survey questionnaires, interviews) measures.	NA	*Awareness*
Behavioural changes among tenants could lead to more sustainable lifestyles	NA	*Action*
Standardised		
Improved citizen participation (Selected 3 times)	No. of citizens engaged through the Pilot activities	*Intention*
Social Cohesion	NA	*Awareness*
Long-term non-GHG Impacts (Customised according to city/project)		
Participants to project activities	Number	*Awareness*
No. of Projects	No. of citizens engaged through the Pilot activities No. of newsletters and/of press releases produced.	*Awareness*
No. of participants	No. of public officers trained through the Pilot activities total # of counseled activities	*Intention*
% of citizens participating in energy-related initiatives	%	*Action*
Increase of stable agreements & collaborations with relevant energy stakeholders in cities	n° of protocols, pathways, collaborations, memoranda of understanding involving relevant stakeholders	*Awareness*

Online Platform Visitors	Total # of Visitors	*Awareness*
Social spectrograph	Socially weighted average distance from 'aware'	*Action*
Accelerated, socially peaceful change towards NZC Broader acceptance of solutions Social spectrograph	Socially weighted average distance from 'aware'	*Intention*
Climate impact per capita of consumption	t CO2 equivalents / year	*Action*
Successful operational changes made	Number as identified	*Intention*
Number community engagement activities	Number as identified	*Awareness*
Number of REC triggered by the project	Number of REC triggered by the project	*Intention*
Feedback of the organised events, trainings & webinars for the Climate team	Number of organised events, trainings and webinars Number of participants	*Intention*
Observable changes in the behaviour of citizens towards climate neutrality	Indicators different per type of emission, measured using existing general tools	*Action*
Awareness & ability to work across silos, formalised changes in policy, governance, organisational structure, budgets, etc.	Self-assessment Review of changes in yearly reports, number of climate instigated organisational changes, implemented recommendations from project	*Intention*
Awareness and ability to work across silos, formalised changes in policy, governance, organisational structure, budgets, etc. (Selected 2 times)	NA	*Intention*

Based on the frequency of indicator selections linked to behavioural change, a plausible explanation for the disproportionate difference between standardised and customised indicators is the absence of a defined "Behavioural Change" indicator within the list of standardised indicators provided to cities, while cities perceive the need to monitor behavioural change of various actors. This observation highlights an emerging trend in climate interventions being undertaken by cities, which increasingly aim to address behavioural change at a systemic level. It also underscores the potential for the NZC project standardised indicator framework to integrate behavioural change indicators, drawing on the trend of customised indicator selections made by cities, and aligning them with the stages of behavioural change as outlined by the Transtheoretical Model of Change.

On the other hand, Indicators such as *"(%) Share of citizens with eco-friendly behaviours"* and *"Improved citizen participation"* are primarily associated with the early stages of behavioural change (Awareness and Intention). Conversely, a smaller proportion of indicators, such as *"Observable changes in the behaviour of citizens towards climate neutrality"* and *"Behavioural changes among tenants could lead to more sustainable lifestyles"*, correspond to the later stages of behavioural change (Action and Maintenance). This highlights that the prevailing focus of climate actions related to behavioural change, whether directly or indirectly, is on raising awareness or fostering intention (e.g., through participation and engagement) rather than on sustaining action and long-term maintenance of pro-environmental behaviour.

This categorisation offers cities a framework to assess the stage of behavioural change at which their interventions are likely to be effective, ensuring alignment with their intended objectives for systemic behavioural transformation. Additionally, the adaptation of the TTM for assessing behavioural change in climate transitions is a novel contribution to the transdisciplinary literature on behavioural change in systemic transformations, that extends beyond the focus on the individual.

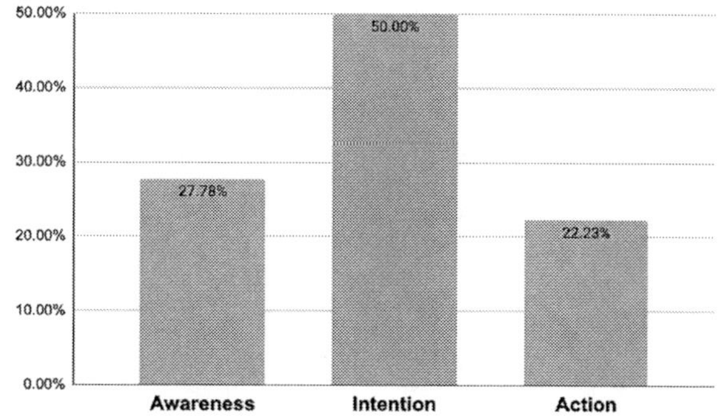

Figure 2. Bar chart showing the distribution of indicators selected by cities across the stages of behavioural change

6. CONCLUSION

This study presents novel insights on the assessment of behavioural change in systemic transformations by categorising indicators of climate interventions chosen by European cities for climate action monitoring and evaluation, based on an adaptation of the Transtheoretical Model of Change (TTM). The indicator selection of the 39 cities provides emerging evidence of how cities are seeking to monitor the behavioural change of citizens within the context of climate neutrality. The classification of indicators into awareness, intention and action supports cities in assessing the implications of planned climate actions on the behaviour of their citizens, helps shape climate policy that incorporates behavioural insights (Rutter et al., 2017), and gauges how cities are positioned in their journey to foster long-term behavioural change of citizens within systemic transformations (Newell et al., 2021). This enables cities to develop not only soft climate adaptation strategies (Sovacool, 2011) but also foster deep scaling of behavioural change through hard and soft policy measures or normative influence.

Hence, this study contributes a timely and pragmatic perspective to the emerging trans-disciplinary body of literature while addressing the crucial needs of cities which are striving towards climate neutrality. However, it's important to mention that this study does come with certain limitations: the NZC project is based in the European context; with appropriate adaptation, the categorisation and insights can be useful to study a wider context of cities globally.

Additionally, future directions of research may expand on how the level of preparedness of cities to address behavioural change correlates with the stages of behavioural change they address through their climate actions. Lastly, the proposed categorisation may be further expanded to assess the degree of dependence (for the success) of climate actions on the various stages of behavioural change of citizens.

ACKNOWLEDGEMENT

This research was funded by the EU, grant number 101139652—SGA2-NZC, NetZeroCities.

REFERENCES

Akenji, L., Lettenmeier, M., Koide, R., Toivio, V. and Amellina, A., (2019). 1.5-degree lifestyles: Targets and options for reducing lifestyle carbon footprints. Hayama: Institute for Global Environmental Strategies. Available at: https://doi.org/10.57405/iges-6719 [Accessed 28 Mar. 2025].

Bernauer, T., (2013). 'Climate change politics', Annual Review of Political Science, 16(1), pp.421–448. Available at: https://doi.org/10.1146/annurev-polisci-062011-154926.

Carter, N., Bryant-Lukosius, D., DiCenso, A., Blythe, J. and Neville, A.J., (2014). 'The use of triangulation in qualitative research', Oncology Nursing Forum, 41, pp.545–547.

Chaudhary, N., Hawkins, P. and Palavicino, C., (2022). NetZeroCities theory of change. NZC Deliverable D2.14.
Cialdini, R.B., Reno, R.R. and Kallgren, C.A., (1990). 'A focus theory of normative conduct: Recycling the concept of norms to reduce littering in public places', Journal of Personality and Social Psychology, 58(6), pp.1015–1026. Available at: https://doi.org/10.1037/0022-3514.58.6.1015.
European Commission, (2021). European Missions – 100 Climate-Neutral and Smart Cities by 2030. Available at: https://netzerocities.eu/.
European Commission Joint Research Centre, (2025). Unlocking the full potential of behavioural insights for policy: From influencing the individual to shaping the system. Luxembourg: Publications Office of the European Union. Available at: https://data.europa.eu/doi/10.2760/7367599 [Accessed 28 Mar. 2025].
Gifford, R. and Nilsson, A., (2014). 'Personal and social factors that influence pro-environmental concern and behaviour: A review', International Journal of Psychology, [online] Available at: https://doi.org/10.1002/ijop.12034.
Haselton, M.G., Nettle, D. and Murray, D.R., (2015). 'The evolution of cognitive bias', in Buss, D.M. (ed.) The handbook of evolutionary psychology. Hoboken: John Wiley & Sons, pp.1–20. Available at: https://doi.org/10.1002/9781119125563.evpsych241.
Hertwich, E.G. and Peters, G.P., (2009). 'Carbon footprint of nations: A global, trade-linked analysis', Environmental Science & Technology, 43(16), pp.6414–6420. Available at: https://doi.org/10.1021/es803496a.
Hirano, Y., Ihara, T. and Yoshida, Y., (2016). 'Estimating residential CO_2 emissions based on daily activities and consideration of methods to reduce emissions', Building and Environment, 103, pp.1–8. Available at: https://doi.org/10.1016/j.buildenv.2016.02.021.
Inman, D. and Jeffrey, P., (2006). 'A review of residential water conservation tool performance and influences on implementation effectiveness', Urban Water Journal, 3(3), pp.127–143. Available at: https://doi.org/10.1080/15730620600961288.
IPCC, (2022). Global warming of 1.5°C: IPCC special report on impacts of global warming of 1.5°C above pre-industrial levels in context of strengthening response to climate change, sustainable development and efforts to eradicate poverty. 1st ed. Cambridge: Cambridge University Press. Available at: https://doi.org/10.1017/9781009157940.
Ivanova, D., Stadler, K., Steen-Olsen, K., Wood, R., Vita, G., Tukker, A. and Hertwich, E.G., (2020). 'Quantifying the potential for climate change mitigation of consumption options', Environmental Research Letters, 15(9), 093001. Available at: https://doi.org/10.1088/1748-9326/ab8589.
Kahneman, D., (2012). Thinking, fast and slow. Choice Reviews Online, 49(10), pp.49-5972–49-5972. Available at: https://doi.org/10.5860/CHOICE.49-5972.
Mitrega, M., Klézl, V. and Spáčil, V., (2022). 'Systematic review on customer citizenship behavior: Clarifying the domain and future research agenda', Journal of Business Research, 140, pp.25–39. Available at: https://doi.org/10.1016/j.jbusres.2021.11.029.
Mondal, R., Bresciani, S. and Rizzo, F., (2024). 'What cities want to measure: Bottom-up selection of indicators for systemic change toward climate neutrality aligned with Sustainable Development Goals (SDGs) in 40 European cities', Climate, 12(3), 41. Available at: https://doi.org/10.3390/cli12030041.
Neumann, H.M., et al., (2022). Comprehensive indicator framework. NetZeroCities Deliverable 2.4.1. Vienna: AIT. Available at: https://netzerocities.eu/results-publications/.
Newell, P., Twena, M. and Daley, F., (2021). 'Scaling behaviour change for a 1.5-degree world: Challenges and opportunities', Global Sustainability, 4, e22. Available at: https://doi.org/10.1017/sus.2021.23.
Prochaska, J.O., Norcross, J.C. and DiClemente, C.C., 1994. Changing for good. 1st Collins paperback ed. New York: William Morrow.
Repo, P. and Matschoss, K., (2019). 'Social innovation for sustainability challenges', Sustainability, 12(1), 319. Available at: https://doi.org/10.3390/su12010319.
Rizki, M., Joewono, T.B. and Susilo, Y.O., (2024). 'Exploring levels of adoption of multi-function transport apps: Transtheoretical model of change on the customer journey of Transport-SuperApp (TSA) users', Communications in Transportation Research, 4, 100125. Available at: https://doi.org/10.1016/j.commtr.2024.100125.
Rutter, H., Savona, N., Glonti, K., Bibby, J., Cummins, S., Finegood, D.T., Greaves, F., Harper, L., Hawe, P., Moore, L., Petticrew, M. and White, M., (2017). 'The need for a complex systems model of evidence for public health', The Lancet, 390(10112), pp.2602–2604. Available at: https://doi.org/10.1016/S0140-6736(17)31267-9.
Sabel, C.F. and Victor, D.G., (2017). 'Governing global problems under uncertainty: Making bottom-up climate policy work', Climatic Change, 144(1), pp.15–27. Available at: https://doi.org/10.1007/s10584-015-1507-y.
Schwartz, S.H., (1977). 'Normative influences on altruism', in Berkowitz, L. (ed.) Advances in experimental social psychology. 1st ed. New York: Academic Press, pp.221–279.
Sovacool, B.K., (2011). 'Hard and soft paths for climate change adaptation', Climate Policy, 11(4), pp.1177–1183. Available at: https://doi.org/10.1080/14693062.2011.579315.
United Nations, (2023). The Paris Agreement. United Nations Climate Change. Available at: https://unfccc.int/process-and-meetings/the-paris-agreement.
Vries, H.D., Mesters, I., van de Steeg, H. and Honing, C., (2005). 'The general public's information needs and perceptions regarding hereditary cancer: An application of the Integrated Change Model', Patient Education and Counselling, 56(2), pp.154–165. Available at: https://doi.org/10.1016/j.pec.2004.01.002.

STANDARDIZATION OF SMART CITY SENSOR NETWORKS: USE-CASE- SPECIFICITY AND DATA FORMATS AS KEY ASPECTS

Johannes Sautter[1], Rebecca Kreuz[2], Frederic Schubert[1], Ekaterina Dobrokhotova[1], Andreas Helsper[1], Philipp Lämmel[3] and Adrian Föhl[4]
[1]Fraunhofer Institut für Arbeitswirtschaft und Organisation IAO
Nobelstr. 12, 70569 Stuttgart, Germany
[2]bwcon research gGmbH
Seyfferstraße 34, 70197 Stuttgart, Germany
[3]Fraunhofer Institut FOKUS
Kaiserin-Augusta-Allee 31, 10589 Berlin, Germany
[4]BABLE GmbH
Reuchlinstrasse 27, 70176 Stuttgart, Germany

ABSTRACT

As smart city evolves rapidly, standardization is increasingly recognized as a technical, political, and strategic tool for fostering long-term resilience and innovation. In-situ sensors capture a variety of data that is crucial for urban governance and data-driven decision-making. This study examines the necessity, form, and challenges of standardizing sensors within the context of smart city development. Using a qualitative methodology that incorporates literature reviews and guided interviews, the research aims to understand the current situation and the perspectives of municipalities and sensor manufacturers, while identifying areas of common ground. The results demonstrate a growing need for standardization to ensure interoperability and optimize the use of sensor technologies. Furthermore, it is revealed that many municipalities are reluctant to implement new smart city use cases due to needed expertise and personnel, even with technical standards in place. Standardization could contribute to the further establishment of smart cities in general and IoT-sensors in particular.

KEYWORDS

Standardization, Sensor Networks, Smart City, Use Cases, Data Interoperability

1. INTRODUCTION

Smart city sensors, which this study defines as in-situ sensors typically operated by local authorities, have become essential for developing intelligent urban environments (Gracias et al., 2023). These sensors collect various types of information, including air quality, temperature, noise levels, traffic flow, and parking availability (Enlund et al., 2022). One initiative that promotes their use is the European Smart Cities Marketplace, formerly the European Innovation Partnership on Smart Cities and Communities (EIP-SCC). Its goal is to create an open marketplace for smart city solutions, particularly in transportation, energy, and ICT (Europa.eu, 2019), and it has encouraged the development of various sensor-based projects in European cities. For example, in Tallinn, Estonia, data on urban noise is collected (Environment, 2023), and in Frankfurt am Main, Germany, sensors have been deployed to monitor air quality and road conditions (e.g., black ice detection) as part of the SMART.sensor project (Smart City Factory, 2024). Similarly, Dresden launched the KLIPS project, which involves installing 300 temperature and humidity sensors throughout the city to address challenges related to urban overheating and climate change (Smart City Dresden, 2022).

The smart city sensor landscape is marked by significant diversity, with different technologies and applications being implemented across urban contexts. For instance, sensor data platforms are essential for processing and visualizing large amounts of data, which facilitates real-time traffic updates and accurate environmental monitoring (Enlund et al., 2022). However, as various sensor applications evolve, there is an increasing need for standardization and interoperability, especially in smaller municipalities and rural areas, in order to realize the full potential of integrated urban and rural environments. Despite some existing frameworks, such as the Smart City Interoperability Reference Architecture (SCIRA) and the European Interoperability Framework for Smart Cities, challenges in data privacy, backward compatibility, and stakeholder engagement persist (U.S. Department of Homeland Security, 2024).

The German Institute for Standardization (DIN) highlights the importance of coordinating standards at the national, European, and international levels. Currently, around 140 relevant standards exist (DIN SPEC 91387, 2020). The German Standardization Roadmap for Smart Cities emphasizes the need for efficient communication processes to enhance economic performance, quality of life, and resource efficiency. Furthermore, the Digital Services Act (DSA) aims to promote innovation while safeguarding consumer rights and fostering transparency and accountability in online platforms.

To address the challenges of smart city initiatives, such as resource constraints for smaller municipalities and institutional inertia for larger ones, the OUPENBW[1] project focuses on the standardization of smart city sensors. This paper explores the adoption of existing standards, the role of data management, and the need for standards linked to specific use cases, such as waste management and parking systems. The project aims to improve integration across urban organizational structures, enhance the user experience, and increase acceptance of sensor systems by demonstrating their ease of use and usefulness.

2. STATE OF THE ART

Depending on individual requirements such as range, energy efficiency, cost, and data throughput, smart city sensor networks use different wireless technologies and network protocols, often in combination. Short-range wireless technologies, such as Wi-Fi, Bluetooth, and Zigbee, are primarily used in local sensor systems, like those found in smart homes. They offer high data rates and low latency but have a limited range of less than 1 km. Mobile radio technologies, such as LTE and 5G, as well as special IoT extensions, such as LTE-M, offer greater range, high data rates, low latency, and broad existing network coverage. Additionally, they use licensed frequency bands that ensure reliable data transmission. Low Power Wide Area Networks (LPWANs), such as LoRaWAN, Mioty, Sigfox, and NB-IoT, are designed for energy-efficient applications that communicate over long ranges at low data rates and frequencies. They enable years of low-maintenance sensor operation using batteries or solar energy and function on both license-free and license-required frequency bands. (Abadía et al., 2022)

Sensors are the most prevalent type of device in the Internet of Things (IoT), though other end devices, such as actuators, tags, and smart meters, also play an important role. Sensors are used to record physical, chemical, or biological conditions qualitatively or quantitatively and transform them into electrical signals that can be processed further. Actuators convert electrical signals, typically from control and regulation processes, into physical actions. Tags uniquely identify objects and people and provide information about them, such as their location and access rights. (Bauer et al., 2021)

The application fields for sensor technologies are also very diverse. Some of the most important fields include governance, housing and infrastructure, mobility, business, industry, energy, the environment, and healthcare. To address these diverse and extensive needs, sensor network architectures typically employ three- or five-layer models, integrating edge, fog, and cloud computing. One of the biggest challenges of implementing sensors across all use cases is ensuring the interoperability of heterogeneous systems. Numerous manufacturer-specific, isolated solutions hinder data exchange and result in high operating costs.

Other challenges include data protection and cybersecurity, both of which are essential for the social acceptance of sensor applications in cities. Open standards and platforms can address these issues, as well as ensure the scalability and long-term maintainability of sensor infrastructures necessary for sustainably managing growing urbanization and the digitization of municipal infrastructures (Bellini et al., 2022).

[1] OUPEN SmartCity made in BW Plug&Play Systeme für GreenTech in Kommunen (Acronym: OUPENBW)

Furthermore, coordinated standardization involving standardization organizations, industry, and the public sector can generate economies of scale and strengthen the technological sovereignty of European cities. This is relevant not only on a local level, but also in the context of challenging overarching European initiatives, such as the Green Deal and the Digital Decade. Therefore, standardization is increasingly recognized as a technical, political, and strategic tool for fostering long-term resilience and innovation in urban areas. (Albouq et al., 2022; European Commission, n.d.) DIN SPEC 91357 establishes an open urban platform architecture for German smart cities. DIN SPECs 91367, 91387, 91607, and 91377 focus on real-time mobility data, digital transformation, urban digital twins, and interoperability protocols. ISO 37107:2019, ISO 37120:2018, and ISO 37122:2019 provide frameworks and indicators for evaluating and advancing smart, sustainable communities globally.

3. METHOD

In order to investigate the research questions, "What requirements exist for standardization?" "What is the current status of municipalities and sensor manufacturers regarding smart city sensors?" and "What are the potential benefits of additional standards?," a qualitative research design was chosen. An enhancing literature review aimed to assess the current state of research as well as key terms, concepts and research gaps.

Guided interviews are used to gain practical insights and opinions from experts or those affected. This method is particularly suitable for gathering detailed information on specific topics because it combines open-ended questions with a clear structure. The interviews were conducted with representatives of sensor manufacturers as well as with municipal decision-makers in order to shed light on the status quo and current developments in the sensor sector from different perspectives. Participants were selected through purposive sampling to ensure relevance to the research questions. Selection criteria included professional involvement in smart city projects, affiliation with either municipalities or sensor manufacturers, and expertise in sensor deployment, standardization, or strategy. Efforts were made to include a range of perspectives from organizations of different sizes and roles. While the sample is not representative, it was designed to capture diverse, practical insights.

Concise and representative interview excerpts were selected for the article to provide practical examples that support the theoretical findings from the literature review. To ensure the confidentiality of the participants, these quotes were presented anonymously. Table 1 provides an overview of the anonymized interviewees from municipalities and sensor manufacturers in Germany. The cities are classified according to the NUTS categories.[2]

Table 1. Overview of interviewees

No. of Interview	Position of interviewee	Type of the city or enterprise
1	Senior Manager IT & Digital Solutions	Middle city
2	Lead Smart City department, Deputy Head of Digitalization	Small city
3	Digitalization department, Smart City	Small city
4	Lead Data Science department	Middle city
5	Head of Climate Protection Unit	Small city
6	CEO	Enterprise in ownership of Middle city
7	Manager Embedded Engineering	IT consulting company
8	Business Development Manager	Supplier of measuring devices and solutions for industrial process technology
9	Founder & CEO	Start-up, sensor developer

[2] NUTS - Nomenclature of territorial units for statistics [online] https://ec.europa.eu/eurostat/web/nuts/ [Accessed 18 March 2025]

4. ASPECTS OF STANDARDIZATION

In the context of smart cities and digitalization in Germany, the amount of knowledge municipalities have about sensors varies significantly. Some municipalities are just starting out, while others, typically large cities, have built extensive expertise with the help of many people. Small cities with limited financial and human resources tend to wait for the longest for technologies to become more established, standardized, and user-friendly. Although the surveyed companies recognize the potential of standardizing smart city solutions, they face challenges in their practical implementation.

4.1 Desire for More Smart City Governance

There are general statements from the municipalities on smart cities, such as the idea that development should start top-down, at the federal level, to enable better scaling. Furthermore, there are calls for the state to move forward more consistently with greater governance and less federalism and to break away from experimentalism and establish binding standards. (Colangelo & Borgogno, 2023).

A central problem is that "90%" of cities do not build their own sensors but instead put them out to tender (interview no. 4). In this context, the concept of an "IoT Ready!" certification is introduced that could provide cities with easy access to the necessary information to help them become more independent (interview no. 4). Another frequently mentioned point is that there are "far too many options on the market," including different providers, systems, and data formats (interview no. 5).

Many companies already offer cloud-based or mobile-supported open-source solutions which, from their point of view, should enable simple integration.

Therefore, it is important for the state to clearly recommend what should be used. A "list of requirements" (interview no. 5) for sensors would be helpful to ensure strict quality criteria are met, such as data protection and open data (Schnabel et al., 2020).

4.2 "Trial and Error" Necessary as Only Technical Standards Exist

Currently, only technical standards exist, which means that "there is nothing out of the box, try and error; there is no solution that can be used for every city" (interview no. 1). Standardization is sought more at a platform level, but there is no comprehensive platform that covers all needs. It is pointed out that "many are pushing in different directions" (interview no. 2), and it is unclear what specifically should be standardized. One sensible approach would be, for example, standardizing the link to the data catalog containing the data product created by the sensor network defining technical metadata and access rights (Korte et al., 2019).

The future of smart city solutions lies in "a lot of transferability" (interview no. 2), so that cities can simply move solutions from city A to city B. Consolidation would be desirable to break up existing specialist silos where governance structures are lacking. Regarding specific standards at sensor manufacturers, the wish was expressed that "something like this comes from the top - the compulsion comes from the top down" (interview no. 2).

The companies surveyed recognize the potential of standardizing smart city solutions, but at the same time see challenges in the practical implementation. "A central platform makes sense if information from a large number of sources is to be combined and made available to different users" (interview no. 8).

A functioning certification requires a "seal from the public authority" (interview no. 5), accompanied by an openly designed proof of concept to facilitate easier use. Even if software was more expensive, the cities would be prepared to pay for the seal. The current API standard, to which everyone should adhere, is also relevant.

4.3 Sensor Maintenance Costs Resources

In addition to organizational hurdles, there are also technical challenges, such as the standardization of data formats or the maintenance of sensors. "They are tools that need to be maintained" (interview no. 9). The sensor manufacturers emphasize that in the long term, stronger strategic cooperation with local authorities is necessary in order to successfully implement standardization approaches and leverage the benefits of interoperable solutions in administration.

4.4 The Ideal Standard

Several requirements are named with regard to an ideal standard: It should be "Open Data capable" (interview no. 5), it should make sense for municipalities to be able to choose between different licenses, and clear specifications must be made regarding data formats. Data protection and readability for databases such as mobilithek[3], MobiDataBW[4] or European Data Portal[5] are also important in order to ensure universal use.

To enable full certification, it is important to "explain a lot! Why you take what" (interview no. 5) to lower the barrier to entry. Wording is often a hurdle, which is why explanations for dummies should be provided. The transfer process, especially in larger cities, costs a lot of human resources and requires a lot of human resources, knowledge, and time.

Overall, a "higher degree of standardization" (interview no. 6) is desired in the area of interfaces, particularly with regard to data models and how data is processed. However, it remains difficult to create concrete specifications for specific systems.

4.5 Acceptance and Penetration in Cities Vary Greatly

Although open standards and central platforms are recognized as means of achieving interoperability and increasing efficiency, municipalities often hesitate to integrate new technologies. "Penetrating municipal administrations with a single solution is quite difficult" (interview no. 7). Many companies already offer cloud-based or mobile-based open source solutions that are designed to enable simple integration. Although the solutions are quite easy to use from the perspective of the sensor-producing companies, acceptance within municipal administrations varies greatly: while specialized departments welcome application-oriented solutions, there is often a lack of an overarching strategy for using a central platform and the necessary know-how. Added to this are concerns about data protection when processing personal data and complex administrative processes. "I came up with over 25 standards and certifications that our solution has to meet" (interview no. 7).

4.6 Use Cases of Sensor Infrastructure

Sensors have many applications in urban environments. They help improve quality of life and increase the efficiency of city administration.

Weather stations that monitor weather conditions as well as temperature and soil moisture sensors are used particularly frequently to optimize environmental conditions (interview no. 1).

In the area of waste and environmental management, digital waste containers, sensors in mobile toilets and temperature detection systems in the recycling center are used to identify fire hazards at an early stage (interview no. 1). Leakage sensors and climate monitoring help to use water resources efficiently and identify problems in good time (interview no. 2). In addition, parking space monitoring systems are being developed to optimize the use of e-charging stations and disabled parking spaces (interview no. 2).

Sensors for monitoring pedestrians and irrigation sensors in parks improve environmental conditions in urban areas (interview no. 3). Special climate monitoring sensors support the analysis of urban climates and air quality (interview no. 4).

Another example is the digital pedestrian counter, which measures the number of people passing by an object. This data is used by the specialist department for tourism, marketing and events to evaluate events. "Solution provides data in real time, every 60 seconds" (interview no. 7).

Another important area of application is flood management. A level sensor is used here, which monitors the water levels and issues a warning if a certain limit value is exceeded. This allows for the initiation of further steps in a timely manner. "We use a web-based interface and, for example, e-mail communication for alerting" (interview no. 8).

[3] https://mobilithek.info/
[4] https://www.mobidata-bw.de/
[5] https://data.europa.eu/

In addition, the monitoring of drought stress in tree crowns is also supported by sensors. Such sensors help to monitor the drought stress in tree crowns in order to optimize the irrigation of young trees in a targeted manner. "We hang the sensor at the top of the tree, so it's not a ground sensor, as is usually the case" (interview no. 9).

These applications demonstrate how sensors can increase the efficiency and effectiveness of urban services while providing valuable information for urban planning and management.

The potential of sensor-based use cases varies depending on a municipality's characteristics, especially its population and size. For instance, sensors designed to optimize parking and traffic flow are more beneficial in large cities than in small towns and villages because of their comparatively low traffic density. Furthermore, geographical location can lead to differences in application potential. For example, the usefulness of water level and black ice sensors depends on a municipality's topography and weather conditions. In turn, monitoring safety in buildings and public spaces is an example of a relevant use case for every municipality, although its importance varies depending on the local crime rate. (Interviews conducted as part of the public research project “LoRaWAN Labs@BW.”)

4.7 Sensor Interpretation and Fusion

"A central platform makes sense if information from a large number of sources is to be summarized and made available to various users" (interview no. 8). Currently, numerous municipalities are developing urban data platforms for this purpose. However, one interviewee stated that local authorities and the relevant specialist departments sometimes still lack the necessary expertise when it comes to interpreting sensor data in order to draw the right conclusions. "People underestimate how complex data analysis can be. [...] You need someone to contextualize this data, and this context is of course provided on the one hand by the data sources, i.e. many different sensors, but above all by the expert" (interview no. 9).

It is important to clarify what expertise is needed on-site. This may include technical skills for the establishment and operation of networks and sensors, as well as subject matter expertise on topics such as climate protection and mobility. Recognizing these distinctions is crucial to prevent IoT experts from suddenly becoming climate protection experts.

Sensor fusion, or sensor data fusion, refers to the merging of measurement data from different sensors to improve the information quality of a parameter to be determined (Human, 2024). "For example, the central municipal platform collects the measurement data for the water quality of a bathing lake, simultaneously obtains the data from the German Weather Service and then merges this information into a recommendation for the lifeguard" (interview no. 8).

One of the companies surveyed sees the centralization of data as a way to facilitate sensor fusion. A common platform would enable the comparison of different sensor data and reduce the administrative burden. However, another interviewee emphasizes that a central platform model does not make sense across the board. Instead of a universal "pope platform", it is more important to have networked, specialized platforms that are used depending on the area of application. This supports an open, heterogeneous IoT ecosystem in which different systems are interoperable. "Heterogeneous means that you should not claim that your own platform is the sole and only usable platform. [...] There will not be one single ‘pope platform’" (interview no. 8).

5. LIMITATIONS

The use of a qualitative methodology entails the limitation that the findings are not representative of all municipalities within a given region. The selection of interview participants may also introduce certain biases. Furthermore, the diversity of technologies and use cases complicates the comparability of the results. Finally, the focus on specific regional contexts limits the transferability of the findings to other geographical areas.

6. CONCLUSION AND OUTLOOK

The paper discloses a discrepancy in resources and knowledge between large and small cities: While larger cities often have extensive knowledge, smaller municipalities are often dependent on the standardization and establishment of such technologies. They have fewer staff and consequently less digital expertise. This forces them to rely heavily on external service providers, which presents the challenge of allocating the necessary financial resources for comprehensive outsourcing. Additionally, they may struggle to allocate internal administrative resources for tasks that cannot be outsourced. These tasks include providing property lists for network planning, drafting tenders, and obtaining permits.

The reluctance of many municipalities to introduce new technologies is often due to a lack of expertise and data protection concerns. Clear governance from the government side is crucial, to drive forward the standardization and use of smart city solutions. In addition, the development of a common platform for data integration is considered essential to increase the efficiency of urban administration. The paper suggests that standards could contribute to the further establishment of smart city in general and IoT-sensors in particular. However, it is also important to focus on the specific subject of these standards to ensure they effectively address the needs of different municipalities and use cases.

As a next step, a workshop with municipalities and enterprises is planned. Also, a practitioner study enhancing the corporate perspective is planned. The advancing digitalization of urban spaces presents cities with challenges in the areas of interoperability, scalability, and cost-effectiveness. Open standards and innovative test environments enable the seamless integration of different technologies, thus laying the foundation for smart, connected cities. Modern approaches, such as the use of cloud technologies and automated processes, help reduce implementation barriers and strengthen digital sovereignty by keeping control over data and systems in the hands of municipalities. Close collaboration between public administrations, technology providers, and research institutions is crucial for developing practical and future-proof solutions. In the long term, these strategies can help improve urban quality of life, generate economic stimulus, and pave the way to a sustainable, digital future.

ACKNOWLEDGEMENTS

The project "OUPEN Smart City made in BW Plug&Play Systeme für GreenTech in Kommunen" (OUPENBW) aims for creating a testbed for smart city standards in order to foster replication and up-scaling of wellbeing, sustainability and digitalization. The project is led by BABLE, funding number is BW1_5058/04.

REFERENCES

Abadía, J.J.P., Walther, C., Osman, A. and Smarsly, K. (2022). A systematic survey of Internet of Things frameworks for smart city applications. Sustainable Cities and Society, 83, p.103949. DOI: https://doi.org/10.1016/j.scs.2022.103949.

Albouq, S. S., Sen, A. A. A., Almashf, N., Yamin, M., Alshanqiti, A. & Bahbouh, N. M. (2022). A Survey of Interoperability Challenges and Solutions for Dealing With Them in IoT Environment. IEEE Access, 10, 36416–36428. DOI: https://doi.org/10.1109/access.2022.3162219

Bauer, M., Sanchez, L. and Song, J. (2021). IoT-Enabled Smart Cities: Evolution and Outlook. Sensors, 21(13), p.4511. DOI: https://doi.org/10.3390/s21134511.

Bellini, P., Nesi, P. and Pantaleo, G. (2022). IoT-Enabled Smart Cities: A Review of Concepts, Frameworks and Key Technologies. Applied Sciences, 12(3), p.1607. DOI: https://doi.org/10.3390/app12031607.

Colangelo, G. & Borgogno, O. (2023). Shaping Interoperability for the Internet of Things: The Case for Ecosystem-Tailored Standardisation. European Journal Of Risk Regulation, 1–16. https://doi.org/10.1017/err.2023.8

DIN SPEC 91387: Communities and digital transformation - Overview of the spheres of activity (DIN SPEC 91387:2020-08) (2020). DOI: https://doi.org/10.31030/3181316.

Enlund, D., Harrison, K., Ringdahl, R., Börütecene, A., Löwgren, J. and Angelakis, V. (2022). The role of sensors in the production of smart city spaces. Big Data & Society, 9(2), p.205395172211102. DOI: https://doi.org/10.1177/20539517221110218.

Environment. (2023). Tallinn - Smart City sensors. [online] Available at: https://environment.ec.europa.eu/topics/urban-environment/european-green-capital-award/inspiration/tallinn-smart-city-sensors_en [Accessed 25 March 2025].

Europa.eu. (2019). Solutions | Smart Cities Marketplace. [online] Available at: https://smart-cities-marketplace.ec.europa.eu/insights/solutions [Accessed 25 March 2025].

European Commission. (n.d.). Europe's Internet of Things Policy. [online] Available at: https://digital-strategy.ec.europa.eu/en/policies/internet-things-policy [Accessed 20 May 2025].

Gracias, J.S., Parnell, G.S., Specking, E., Pohl, E.A. and Buchanan, R. (2023). Smart Cities-A Structured Literature Review. Smart Cities, [online] 6(4), pp.1719-1743. DOI: https://doi.org/10.3390/smartcities6040080.

Human, S. (2024). Was ist Sensorfusion und wo nutzt sie? [online] https://www.industry-of-things.de/sensorfusion-informationsqualitaet-a-642221b9c4bb66f11a7099c69e7047e9/ [Accessed 17 March 2025].

Korte, Tobias, Martin Fadler, Markus Spiekermann, Christine Legner, and Boris Otto. "Data Catalogs - Integrated Platforms for Matching Data Supply and Demand." Fraunhofer Verlag, 2019. https://publica.fraunhofer.de/handle/publica/299505.

Schnabel, F., Heydkamp, C., Schmitz, M., Braun, S., Albrecht, P., Lonien, J., Rothballer, C., Jacobi, N. and Ullrich, P. (2020). Smarte umweltrelevante Infrastrukturen: Anwendungsfelder, Bedarfe, Praxiserfahrung aus kommunaler Sicht. [online] Available at: https://www.umweltbundesamt.de/sites/default/files/medien/479/publikationen/texte_74-2020_sui_abschlussbericht_barrierefrei_03092019.pdf [Accessed 25 March 2025].

Smart City Dresden. (2022). Temperature sensors (KLIPS). [online] Available at: https://smartcity.dresden.de/en/projects/temperature-sensors.

SMART CITY FACTORY. (2024). Smart City Frankfurt - SMART CITY FACTORY. [online] Available at: https://smart-city-factory.com/en/case-frankfurt/ [Accessed 25 March 2025].

U.S. Department of Homeland Security. (2024). ST-Smart City Interoperability Reference Architecture Fact sheet | Homeland Security. [online] Available at: https://www.dhs.gov/archive/science-and-technology/publication/st-smart-city-interoperability-reference-architecture-fact-sheet.

DOES SMART CITY CONSTRUCTION PROMOTE SUSTAINABLE DEVELOPMENT GOALS? A CROSS-COUNTRY EMPIRICAL STUDY

Jinghua Yin, Haiying Song and Chunlan Zhao
Zhejiang International Studies University
No. 299, Liuhe Road, Hangzhou, China

ABSTRACT

To address "urban diseases," such as air pollution, traffic jams, and income gaps, global cities pursue sustainable development goals (SDGs) with digital technologies. However, the uneven impact of smart city construction (SCC) on the three pillars of SDGs remains underexplored. Based on panel data from 105 countries spanning 2014 to 2021, this study employed the entropy method to construct comprehensive indices for SDGs and SCC levels. Further, a bidirectional fixed-effects model was used to empirically analyze the relationship. The results reveal that SCC significantly promotes SDGs. The economic impact of SCCs is the largest, the social impact is moderate, and the environmental impact is the mildest. The Asian SCC model has the greatest influence on SDGs, followed by the European and American models.

KEYWORDS

Smart City Construction, Sustainable Development, American Model, Asian Model, European Model

1. INTRODUCTION

As digital transformation reshapes global urbanization, cities have emerged as both critical agents for achieving the United Nations (UN) sustainable development goals (SDGs) and laboratories for deploying intelligent technologies to reconcile economic growth with ecological boundaries. To confront escalating urban challenges—from resource depletion to climate pressures—over 200 smart city initiatives worldwide leverage Internet of Things (IoT), artificial intelligence (AI), and big data to reimagine urban ecosystems. From New York's resilience planning to Amsterdam's circular economy models, these projects exemplify how technology-driven urban governance seeks to operationalize the 2030 Agenda for Sustainable Development. China's systematic integration of smart cities into its Five-Year Plans further underscores their strategic role in balancing modernization with sustainability.

Sustainable development is defined as development that meets current needs without compromising the ability of future generations to meet their own needs (World Commission on Environment Development, 1987). ISO 37120 quantifies "urban sustainability" as the integrated performance of three core indicator categories: economic, environmental, and social governance. A smart city constitutes a complex adaptive system where information and communications technology (ICT) serves as the digital backbone, enabling multi-stakeholder collaborative governance to holistically optimize six dimensions—economy, environment, quality of life, human capital, mobility, and governance—to ultimately achieve sustainable human-centered development (Hajek, Youssef, and Hajkova, 2022; Park and Yoo, 2023).

While the SDGs provide a universal framework for equitable development, the interplay between smart city development (SCC) and sustainability outcomes remains contested. Current research reveals paradoxical impacts: smart technologies enhance poverty targeting (SDG1-2) through data-driven interventions, yet risk exacerbates inequality via algorithmic bias (Vinuesa et al., 2020). They enable real-time environmental monitoring (SDG6-7, 13-15) but often prioritize corporate efficiency over equitable service access (Parmentola et al., 2022). Such contradictions highlight critical gaps in existing scholarship. First, fragmented case studies dominate the field, lacking cross-regional analyses to compare Asian, European, and American SCC models. Second, empirical evidence remains inconclusive—while Zhu et al. (2020) acknowledge SCC's dual impacts,

Haarstad (2017) argues sustainability often becomes secondary to innovation in European Union (EU) contexts. Third, few studies systematically map how SCC's technological, institutional, and social dimensions interact to advance or hinder specific SDGs.

This study addresses these gaps through three contributions. (1) Mechanistic framework: The framework proposes a unified model that analyzes SCC's multidimensional impacts across economic, social, and environmental SDG pillars. (2) Comparative analysis: The study empirically contrasts SCC outcomes across 105 nations, revealing Asia's policy-driven model are different from Western market-centric approaches in SDG attainment. (3) Policy integration: The study identifies governance levers to align SCC investments with SDG priorities, emphasizing context-sensitive strategies for Global South cities.

By bridging theoretical fragmentation and regional biases, this research reorients the smart city discourse from techno-centric optimism to evidence-based policymaking, offering actionable pathways to harness urban digitalization as a force for equitable sustainability.

2. THEORETICAL FRAMEWORK AND RESEARCH HYPOTHESES

2.1 Uneven Impact of SCC on Different Dimensions of SDGs

SCC primarily relies on ICT, big data, and innovative governance to address various urban challenges; however, there is significant inconsistency between smart and sustainable cities that directly impacts their effectiveness in terms of economic, social, and environmental sustainability. Bibri and Krogstie (2017) highlight that there are fundamental differences in design concepts and objectives between smart cities and sustainable cities. Sustainable cities emphasize design principles focused on environmental protection and social equity, whereas smart cities tend to prioritize the application of modern information technology and efficient solutions, often neglecting deeper design considerations. Due to this discrepancy, smart cities fail to integrate environmental and social needs effectively when pursuing development goals.

Ahvenniemi et al. (2017) further analyzed the current assessment framework for smart cities and found that it places excessive emphasis on ICT and intelligent applications while lacking consideration for environmental indicators. Consequently, smart cities perform relatively well in terms of addressing economic and social issues but their contribution to environmental sustainability is comparatively weak. Smart city assessment systems focus more on social welfare and economic benefits than on environmental protection and ecological balance.

As a result of these inconsistencies, the impact of SCC on the different dimensions of sustainable development varies significantly. It has been noted that the tension between the three pillars of urban development—economic, social, and environmental development—leads to tradeoffs. The technology-driven nature of smart cities often prioritizes economic growth, which leads to the marginalization of social and environmental concerns. For example, certain capital investments may promote economic development but can also result in social inequality and environmental degradation. This trend indicates that the impact of SCC is greatest in the economic domain, whereas its effects on social and environmental aspects are relatively weak.

Yigitcanlar et al. (2018) proposed an input-process-output model framework that highlights the importance of community, technology, and policy in SCC and illustrates how the interaction of these factors influences economic, social, and environmental sustainability. However, SCC often fails to achieve balanced development across these dimensions, with economic gains typically taking priority over social and environmental needs.

Therefore, this paper proposes the following hypotheses:

H1: SCC's impact is largest in the economic dimension, moderate in the social dimension, and smallest in the environmental dimension

2.2 Differential Impact of Various SCC Models on SDGs

Asia, Europe, and the Americas have adopted distinct development models for the global SCC landscape, each exerting varying degrees of impact on sustainable development.

2.2.1 Dominance of the Asian Model

The construction of smart cities in Asia is typically government-led and emphasizes rapid infrastructure development and technological applications. The core of this model lies in achieving swift economic growth through substantial infrastructure investment. For example, smart city projects in China often focus on infrastructure development in areas such as transportation, energy, and urban management (Hu and Zheng, 2021). Although this model may accelerate economic development in the short term, it can exacerbate environmental issues and overlook social needs. As a result of the centralized decision-making process and lack of social feedback mechanisms, technological solutions frequently fail to consider the actual needs of residents adequately, negatively affecting the realization of SDGs (Kitchin, 2014).

2.2.2 Balanced Development in the European Model

In contrast to the Asian model, the European approach emphasizes social and environmental sustainability. SCC in Europe prioritizes multi-stakeholder participation and collaboration, ensuring broad involvement of the government, businesses, and citizens. Resident needs and quality of life are of significant importance in the policymaking process, with a focus on ecological friendliness and social welfare (Hollands, 2020). For example, many European cities are committed to promoting green buildings, renewable energy, and intelligent transportation systems to reduce carbon emissions and ecological footprints (Caragliu and Del Bo, 2019). Although the European model excels in terms of sustainable development, economic growth rates tend to be relatively low.

2.2.3 Market-oriented Model in the Americas

The smart city model in the Americas is primarily citizen centered, emphasizing market-driven approaches and technological innovations. However, efforts toward environmental sustainability in this model are insufficient. Although some cities have taken measures to promote sustainable development, overall policies and practices remain inadequate and environmental issues persist (Bibri and Krogstie, 2017). SCC in the Americas often relies on private sector participation. Although there are numerous channels for resident engagement, improvements are still needed in terms of decision-making transparency and social inclusivity.

Therefore, the following hypothesis is proposed:

H2: SCC models in different regions exhibit heterogeneous impacts on sustainable development.

3. MATERIALS AND METHODS

3.1 Model Setting

The baseline model used for analysis in this study is defined as follows:

$$SDG_{it} = c_0 + \alpha_0 SCC_{it} + \beta_0 control_{it} + u_i + v_t + \varepsilon_{it}. \quad (1)$$

The subscripts i and t represent different countries and years, respectively; *SDG* denotes SDGs; *SCC* represents the level of SCC; and *control* refers to the set of control variables, u_i represents country fixed effects, and v_t represents time fixed effects. Additionally, c_0 is a constant, α_0 and β_0 are regression coefficients, and ε_{it} is a random disturbance term. A significantly positive coefficient of *SCC* indicates that *SCC* promotes sustainable development.

3.2 Variables

Table 1 presents the names, meanings, calculation methods, and data sources of the variables considered in the model described above.

Table 1. Variable definition

Type	Meaning	Variable	Calculation	Data source
Dependent variable	Sustainable Development	*SDG*	The total score calculated by the entropy method.	
Independent variable	Smart City Construction	*SCC*	The total score calculated by the entropy method.	
Control variables	Government Intervention Level	*Gov*	Government expenditure on health and education as a percentage of GDP	United Nations Global Sustainable Development Report
	Upgrading of Industrial Structure	*Str*	Value added of the tertiary sector as a percentage of GDP	World Bank
	Level of Openness	*Open*	Trade volume (imports and exports) as a percentage of GDP	World Bank
	Level of Industrialization	*Ind*	Industrial value added (in current USD)	World Bank

3.2.1 Dependent Variable

We constructed an indicator system encompassing economic, social, and environmental dimensions to measure SDGs (see Table 2).

Economic dimensions (economy). Per-capita GDP was used as an indicator to assess the state of economic development and predict future trends.

Social dimensions (society). The accumulation of human capital is crucial for enhancing social sustainability. The human capital indices of various countries were employed to measure this dimension.

Environmental dimensions (environment). Promoting environmental protection and green low-carbon development contributes to industrial upgrading and resource utilization efficiency. We used carbon dioxide emissions as a negative indicator of environmental sustainability.

We used the entropy method to determine the level of sustainable development across countries. Entropy is a physical measurement unit. The higher the entropy, the more chaotic the data, and the less information it carries, resulting in a smaller utility value and, consequently, a lower weight. In entropy weight analysis, weighting depends only on the discrete degree of the data itself, which reduces the loss of information and tends to be more objective. Entropy weight analysis assigns weights based solely on the degree of dispersion of the data itself, which reduces the degree of information loss, making it more objective.(Bennett and Bowers, 1976).

Table 2. SDG index

Evaluation Object	Indicator	Symbol	Measurement Indicator	Attribute	Weight
SDGs	Economic	Economy	Per-Capita GDP (current USD)	Positive	34.21%
	Social	Society	Human Capital Index	Positive	59.39%
	Natural	Environment	Carbon Dioxide Emissions	Negative	6.40%

3.2.2 Independent Variable

SCC is performed through the application of next-generation information technologies, such as the IoT, cloud computing, and social networks, with the goal of creating urban forms conducive to intelligent innovation. The level of digital development reflects the status of SCC. Therefore, we selected *governance digitalization*, *industrial digitalization*, and *new urbanization* as representative indicators (see Table 3). We used the entropy method to determine weights to assess the level of SCC across countries.

Governance digitalization. This is a comprehensive concept that encompasses political, cultural, economic, social, and environmental aspects, aiming to achieve digital and intelligent governance systems and methods. We utilized the E-Government Development Index to measure governance digitalization, which considers factors such as telecommunications infrastructure and online services, and provides a comprehensive reflection of the state of e-government development in various countries.

Industrial digitalization. This indicator refers to the transformation and upgrading of traditional industries using digital technologies. The core focus is optimizing resource allocation and enhancing production efficiency through data-driven approaches, thereby promoting high-quality economic development. The Digital Market Index is used to represent the level of industrial digitalization, encompassing the exchange of products and services based on internet technologies.

New urbanization. This indicator reflects the progress of urbanization in a country or region, emphasizing coordinated rural-urban development, the interaction between industries and cities, and the concept of harmonious ecological development. This indicator represents the characteristics of new urbanization within the context of SCC.

Table 3. Indicator system for SCC level

Evaluation Object	Indicator	Symbol	Measurement Indicator	Attribute	Weight
SCC	Governance Digitalization	Governance	E-Government Development Index	Positive	31.33%
	Industrial Digitalization	Industry	Digital Market Index	Positive	40.19%
	New Urbanization	Urban	Urban Population (% of total population)	Positive	28.48%

3.2.3 Control Variables

We control government *intervention*, upgrading of industrial structure and openness. We measure the level of government intervention using the proportion of government fiscal expenditures over GDP, and use the proportion of value added by the tertiary sector in each country as an indicator of industrial structure upgrading. We measure the level of industrialization as industrial value added.

3.3 Data Sources and Statistics

We excluded countries with significant missing data and selected a global research sample of 105 countries from 2014 to 2021 covering six regions: Arab states, the Commonwealth of Independent States (CIS), Africa, the Americas, Europe, and the Pacific. The sample countries were categorized into Asian, European, American, and other models based on regional differences, and further classified into high-income, upper-middle-income, lower-middle-income, and low-income groups based on income differences. This sample design considered disparities at both the regional and economic development levels, accurately reflecting global conditions. The data sources used in this study include the World Bank, World Intellectual Property Organization, United Nations Global Sustainable Development Report, United Nations E-Government Database, and Chinese Academy of Social Sciences Global Digital Economy Development Index (TIMG). Linear interpolation was used to fill in missing data. Table 4 presents the descriptive statistics of the relevant variables.

Table 4. Descriptive statistics

Variable	Sample size	Mean	Std.	Min.	Max.
Economy	840	0.141	0.165	0.000	1.000
Society	840	0.633	0.222	0.000	1.000
Environment	840	0.972	0.099	0.000	1.000
SDG	840	2.184	0.553	0.417	3.195
SCC	840	0.532	0.184	0.103	0.927
Gov	840	22.190	6.405	8.000	43.000
Str	840	57.471	9.407	16.775	81.275
Open	840	87.884	56.531	20.723	388.848
Ind	840	26.695	8.084	9.985	70.549

4. RESULTS

Table 5 presents the baseline model estimation results. The inclusion of control variables slightly altered the SCC regression coefficient; nonetheless, its significance in terms of SDGs remained unchanged. SCC significantly promotes sustainable development with a positive impact of 0.756 on SDGs, indicating that SCC enhances sustainability. The coefficients for SCC's impact decrease in the order of 1.088 > 0.656 > −0.026 for the economic, social, and environmental dimensions, indicating the greatest effect on the economic dimension, a moderate effect on the social dimension, and a small effect on the environmental dimension, supporting hypothesis H1.

Table 5. Benchmark model estimation results

Variable	(1) SDG	(2) Economy	(3) Society	(4) Environment
SCC	0.756*** (16.270)	1.088*** (8.498)	0.646*** (27.166)	−0.026** (−1.990)
Control	Controlled	Controlled	Controlled	Controlled
Year Fixed	YES	YES	YES	YES
Country Fixed	YES	YES	YES	YES
N	840	840	840	840
R2	0.440	0.328	0.460	0.151

Note: *, **, and *** denote significance at the 1 %, 5 %, and 10% levels, respectively, and t-values are presented in parentheses.

There are spatial disparities in sustainable development and SCC across regions. This study categorized the SCC models of 105 countries into Asian (Asia-Pacific), European, American, and other models (CIS, Arab, and African regions).

Table 6 indicates that SCC models significantly promote SDG, with Asian, European, and American models exhibiting significance at the 1% level, and the other models exhibiting significance at the 5% level. The impact of SCC on SDGs varies by region, with the Asian model having the greatest impact (1.123), followed by European (0.919), American (0.363), and other models (0.28). These findings support hypothesis H2.

Table 6. Regression results of different SCC patterns

Variable	(1) Asian Model	(2) European Model	(3) American Model	(4) Other Models
SCC	1.123*** (10.461)	0.919*** (17.304)	0.363*** (4.549)	0.287** (2.518)
Control	Controlled	Controlled	Controlled	Controlled
Year Fixed	YES	YES	YES	YES
Country Fixed	YES	YES	YES	YES
N	136	312	176	216
R2	0.673	0.610	0.331	0.305

Note: *, **, and *** denote significance levels at 1%, 5%, and 10%, respectively, and t-values are presented in parentheses.

Figure 1 indicates that different SCC models have varying impacts on different SDG dimensions. The Asian and European models significantly affect all three pillars of SDGs, whereas the American model exhibits significant effects only on the social and environmental dimensions. These findings support hypothesis H3. The influence of the Asian and European models on the economic, social, and environmental dimensions decreases sequentially, reinforcing H2. The Asian SCC model significantly promotes economic development (1.899) and social sustainability (0.771) but leads to environmental degradation (0.047). The European SCC model fosters economic growth (1.701) and social equity (0.575) while enhancing environmental protection (−0.081).

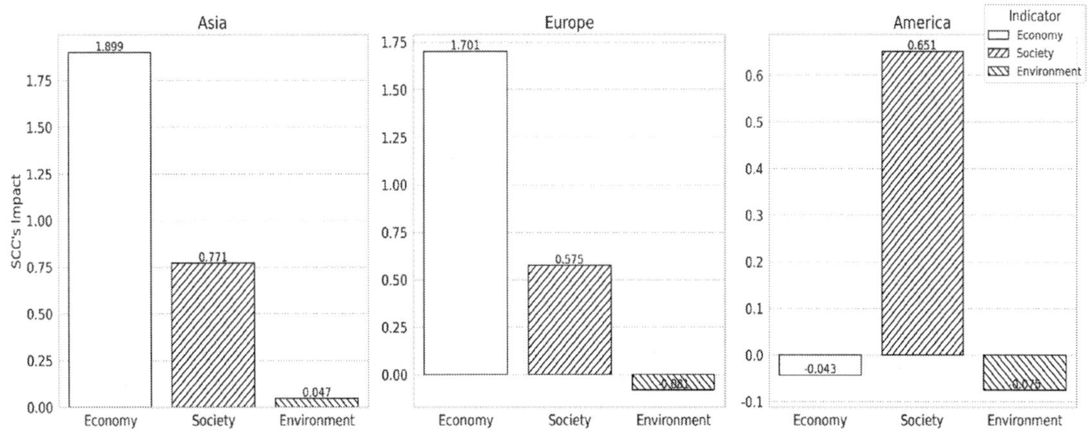

Figure 1. SCC's impacts by continent and indicator

5. DISCUSSION

5.1 Differential Impact of SCC on Various Dimensions of SDG

Our analysis shows that the impact of SCC on the economic dimension is the most significant (1.088), aligning with the findings of Li et al. (2024), who emphasized that technology is a core driver of economic development. Regarding the social dimension, although the impact of smart cities was moderate (0.646), it remained significant. Corsi et al. (2022) demonstrated that smart sustainable cities have the greatest impact on the social dimension; however, public concerns regarding transparency and privacy must be addressed (Lim, Edelenbos, and Gianoli, 2023). The environmental impact of SCC was relatively low in our analysis (-0.026), suggesting that environmental protection may not have received adequate attention in the SCC process. Xu, Wang, and Yu (2022) emphasized that economic incentives may neglect of environmental protection, particularly in economically advanced cities (Kumar, Dutta, and Gupta, 2022).

SCC significantly boosts the economy (capital accumulation) but has minimal impact on environmental improvement (natural capital). This confirms the dominance of the "weak sustainability" theory in the context of smart cities: technology-driven growth takes precedence over the maintenance of ecological boundaries. The results challenge the optimistic assumption that smart cities can "automatically achieve sustainability" and support the core argument of the "strong sustainability" theory: economic growth cannot automatically compensate for the depletion of natural capital, and institutional constraints (such as environmental policies in Europe) are needed to ensure that the environmental dimension is not sacrificed.

5.2 Asian SCC Environmental Sustainability Issues

SCC in Asia significantly enhances economic development (impact coefficient: 1.899) and social sustainability (impact coefficient: 0.771) but also leads to environmental degradation (impact coefficient: 0.047). The effects of SCC differ notably between Asia and Europe. While Asian SCC increases carbon emissions, European SCC tends to reduce emissions. This is because of the inverted U-shaped relationship between SCC and carbon emissions, with many Asian countries still in the early stages of development and yet to reach a turning point (Añón Higón, Gholami, and Shirazi, 2017).

The Asian model (strong government + infrastructure-driven development) is efficient but comes with high environmental costs, revealing the limitations of ecological modernization theory. Without institutional constraints and social participation, mere technological investment may trigger the Jevons Paradox—efficiency improvements inadvertently stimulate total resource consumption (e.g., carbon emissions rise with economic growth). For example, Indonesia's Sleman region faced environmental management challenges during its smart city initiatives (Fridayani and Rifaid, 2019). South Korea's Songdo area promotes economic and social sustainability but also contributes to environmental decline (Kim, 2019). In Malaysia, the low acceptance of environmental goals within smart city frameworks reflects inadequate awareness (Lim et al., 2021). China's smart city policies have shown unstable impacts on green productivity (Yang, Su, and Yu, 2024), and Thailand's local governments often focus excessively on technology and economic growth (Choi and Kenney, 2024).

5.3 European SCC Benefits for Environmental Sustainability

In Europe, SCC enhances not only economic development (impact coefficient: 1.701) and social equity (impact coefficient: 0.575) but also environmental effectiveness (impact coefficient: −0.081). This success is linked to the EU's strong environmental policies, diverse economic structures, and strong public participation.

EU countries have implemented stricter environmental regulations that promote sustainable development and pollution reduction, driving the adoption of green technologies (Gradinaru and Maricut, 2023). In contrast, China's environmental policies face challenges in terms of enforcement and public engagement, particularly within a manufacturing-driven economy. And public participation models also differ significantly. Helsinki emphasizes community involvement and enhances social cohesion through public consultations (Shamsuzzoha et al., 2021), whereas Singapore's top-down approach limits citizen participation despite technological efficiency. These disparities result in varying outcomes in terms of social equity and environmental sustainability.

The European model (strong institutions + public participation) achieves better environmental-economic synergy. European cases support ecological modernization theory, showing that technological innovation (SCC) can achieve economic-environmental win-win outcomes (Gradinaru and Maricut, 2024). under supportive institutions (strict environmental regulations) and pluralistic participation.

5.4 SCC's Different Impacts on North and South America

SCC in America enhances social sustainability (impact coefficient: 0.651) and reduces CO_2 emissions (−0.075); however, its economic impact is less significant. These effects differ between South and North America. Income heterogeneity tests show that SCC in the US and Canada significantly benefits the economic, social, and environmental dimensions. SCC has an insignificant impact on economic growth, primarily because of political and institutional factors affecting development in Central and South American countries which, except for Chile, are largely trapped in the middle-income bracket.

The aforedescribed results highlight the criticality of political economy/institutionalist perspectives. The sustainability outcomes of SCC are highly contingent on the institutional environments in which they are embedded (Mora, et al., 2022), such as citizen participation mechanisms in Europe, centralized decision-making models in Asia, and private-sector dominance in the Americas. This finding verifies the social construction of technology: the "wisdom" and "sustainability" attributes of smart city technology are not inherent to the technology itself but are constructed by the social-institutional environment in which it is applied (Zhai and Gao, 2024).

6. CONCLUSION

This study analyzed a sample of 105 countries from 2014 to 2021, utilized the entropy method to measure sustainable development and SCC, and employed a two-way fixed effects model to investigate the impact of SCC on SDGs. Our key findings are summarized below. (1) SCC significantly enhances sustainable development, with the largest effect on the economic dimension, a moderate effect on the social dimension, and the smallest effect on the environmental dimension. (2) The influence of different SCC models on sustainable development varies in the following order: Asian model > European model > American model > other models. This study not only provides empirical evidence for the optimization of smart city policies but also deepens the theoretical understanding of the transformation of sustainable development by revealing the complex inter-structure among technology, institutions and sustainability. It makes valuable contributions, particularly in reconciling economic growth and ecological boundaries, elucidating the differentiation of regional paths, and integrating the perspectives of social and technological systems.

Based on these findings, we propose the following recommendations. Governments and enterprises should increase investment in emerging technologies, particularly AI, IoT, and big data analysis, to enhance economic efficiency and promote growth. Governments should develop policies to support entrepreneurship in the smart city sector and provide funding and market access to encourage innovation. SCC should enhance public participation by utilizing digital platforms and social media to improve community involvement in planning and implementation, and increase policy transparency. An emphasis on environmental protection is crucial. Learning from successful European environmental policy practices and public participation can enhance sustainability. The North American experience highlights the importance of strengthening education and training to develop a skilled workforce that fosters innovation. Advocating for political reforms and institutional development can improve the investment environment and enhance the economic impact of SCC in Central and South America.

REFERENCES

Ahvenniemi, H., Huovila, A., Pinto-Seppä, I., and Airaksinen, M., (2017). Smart city assessment frameworks: A systematic review. *Sustainable Cities and Society*, Vol. 32, pp. 596–603.

Añón Higón, D. A., Gholami, R., and Shirazi, F., (2017). ICT and environmental sustainability: A global perspective. *Telematics and Informatics*, Vol. 34, pp. 85–95.

Bibri, S. E., and Krogstie, J., (2017). Smart sustainable cities of the future: An extensive interdisciplinary literature review. *Sustainable Cities and Society*, Vol. 31, pp. 183-212.

Bennett S., and Bowers D., (1976). Principal factor analysis. In: An *Introduction to Multivariate Techniques for Social and Behavioural Sciences*. MacMillan Education, London, UK.

Caragliu, A., and Del Bo, C. F., (2019). Smart innovative cities: The impact of smart city policies on urban innovation. *Technological Forecasting and Social Change*, Vol. 142, pp. 373–383.

Choi, Y. J., and Kenney, L., (2024). A conceptual framework to explore considerations of the social implications in internet of things and smart city governance and policy: The case of Thailand. *Policy and Internet*, Vol. 16, pp. 242–271.

Corsi, A., Pagani, 'Rocha e Cruz T. B. R., de Souza, F. F., and Kovaleski, J. L., (2022). Smart sustainable cities: Characterization and impacts for sustainable development goals. *Revista de Gestão Ambiental e Sustentabilidade*, Vol. 11, e20750-e20750.

Fridayani, H. D., and Rifaid, R. J., (2019). Smart city as A Tool is to achieve sustainability city. *Jurnal Ilmiah Tata Sejuta STIA Mataram*, Vol. 5, pp. 308–320.

Gradinaru, G. I., & Maricut, A. C. (2024). The impact of economic behavior environment: A comparative study among EU's countries and China. Journal of the Knowledge Economy, Vol.15, pp.10470-10484.

Haarstad, H., 2017. Constructing the sustainable city: Examining the role of sustainability in the "smart city" discourse. *Journal of Environmental Policy and Planning*, Vol. 19, pp. 423–437.

Hajek, P., Youssef, A., and Hajkova, V., (2022). Recent developments in smart city assessment: A bibliometric and content analysis-based literature review. *Cities*, Vol. 126, p. 103709.

Hollands, R. G., (2020). Will the real smart city please stand up? Intelligent, progressive or entrepreneurial? In K. S. Willis, and A. Aurigi (Eds.), *The Routledge Companion to Smart Cities* (pp. 179–199). Routledge, New York, USA.

Hu, Q., and Zheng, Y., (2021). Smart city initiatives: A comparative study of American and Chinese cities. *Journal of Urban Affairs*, Vol. 43, No. 4, pp. 504-525.

Kim, M. H., (2019). Transition of initiative from ubiquitous city to sustainable smart city by Korea: Focusing on the case of Songdo City. *Sustainability*, Vol. 11, pp. 1–15.

Kitchin, R., (2014). *The Data Revolution: Big Data, Open Data, Data Infrastructures and Their Consequences.* SAGE Publications, Thousand Oaks, USA.

Kumar, A., Dutta, S., and Gupta, S., (2022). Smart cities: A step toward sustainable development. In *Smart Cities* (pp. 1–43). CRC Press, Boca Raton, USA.

Lim, S. B., Malek, J. A., Yussoff, M. F. Y. M., and Yigitcanlar, T., (2021). Understanding and acceptance of smart city policies: Practitioners' perspectives on the Malaysian smart city framework. *Sustainability*, Vol. 13.

Lim, Y., Edelenbos, J., and Gianoli, A., (2023). Dynamics in the governance of smart cities: Insights from South Korean smart cities. *International Journal of Urban Sciences*, Vol. 27, Sup. 1, pp. 183–205.

Lim, Y., Edelenbos, J., and Gianoli, A., (2024). What is the impact of smart city development? Empirical evidence from a smart city impact index. *Urban Governance*, Vol. 4, pp. 47–55.

Mora, L., Deakin, M., Zhang, X., Batty, M., de Jong, M., Santi, P., and Appio, F. P., (2022). Assembling sustainable smart city transitions: An interdisciplinary theoretical perspective. In L. Mora & M. Deakin (Eds.), *Sustainable Smart City Transitions* (pp. 1–27). Routledge, New York. USA.

Parmentola, A., Petrillo, A., Tutore, I., and De Felice, F., (2022). Is blockchain able to enhance environmental sustainability? A systematic review and research agenda from the perspective of sustainable development goals (SDGs). *Business Strategy and the Environment*, Vol 31, No. 1, pp. 194–217.

Park, J., and Yoo, S. (2023). Evolution of the smart city: Three extensions to governance, sustainability, and decent urbanization from an ICT-based urban solution. *International Journal of Urban Sciences*, Vol. 27, Sup. 1, pp. 10–28.

Shamsuzzoha, A., Nieminen, J., Piya, S., and Rutledge, K., 2021. Smart city for sustainable environment: A comparison of participatory strategies from Helsinki, Singapore and London. *Cities*, Vol. 114, p. 103194.

Vinuesa, R., Azizpour, H., Leite, I., Balaam, M., Dignum, V., Domisch, S., and Fuso Nerini, F., (2020). The role of artificial intelligence in achieving the sustainable development goals. *Nature Communications*, Vol. 11, No. 1, p. 233.

World Commission on Environment Development (WCED). (1987). Our common future: 17 special working session.Pearson.

Xu, L., Wang, D., and Du, J., (2022). Spatial-temporal evolution and influencing factors of urban green and smart development level in China: Evidence from 232 prefecture-level cities. *International Journal of Environmental Research and Public Health*, Vol. 19, p. 3939.

Yang, S., Su, Y., and Yu, Q., (2024). Smart-city policy in China: opportunities for innovation and challenges to sustainable development. *Sustainability*, Vol. 16.

Yigitcanlar, T., Kamruzzaman, M., Buys, L., Ioppolo, G., Sabatini-Marques, J., da Costa, E. M., and Yun, J. J., (2018). Understanding "smart cities": Intertwining development drivers with desired outcomes in a multidimensional framework. *Cities*, Vol. 81, pp. 145–160.

Zhai, Y., and Gao, P., (2024). *Smart City Governance with Socio-Technical Systems Theory: A Case Study and Analytical Framework of Shenzhen in China* [Manuscript in preparation].

Zhu, S., Li, D., Feng, H., Gu, T., Hewage, K., and Sadiq, R., (2020). Smart city and resilient city: differences and connections. *Wiley Interdisciplinary Reviews: Data Mining and Knowledge Discovery*, Vol. 10, e1388.

FEDERATED LEARNING WITH TRANSFORMER-BASED REINFORCEMENT LEARNING FOR SMART CITY SECURITY

Laila Alterkawi[1] and Mokhled Altarawneh[2]
International University Kuwait (IUK)
Al-Ardiya, Kuwait

ABSTRACT

The growing integration of devices from the Internet of Things (IoT) in smart city infrastructures creates new opportunities for autonomous decision-making but also introduces significant challenges related to scalability, coordination, and data privacy. Traditional reinforcement learning (RL) models, particularly Multi-Agent Actor-Critic (MAAC) frameworks, offer structured coordination but rely on centralized critics and recurrent networks, limiting their effectiveness in dynamic, large-scale environments. In this work, we propose a novel federated reinforcement learning framework that integrates Decision Transformers (DT) into a multi-agent federated learning (FL) paradigm. Our method replaces conventional policy gradient optimization with trajectory-based learning using self-attention mechanisms, enabling agents' to model long-term dependencies without centralized value functions. By combining DT with FL, our framework supports decentralized training on IoT nodes, preserving privacy and reducing communication overhead. We benchmark our model against MAAC in a mobile edge computing (MEC) simulation environment, demonstrating that Transformer-based RL achieves higher reward efficiency, better scalability, and improved adaptability in non-stationary environments - although with slightly higher variance in early training. These results highlight the potential of Transformer-based RL as a robust alternative for decentralized, security-sensitive smart city applications.

KEYWORDS

Internet of Things, Smart Cities, Multi-Agent Reinforcement Learning, Decision Transformer, Federated Learning, Mobile Edge Computing

1. INTRODUCTION

Rapid advancement of smart city technologies has led to an increasing reliance on Internet of Things (IoT) devices to facilitate urban operations such as transportation, surveillance, and energy management (Sookhak, Tang, He, & Yu, 2019; Al-Fuqaha, Guizani, Mohammadi, Aledhari, & Ayyash, 2015). These environments generate large volumes of distributed sequential data and require intelligent and adaptive systems for real-time decision making. Although reinforcement learning (RL) has shown promise in such applications, traditional models, particularly Actor-Critic and value-based approaches, struggle with capturing long-term temporal dependencies and often rely on inefficient trial-and-error exploration (Silver, 2017 ;Sutton, R. S. & Barto, A. G., 2018). In addition, the centralized nature of their training pipelines introduces communication bottlenecks, scalability issues, and privacy risks, which are especially problematic in distributed IoT networks.

Multi-Agent Actor-Critic (MAAC) methods were introduced to improve coordination among agents in complex environments. However, MAAC still relies on a centralized critic, which limits scalability and creates a single point of failure. In addition, MAAC models typically use recurrent neural networks to capture temporal context, which are prone to vanishing gradient issues and struggle with long-range dependencies. These limitations hinder MAAC's effectiveness in dynamic, high-dimensional smart city applications where both privacy and long-horizon reasoning are critical.

Recent breakthroughs in deep learning, particularly Transformer-based architectures, have reshaped the landscape of sequential decision-making. Transformers leverage self-attention to model long-term dependencies more effectively than RNNs or LSTMs (Vaswani, Ashish, et al., 2017; Chen, 2021). Decision Transformers (DT), which frame RL as a sequence modeling problem, enable trajectory-based learning without relying on value functions or dynamic programming. Unlike traditional RL methods that require explicit state-action mappings, DTs generate actions conditioned on return-to-go estimates, offering better sample efficiency and generalization (Chen, 2021; Kapturowski, 2019).

In this paper, we propose a novel framework that integrates Decision Transformers within a Federated Learning (FL) paradigm to address the limitations of MAAC and other centralized MARL methods. Our approach enables multiple agents to collaboratively learn policies while keeping data local, thereby enhancing privacy and reducing communication overhead. By removing the dependence on a centralized critic and replacing recurrent components with self-attention, our model improves scalability, robustness, and adaptability in real-world smart city scenarios.

This methodology tackles three critical challenges in existing MARL-based smart city applications. First, it improves data efficiency by using sequential pattern recognition to reduce exploration requirements and speed up convergence (Chen, 2021). Second, it offers horizontal scalability, allowing seamless integration of new agents in large IoT systems without retraining from scratch (Parisotto, 2020). Third, it increases resilience to adversarial perturbations, as attention-based models are more robust to localized attacks and model drift (Carlini & Wagner, 2017).

The remainder of this paper is structured as follows. Section 2 reviews the existing literature on reinforcement learning in smart city security, with an emphasis on MAAC and Transformer-based RL models. Section 3 introduces our proposed Federated Learning framework with Decision Transformers. Section 5 describes the experimental setup, including training environments and evaluation metrics. Section 6 presents a comprehensive analysis of the results, comparing our method with MAAC. Finally, Section 7 concludes the paper and outlines future research directions.

2. RELATED WORK

2.1 Multi-Agent Reinforcement Learning

The integration of Reinforcement Learning in smart city security has spurred numerous advances. Traditional Actor–Critic models serve as foundational sequential decision-making tools but often struggle with capturing long-term dependencies and incur high computational costs due to extensive exploration (Silver, 2017; Sutton, R. S. & Barto, A. G., 2018). Hierarchical and multi-agent RL extensions–such as Multi-Agent Actor–Critic (MAAC) (Iqbal, Shariq & Sha, Fei, 2019), improve cooperation and scalability but depend on centralized training, which increases communication overhead and raises privacy concerns in expansive IoT networks.

Multi-Agent Reinforcement Learning (MARL) techniques have been systematically reviewed by Buşoniu et al. (Buşoniu, Lucian, et al., 2010) who categorize approaches into *value-based methods* (e.g., Q-learning), *policy-gradient techniques*, and *actor-critic architectures*. While effective in small-scale scenarios, these methods face critical challenges in scalability, coordination, and adaptability when deployed across heterogeneous, dynamic IoT infrastructures.

2.2 Decision Transformer and Limitations in Federated Settings

Decision Transformers (DT) reframe RL as a sequence modeling problem, treating state–action–reward trajectories as input tokens to a Transformer encoder–decoder (Chen, 2021). This trajectory-based formulation eliminates dynamic programming and excels in offline RL. However, DTs assume centralized access to large, pre-collected datasets and lack mechanisms for online or decentralized learning, making them unsuitable for federated environments. Their single-agent focus further limits coordination among multiple IoT nodes, and the absence of federated update protocols leads to prohibitive communication costs in bandwidth-constrained networks.

2.3 Transformer-Based RL with Federated Learning

Transformer architectures leverage self-attention to capture long-range dependencies and improve sample efficiency (Vaswani, Ashish, et al., 2017; Chen, 2021). Prior work has demonstrated FL integration with Transformers in recommendation systems (Zhang, J., et al., 2023) and personalized federated self-attention (FedTP) to mitigate client heterogeneity (Li, R., et al., 2022). In smart city security, converging RL, blockchain, and IoT has shown promise for robust decentralized systems (Alazab, M., et al., 2023) and deep RL has been applied to multi-UAV coordination for surveillance (Yun, H., et al., 2022).

3. METHODOLOGY

We address the core limitations of conventional Multi-Agent Reinforcement Learning (MARL) specifically Multi-Agent Actor-Critic (MAAC), by embedding a Decision Transformer (DT) within a Federated Learning (FL) loop. Our hybrid framework empowers each IoT agent to model long-term dependencies, scale horizontally, and preserve data privacy:

- **Self-Attention for Sequential Dependencies.** Instead of RNNs or LSTMs, each agent uses a Decision Transformer to process its entire state-action-reward trajectory via multi-head self-attention. This captures long-range temporal context without vanishing gradients, dramatically improving sample efficiency.
- **Federated Averaging for Communication Efficiency.** Agents never share raw observations or gradients. After local training on their DT parameters, they send only encrypted weight deltas to a central FL server and receive back the federated average. This slashes per-round communication overhead and removes single-point data-leakage risks.
- **Decentralized Critic Replacement.** MAAC's centralized critic creates coordination bottlenecks and requires constant access to other agents' data. In our design, the critic logic is fully absorbed into each local DT block, so policy evaluation and update happen on-device using only local information and the latest global model.
- **Scalability and Robustness.** New agents can join at any time by downloading the latest global model and contributing their own updates, with no change to the server architecture. Eliminating a central critic also removes a single point of failure, making the system resilient to node outages or communication dropouts.

In practice, each training round proceeds as follows: first, every agent collects a batch of recent state-action-reward trajectories and embeds them via a shared dense layer. Next, agents process these embeddings through their Decision Transformers to predict the next action conditioned on target returns. Once local updates are computed, agents upload only their model weight deltas to the FL server, which aggregates them via federated averaging. The server then distributes the new global model back to all agents. This cycle repeats until convergence, ensuring that each agent benefits from collective experience without ever revealing its private data.

Extensive simulations in smart-city scenarios show that our federated DT framework converges faster, achieves higher decision accuracy, and handles dynamic node participation far more gracefully than MAAC, demonstrating its promise for large-scale, privacy-sensitive urban IoT deployments.

4. COMPARATIVE ANALYSIS WITH MAAC

To evaluate the effectiveness of our proposed framework, we conduct a comprehensive comparison against the Multi-Agent Actor-Critic (MAAC) algorithm, a widely adopted baseline in Multi-Agent Reinforcement Learning (MARL). MAAC employs centralized critics to support cooperative learning across agents, but this architecture introduces coordination bottlenecks, privacy risks due to centralized data exchange, and limited capacity to model long-term temporal dependencies.

Our approach addresses these limitations by replacing centralized critic architectures with Decision Transformers embedded within a Federated Learning (FL) paradigm. This decentralized design allows agents to learn coordinated behavior through shared self-attention-based policies while keeping raw data local, enhancing both scalability and privacy preservation.

Table 1 summarizes the conceptual and architectural differences between MAAC and our method across several dimensions relevant to MARL systems.

Table 1. Comparison of MAAC and Transformer-Based RL with Federated Learning

Aspect	MAAC (Multi-Agent Actor-Critic)	Decision Transformer + FL (Proposed)
Policy Learning	Uses actor and centralized critic for cooperative learning	Uses trajectory-based sequence modeling; conditioned on return-to-go
Sequential Dependencies	Limited to short-term temporal context via RNNs or LSTMs	Captures long-term dependencies via self-attention
Training Paradigm	Centralized critic aggregates joint agent data during training	Decentralized learning via federated model weight updates
Coordination Strategy	Agents coordinate through shared critic gradients	Agents coordinate via shared attention-based policies
Scalability	Centralized critic creates training bottlenecks in large networks	Scales easily across edge nodes in heterogeneous environments
Privacy	Raw agent data may be shared with central critic	Raw data remains local; only model weights are shared
Communication Overhead	Frequent gradient synchronization with critic	Lightweight periodic model update transfers
Robustness	Sensitive to critic failure or communication issues	Fully decentralized; resilient to node or server failures

To further highlight the system-level improvements, Table 2 contrasts the overall design characteristics of the original MAAC-based architecture with our updated federated transformer-based framework.

These comparisons highlight the architectural and operational advantages of our framework over MAAC, particularly in terms of decentralized coordination, sequence modeling capacity, communication efficiency, and fault tolerance in real-world smart city environments.

Table 2. System comparison: MAAC (centralized) vs. Transformer RL with FL (decentralized)

Aspect	Original System (MAAC Centralized)	Updated System (Decision Transformer + FL)
Data Flow	Raw data sent to centralized critic for global value estimation	Only local model updates exchanged; raw data stays local
Policy Update	Critic gradients broadcasted to update actors	Global model aggregated via federated averaging
Sequence Handling	Short-term temporal context handled via RNNs	Long-term temporal patterns captured via self-attention
Exploration Strategy	Trial-and-error exploration with critic-guided updates	Trajectory-based learning conditioned on return-to-go
Scalability	Central critic becomes a bottleneck in large settings	Horizontally scalable across many agents

Aspect	Original System (MAAC Centralized)	Updated System (Decision Transformer + FL)
Privacy	Exposes raw observations during training	Privacy-preserving by design (no raw data exchange)
Communication Overhead	High due to frequent data and gradient sync	Low; periodic lightweight model updates only
Robustness	Vulnerable to critic or communication failures	Resilient to failures in nodes or aggregator

In the following sections, we detail our experimental setup, evaluation metrics (decision accuracy, convergence speed, communication overhead, and security robustness), and comparative results against MAAC baselines.

5. EXPERIMENTAL SETUP

To ensure a fair and rigorous evaluation, both models are trained under identical environmental conditions, maintaining consistency in hyperparameters, training episodes, and computational resources. The MAAC framework is implemented using the publicly available code from the original paper's GitHub repository[1], which serves as a benchmark for comparison. The Transformer-Based RL model is adapted to match the same experimental conditions, enabling direct assessment of its performance against the Actor-Critic baseline.

To ensure consistent evaluation and reproducibility, both models were executed on the same computational setup. The experiments were conducted on an Apple Mac system equipped with an Apple M1 Max processor, 64GB Unified RAM, and running macOS Sequoia 15.3.1. The software environment included Python 3.9, TensorFlow 2.9, and PyTorch 2.0, providing a stable and efficient platform for training and evaluating both reinforcements learning models. The training procedure follows a structured workflow to ensure a fair and rigorous evaluation of both models. The MAAC model is executed using the online resources, while the Transformer-Based RL model is trained using a modified implementation that integrates Decision Transformers within a Federated Learning framework.

To ensure a fair comparison, both models are trained using the same hyperparameter configurations, as summarized in Table 3. These parameters include learning rates, discount factors, batch sizes, and training durations, ensuring that performance differences stem from architectural variations rather than experimental inconsistencies.

Table 3. Experimental Parameters for MAAC and Transformer-Based RL Training

Parameter	MAAC	Transformer RL
Number of Agents	10	10
Training Episodes	50,000	50,000
Batch Size	64	64
Discount Factor (γ)	0.99	0.99
Learning Rate (Actor)	0.0005	0.0005
Learning Rate (Critic)	0.0005	0.0005
Optimizer	Adam	Adam
Exploration Strategy	ϵ-greedy (decay from 1.0 to 0.1)	ϵ- greedy (decay from 1.0 to 0.1)
Target Update Frequency	Every 100 steps	Every 100 steps
Communication Frequency (FL)	N/A	Every 10 episodes
Replay Buffer Size	1,000,000	1,000,000
Environment Type	MEC Simulation	MEC Simulation

[1] Original MAAC implementation available at: https://github.com/shariqiqbal2810/MAAC

6. RESULTS AND DISCUSSION

The experiments are designed to evaluate how each model processes sequential decision-making tasks while adapting to dynamic environments. Decision accuracy is assessed by the models' ability to select optimal actions over multiple trials, while training convergence speed is measured by the stabilization rate of each policy.

To analyze the differences between the two approaches, we present graphical illustrations of their respective learning behaviors and decision-making patterns. Figure 1 compares agent positioning and decision-making strategies between the Transformer-Based RL and MAAC models. The Transformer-Based RL model exhibits a more distributed approach, where decision points are scattered across the environment, indicating a wider exploration strategy. In contrast, the MAAC framework demonstrates a structured agent placement in well-defined decision zones, emphasizing cooperative learning among multiple agents. Comparison of agent distribution strategies. Transformer RL promotes broader exploration while MAAC clusters agents in tighter formations.

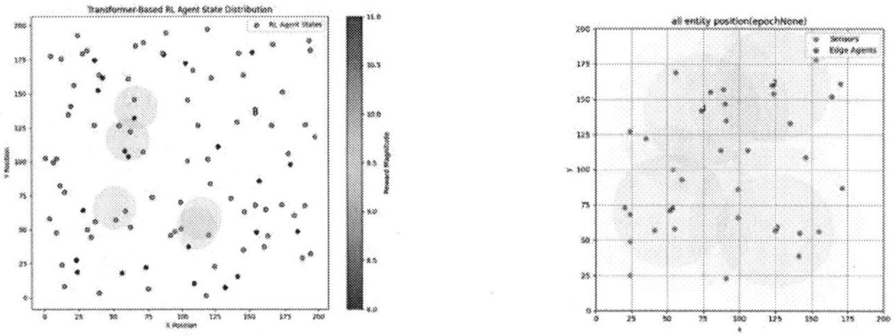

(a) Transformer-Based RL (b) MAAC

Figure 1. Comparison of agent distribution strategies. Transformer RL promotes broader exploration while MAAC clusters agents in tighter formations

The comparison between the Multi-Agent Actor-Critic (MAAC) and Transformer-Based RL models reveals significant differences in agent clustering, learning efficiency, and adaptability:

1. Agent Clustering and Density: Figure 2 (left) shows that MAAC maintains a denser agent distribution with a lower average pairwise distance. This structured behavior promotes cooperative learning and stability but limits flexibility in dynamic environments. In contrast, Transformer-based RL results in a more dispersed agent distribution, indicating broader exploration and adaptability.

2. Learning Efficiency and Reward Distribution: Figure 2 (right) presents the reward distributions. The MAAC model exhibits a more stable reward distribution, with most values concentrated in the 10-12 range. This reflects a consistent, structured learning approach. In contrast, Transformer-based RL achieves higher rewards, exceeding 22, but with a wider variance. This indicates greater exploration capability, but also a less stable learning process.

3. Decision-Making Adaptability: MAAC optimizes collaborative agent interactions, making it effective for structured, cooperative tasks. However, its reliance on predefined agent roles limits its ability to handle dynamic environments. Transformer-based RL, leveraging federated learning, learns state-action sequences dynamically and reduces communication overhead by sharing model updates instead of raw data. This makes it more suitable for decentralized, evolving environments like smart cities, where adaptive decision-making is crucial.

4. Scalability and Performance Comparison The comparison between Transformer-Based RL and MAAC, as illustrated in Figure 3, highlights the scalability of both models in multi-agent environments. The mean episode rewards across different agent counts indicate that the Transformer-Based RL model consistently achieves higher rewards with increased agent numbers, demonstrating superior adaptability to complex interactions. MAAC, while stable, exhibits a gradual decline in performance as the number of agents increases, suggesting that its structured cooperative learning struggles to generalize effectively in highly dynamic environments.

The Transformer-Based RL model, leveraging self-attention mechanisms and sequential decision modeling, maintains robust learning efficiency even as agent complexity grows. These results reinforce the suitability of Transformer-Based RL for large-scale smart city applications, where the number of interconnected IoT agents can dynamically vary. In contrast, MAAC's stability in small-scale settings may still be beneficial for tasks requiring structured collaboration but falls short in decentralized and evolving agent-based systems.

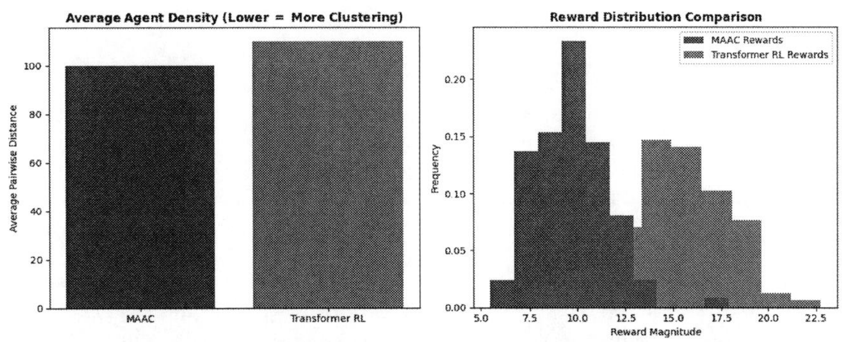

Figure 2. Comparison of MAAC and Transformer RL. (Left) Average pairwise distance among agents shows that Transformer RL maintains a more distributed layout, promoting adaptive behavior. (Right) Reward distribution indicates that Transformer RL achieves higher peak rewards with greater variance, while MAAC yields more stable but lower performance

Scalability comparison between Transformer-Based RL and MAAC across varying agent counts. Transformer RL maintains high reward efficiency as the number of agents increases, while MAAC performance degrades due to centralized critic bottlenecks and coordination challenges.

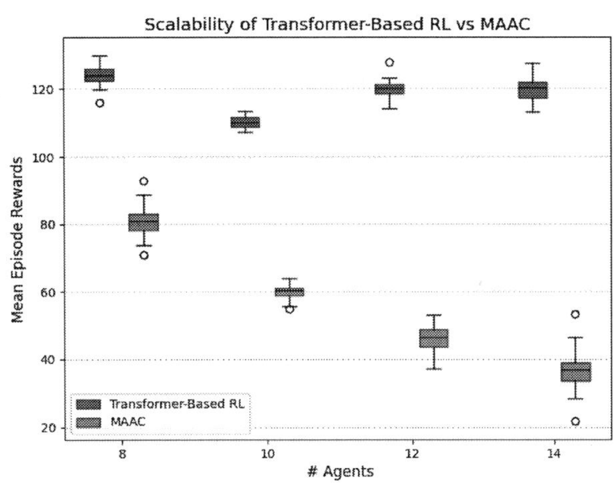

Figure 3. Scalability comparison between Transformer-Based RL and MAAC across varying agent counts. Transformer RL maintains high reward efficiency as the number of agents increases, while MAAC performance degrades due to centralized critic bottlenecks and coordination challenges

Comparison of MAAC and Transformer RL. (Left) Average pairwise distance among agents shows that Transformer RL maintains a more distributed layout, promoting adaptive behavior. (Right) Reward distribution indicates that Transformer RL achieves higher peak rewards with greater variance, while MAAC yields more stable but lower performance.

In summary, MAAC excels in cooperative stability, maintaining consistent performance within structured environments and small-scale multi-agent settings. However, as demonstrated in the scalability comparison, its performance declines as the number of agents increases, indicating limitations in handling dynamic and large-scale environments. In contrast, Transformer RL demonstrates greater adaptability, effectively scaling to higher agent counts while maintaining superior rewards. Despite exhibiting increased variability, its ability to leverage sequential decision modeling allows it to generalize better in decentralized smart city applications. This highlights the trade-off between structured multi-agent reinforcement learning, which ensures stability in well-defined cooperative tasks, and transformer-based trajectory optimization, which offers robustness and scalability in evolving agent-based systems.

7. CONCLUSION

This study presented a comparative analysis between the Multi-Agent Actor-Critic model and Transformer-based RL framework within a Federated Learning paradigm for smart city applications. Experimental results demonstrated that MAAC effectively optimizes cooperative decision-making by maintaining structured agent placement and stable reward distribution but struggles with scalability in dynamic environments. Transformer-based RL, on the other hand, achieves higher adaptability and reward efficiency through self-attention mechanisms, though with increased variability.

While MAAC is well-suited for structured, multi-agent tasks requiring coordination, Transformer-based RL excels in decentralized, evolving environments. Future research could explore hybrid approaches that integrate the cooperative stability of MAAC with the adaptability of Transformer RL, allowing for enhanced performance in both structured and dynamic environments. Additionally, optimizing Transformer RL to reduce reward variance while maintaining its high learning efficiency could improve the stability of decision-making in real-world deployments.

A crucial direction for future research is enhancing the security resilience of Transformer-based RL in federated settings. While Federated Learning provides privacy benefits by preventing direct data sharing, it remains vulnerable to adversarial attacks, model poisoning, and inference risks. Investigating secure aggregation mechanisms, adversarial defense techniques, and robust anomaly detection within Transformer RL could further strengthen its applicability in smart city security.

REFERENCES

Kan Zheng. (2021). Federated Reinforcement Learning: Techniques, Applications, and Open Challenges. *ArXiv, abs/2108.11887*(https://api.semanticscholar.org/CorpusID:237303858).

Ryan Lowe, Yi Wu, Aviv Tamar, Jean Harb, Pieter Abbeel, & Igor Mordatch. (2017). Multi-Agent Actor-Critic for Mixed Cooperative-Competitive Environments. *31st International Conference on Neural Information Processing Systems*, (pp. 6382–6393). Long Beach, California, USA.

Buşoniu, Lucian, Babuška, Robert, & Babuška, Robert. (2010). Multi-agent Reinforcement Learning: An Overview. In Srinivasan, Dipti, Jain, Lakhmi C., & 1 (Ed.), *nnovations in Multi-Agent Systems and Applications* (pp. 183--221). Berlin, Heidelberg: Springer Berlin Heidelberg.

Iqbal, Shariq, & Sha, Fei. (2019). Actor-attention-critic for multi-agent reinforcement learning. *International Conference on Machine Learning (ICML).*

Zhang, J., Wang, L., & Li, M. (2023). Transformer-enhanced federated learning for personalized recommendation. *IEEE Transactions on Neural Networks and Learning Systems.*

Vaswani, Ashish, Shazeer, Noam, Parmar, Niki, & Uszkoreit, Jakob . (2017). Attention is All you Need. *Advances in Neural Information Processing Systems. 30.* Curran Associates, Inc.

Silver, D. (2017). Mastering the game of Go with deep neural networks and tree search. *Nature.*

Li, R., Zhao, Y., & Wang, X. (2022). FedTP: Transformer-based personalized federated learning framework for heterogeneous clients. *IEEE Transactions on Artificial Intelligence*.

Alazab, M., Gupta, B., & Kim, T. (2023). Machine learning and blockchain for smart city security: Challenges and opportunities. *Computers & Security*.

Yun, H., Liu, X., & Zhang, J. (2022). Cooperative deep reinforcement learning for multi-UAV control in smart city surveillance. *IEEE Internet of Things Journal*.

Sutton, R. S., & Barto, A. G. (2018). *Reinforcement Learning: An Introduction.* MIT Press.

Chen, L. (2021). Decision Transformer: Reinforcement Learning via Sequence Modeling. *Proceedings of NeurIPS*.

Kapturowski, S. (2019). Recurrent Experience Replay in Distributed Reinforcement Learning. *Proceedings of ICLR*.

Sookhak, M., Tang, H., He, Y., & Yu, F. R. (2019). Security and Privacy of Smart Cities: A Survey, Research Issues and Challenges. *IEEE Communications Surveys & Tutorials, 21*, 1718-1743.

Al-Fuqaha, A., Guizani, M., Mohammadi, M., Aledhari, M., & Ayyash, M. (2015). Internet of Things: A Survey on Enabling Technologies, Protocols, and Applications. *IEEE Communications Surveys & Tutorials, 17*, 2347-2376.

Parisotto, E. (2020). Stabilizing transformers for reinforcement learning. *Proceedings of the 37th International Conference on Machine Learning*. JMLR.org.

Carlini, N., & Wagner, D. (2017). Towards Evaluating the Robustness of Neural Networks. *2017 IEEE Symposium on Security and Privacy (SP)*, (pp. 39--57).

FALL TRAJECTORIES IN OLDER ADULTS: PATTERNS, PREDICTORS AND OUTCOMES FROM A 6-YEAR FOLLOW-UP STUDY

Amadou M. Djiogomaye Ndiaye[1, 5], Michel Harel[1,2,7], Laurent Billonnet[6] and Achille Tchalla[1, 3,4, 5]

[1]Laboratoire Vie-Santé, UR 24 134, Faculté de Médecine, 87025 Limoges, France,
[2]INSPÉ de Limoges, Université de Limoges, 87036 Limoges Cedex, France ; UMR5219 UPS
[3]Service de Médecine Gériatrique, CHU de Limoges Dupuytren, Limoges, France,
[4]Pôle Hospitalo-Universitaire de Gérontologie clinique, Service de Médecine Gériatrique, Unité de Prévention de Suivi et d'Analyse du Vieillissement, CHU de Limoges Dupuytren, Limoges, France
[5]Chaire d'excellence chez Fondation Partenariale de l'Université de Limoges I Institut Omega Health, Laboratoire VieSanté, UR 24134 Vieillissement Fragilité Prévention e-Santé
[6f]XLIM-SRI - Systèmes et Réseaux Intelligents (444315) - XLIM - 123, avenue Albert Thomas, 87060 Limoges Cedex, France
[7]Institut de Mathématiques de Toulouse, UMR CNRS 5 219, Université Paul Sabatier, Toulouse 31 062, France

ABSTRACT

Background: Falls may occur as unpredictable events or follow patterns that indicate modifiable risk factors and predict adverse outcomes. Understanding these trajectories, risks, and predictors can help clinicians implement effective preventive measures. We hypothesize that distinct clinical trajectories of fall progression can be identified along with their baseline predictors.
Methods: We studied 954 community-dwelling older adults, aged 60 and older, who were followed prospectively for six years. Participants were evaluated by the UPSAV unit (Unité de Prévention, de Suivi et d'Analyse du Vieillissement) at home. Data collected included cardiovascular risk factors, fall occurrences, socio-environmental characteristics, and a comprehensive geriatric assessment summary score. Fall trajectories were identified using a Gaussian Mixture Model in Python, and multinomial logistic regression was applied to determine the predictors of each trajectory.
Results: We identified three distinct fall trajectories: No-Fallers With Fall History (47.9%), Chronic Fallers (37.3%), and Recurrent Fallers (14.8%). Predictors of Chronic Fallers included the age categories 60-80 years (OR 2.47, 95% CI [1.44 – 4.24]) and 80-90 years (OR 1.72, 95% CI [1.06 – 2.78]), as well as usufruct housing (OR 1.84, 95% CI [1.19 – 2.85]) and IADL <8 (OR 1.64, 95% CI [1.20 – 2.24]). Predictors of Recurrent Fallers included female sex (OR 2.00, 95% CI [1.24 – 3.24]), no social activity (OR 1.97, 95% CI [1.09 – 3.55]), MMSE (between 20 and 24) (OR 1.60, 95% CI [1.00 – 2.57]), and MMSE <10 (OR 1.87, 95% CI [1.09 – 3.20]).
Conclusions: Falls may occur in No-Fallers With a Fall History at discrete time intervals or present as chronically increasing or recurring events, which carry a higher risk of adverse outcomes. Women, individuals with multiple falls, and depressive symptoms should be prioritized for preventive measures. Identifying fall trajectories and their predictors will enable specialists to develop targeted prevention strategies, improving fall management in older adults.

KEYWORDS

Fall, Older Population, Mathematical Modeling, Prevention, IADL, MMSE

1. INTRODUCTION

According to the World Health Organization (WHO), older individuals are those aged ≥ 60 years (WHO 2022). The National Institute of Statistics and Economic Studies (INSEE) estimates that one in three individuals in France will be aged ≥ 60 years by 2060, compared to one in four individuals in 2021 (Insee 2016).

Older age is characterized by a progressive cycle of aging, which manifests as gradual degeneration of functional capacities and an increased risk of falling (ROF) (Moylan and Binder 2007; Thierry 2010). Falls, which are the leading cause of mortality in this age group, can lead to a loss of independence, with social, physical, and physiological consequences. In France, there were 2 million falls among individuals aged > 65 years in 2022, resulting in > 100,000 hospitalizations and 10,000 deaths. The fear of falling causes older individuals to avoid walking and climbing stairs. Although this fear protects against falls, it promotes sedentary behavior that leads to weakness over time (Park 2018).

Recent studies have sought to provide concrete evidence on the causes and mechanisms of falls, underscoring fall prevention as a critical issue in geriatrics and public health. While some falls occur as random, unpredictable events, others follow distinct patterns linked to modifiable risk factors and adverse outcomes. Understanding these fall trajectories and associated risk factors can aid clinicians in developing more effective, targeted prevention strategies (Moylan and Binder 2007).

To contribute to this growing body of research, we analyzed a longitudinal database from the UPSAV unit (Unité de Prévention, de Suivi et d'Analyse du Vieillissement) at home, which systematically collected fall data from older adults living at home over a six-year period. Using this database, we identified subgroups with distinct fall trajectories and examined their baseline characteristics to better understand the factors influencing fall risk.

2. MATERIALS AND METHODS

2.1 Study Participants

Our study examines a six-year dataset collected through multiple visits by UPSAV unit (Unité de Prévention, de Suivi et d'Analyse du Vieillissement) at home. The UPSAV team consists of nurses, geriatricians, and other healthcare practitioners. Each patient underwent an initial visit, followed by a second visit six months later. After the second visit, annual follow-ups were conducted for up to six years, provided the patient remained in their home (Tchalla et al. 2023).

The study includes men and women aged 60 and older. To be eligible, participants had to meet the following criteria:

- Provide written informed consent, either personally or through a legal representative.
- Not be enrolled in a clinical trial that modifies their standard medical management.
- Not have progressive pathologies that could significantly affect short-term prognosis.
- Not reside in a long-term care unit or a nursing home.
- Be covered by social security at 100%.

2.2 Falls and Clinical Outcomes Assessments

During the Follow-up, a fall was defined as unintentionally coming to rest on the ground or other lower level not as a result of a major intrinsic event (e.g., myocardial infarction, stroke, or seizure) or an overwhelming external hazard (e.g., hit by a vehicle) ("The Prevention of Falls in Later Life. A Report of the Kellogg International Work Group on the Prevention of Falls by the Elderly" 1987). Each patient underwent a Comprehensive Geriatric Assessment (CGA) and received a personalized care plan. The CGA is a multidimensional and standardized approach designed to enhance clinical practices in the care of older adults through a comprehensive health assessment (Welsh, Gordon, and Gladman 2014).

2.3 Covariates

Covariates included cardiovascular risk factors, fall occurrences, socio-environmental characteristics, and the comprehensive geriatric assessment (CGA) summary score.

Socio-environmental characteristics assessed in the home included gender, age, previous profession, education level, family situation, lifestyle, number of children, housing conditions, presence of an elevator, long-term illness status, health insurance coverage, leisure activities, social activity, human assistance, and pet ownership.

Cardiovascular risk factors considered were hypertension, diabetes, dyslipidemia, and tobacco use.

The CGA summary score encompassed multiple functional and cognitive assessments, including the Activities of Daily Living (ADL), Instrumental Activities of Daily Living (IADL), Mini-Mental State Examination (MMSE), pathological clock test, pathological fluency test, Geriatric Depression Scale (GDS), Mini Nutritional Assessment (MNA), Body Mass Index (BMI), Short Physical Performance Battery (SPPB), and pathological one-legged support test.

2.4 Data Analysis

A descriptive analysis was conducted to provide an overview of the recorded study variables. Additionally, a figure illustrates the proportions of fallers and non-fallers.

To identify clinically distinct trajectories of falls, we applied a Gaussian Mixture Model (GMM) in Python (Scikit-learn, n.d.). The metric for defining trajectories was the number of years in the study. Falls were classified into three categories based on the number of falls reported at each visit:

- No fall: no falls reported.
- Falls: a single fall within a year.
- Recurrent falls: at least two falls within a year.

The final model was selected using a combination of the Bayesian Information Criterion (BIC)-which balances model complexity and fit quality-and by ensuring that each estimated trajectory group represented at least 5% of the study population (Nagin and Odgers 2010). A multinomial logistic regression was then applied to identify the best predictors for each trajectory.

3. RESULTS

3.1 Participants

A total of 1,648 individuals met the study inclusion criteria, of whom 954 received a second visit. Table 1 presents the socio-environmental and health characteristics of the study sample at baseline. Among the included older adults, 664 (70%) were female and 290 (30%) were male. Additionally, 51% had hypertension, and only 18% engaged in social activities. The mean age of participants was 83 ± 6 years. Regarding fall occurrences, 502 participants (53%) had experienced a fall in the past six months. In terms of housing conditions, 59% were homeowners, while 27% lived in rental or free accommodations, and 14% resided under usufruct arrangements.

Table 1. Overview of Baseline Characteristics

Features of the study	**Total sample (N = 954)** **N (%)**
Woman	664 (70)
Age, mean ± SD, years	83 ± 6
Fell in past six months	502 (53)
Hypertension	485 (51)
Housing	
Owner	561 (59)
Tenant / Free accommodation / Other	258 (27)
Usufruct	135 (14)
Social activity	172 (18)
GDS, mean ± SD	3 ± 4
IADL, mean ± SD	6 ± 2
MMSE, mean ± SD	25 ± 5

Figure 1 shows that 36% of the study population experienced a single fall, 17% had recurrent falls (at least twice a year), and 46% did not fall during the study period. The distribution of falls in our study aligns with findings from the literature, which indicate that among individuals aged 80 years and older, one in two experiences a fall.(Grundstrom, Guse, and Layde 2012).

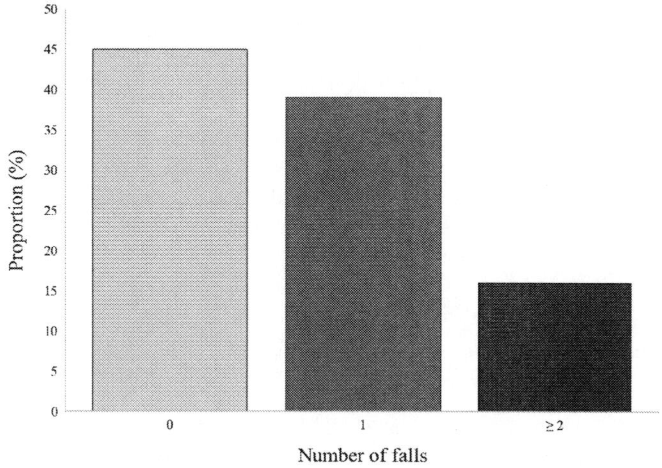

Figure 1. Distribution of fall classes (n= 954)

3.2 Determining Falls Trajectories Over 6 Years

To identify the fall trajectories of participants, we applied clustering methods that group individuals with similar profiles within the study population. One such method is the Gaussian Mixture Model (GMM), which we implemented using Python's Scikit-learn library (Scikit-learn, n.d.). GMM is a probabilistic clustering approach that assumes the data originates from a mixture of multiple Gaussian distributions. Unlike K-means clustering, which assigns each data point to only one cluster, GMM estimates the probability that each data point belongs to multiple clusters, allowing for softer, more flexible classifications. The model is trained using the Expectation-Maximization (EM) algorithm, which iteratively optimizes cluster assignments by estimating the most likely distribution parameters, thereby refining the trajectory classification (Moon 1996).

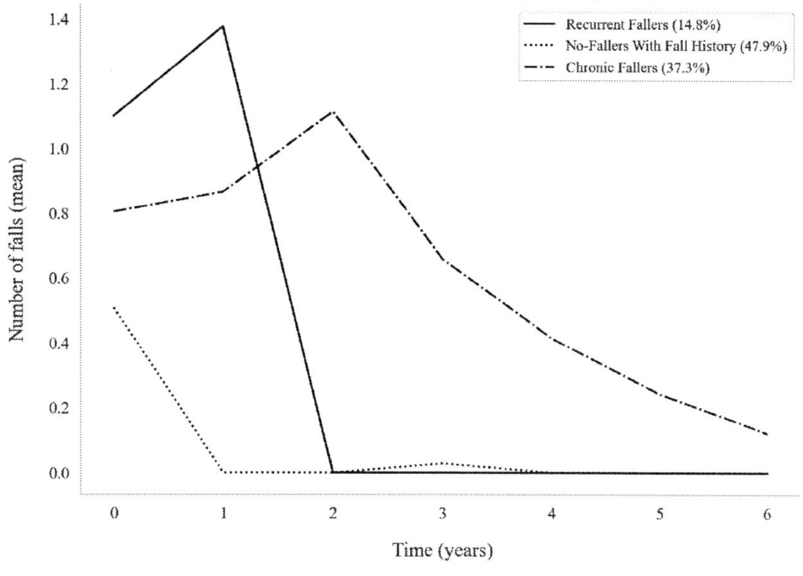

Figure 2. Patterns of Falls Trajectories over 6 years in Older Adults

Three group model performed well, each of the trajectories is constituted by at least five percent of the population study according to the Nagin criteria (Nagin 2014). The distinct trajectories were No-Fallers With Fall History (47.9%; n = 457), Chronic Fallers (37.3%, n = 356) and Recurrent Fallers (14.8%; n = 141).

We can notice in that the trajectory of:

- No-Fallers With Fall History is the most steady trajectory, its trajectory starts with a low fall rate at year 0. Falls gradually decline to nearly zero by year 1 and remain at zero in subsequent years. Patients in this trajectory falls between a year three but not a lot at the point to be in the other trajectories of fallers.
- Chronic Fallers begins with a moderate number of falls at year 0. Falls increase slightly, peaking around year 2. A steady decline follows, but the fall rate does not reach zero by year 6. In this trajectory we can find most of the fallers in our study.
- Recurrent Fallers exhibits initially, a high mean number of falls at year 0. Falls sharply increase, peaking at year 1. A drastic decline follows, reaching nearly zero by year 2, where it remains stable.

3.3 Predictors of Falls Trajectories

Multinomial logistic regression revealed that specific baseline characteristics predicted membership within each of the two fallers trajectory groups as compared to the No-Fallers With Fall History trajectory group are shown in Table 2. Predictors of Chronic Fallers were Usufruct, IADL <8, the categories age 60-80 and 80-90. Predictors of Recurrent Fallers were Women, No Social activity, MMSE (between 20 and 24) and MMSE < 10.

4. DISCUSSION

In this study, we examined the course of falls in community-dwelling older adults. Falls represent a significant concern in this population, with 53% of participants experiencing at least one fall in the past six months, underscoring the high prevalence of fall-related risks. Additionally, 51% of individuals have hypertension, reflecting the substantial burden of cardiovascular conditions among older adults. Regarding housing conditions, 59% of participants are homeowners, while 27% reside in rental or free accommodations, and 14% live under usufruct arrangements. These differences in living conditions may influence environmental risk factors for falls, such as home modifications, accessibility, and safety measures. Furthermore, only 18% of participants engage in social activities, suggesting a potential risk of social isolation, which has been linked to increased frailty, decreased physical activity, and higher fall incidence. Functional and cognitive assessments provide additional insights into the study population. The mean GDS (3 ± 4) indicates a relatively low level of depressive symptoms overall, though some participants exhibit signs of psychological distress. The IADL (6 ± 2) suggests moderate functional independence, with some individuals requiring assistance in instrumental activities of daily living. Additionally, the MMSE (25 ± 5) reflects mild cognitive impairment on average, with variability among participants, potentially affecting their ability to recognize and mitigate fall risks.

We found three clinically distinct trajectories of falls, which we have labeled: No-Fallers With Fall History, Chronic Fallers, and Recurrent Fallers. These distinct trajectories highlight the influence of health interventions, aging, and medical conditions on fall risk patterns. We identified several predictors of these trajectories, including usufruct housing status, IADL <8, age categories 60-80 and 80-90, female sex, lack of social activity, and MMSE (between 20 and 24) or <10.

The trajectory of No-Fallers With a Fall History maintains a very low fall risk, which declines to zero, indicating recovery or a stable low-risk state. Experiencing two or more falls within a year marks the onset of physical decline in older adults, making the early identification of key risk factors essential for prevention. In geriatrics, a patient with recurrent falls is often classified as a chronic faller. The Chronic Fallers trajectory of our study is characterized by a persistently higher fall rate, peaking at year 2, followed by a gradual decline, suggesting a prolonged but decreasing risk. In contrast, Recurrent Fallers experience an initial surge in falls, followed by an abrupt drop, likely due to interventions or external factors that reduce their risk after a high-risk period. The yearly intervention of UPSAV appears to contribute to the reduction of fall rates across all trajectories. If we had a control group of older adults who did not receive UPSAV interventions, we could compare the outcomes between the two groups and assess the true impact of UPSAV in preventing falls among older individuals.

Despite these insights, the study has certain limitations, including sample size constraints and participant recruitment being limited to France. Future research should consider larger, more diverse populations to enhance the generalizability of these findings. Additionally, incorporating a randomized controlled trial (RCT) design could provide stronger evidence of intervention effectiveness.

Table 2. Results of multinomial logistic regression: predictive factors

	Chronic Fallers			Recurrent Fallers		
Characteristic	**OR[1]**	**95% CI[1]**	**p-value**	**OR[1]**	**95% CI[1]**	**p-value**
Gender						
Man	—	—		—	—	
Woman	1.12	0.80 – 1.56	0.5	2.00	1.24 – 3.24	0.005
Age						
Over 90	—	—		—	—	
60-80	2.47	1.44 – 4.24	<0.001	0.88	0.47 – 1.67	0.7
80-90	1.72	1.06 – 2.78	0.027	0.82	0.48 – 1.41	0.5
Hypertension						
No	—	—		—	—	
Yes	0.53	0.39 – 0.72	<0.001	0.76	0.51 – 1.14	0.2
Lifestyle						
With a partner	—	—		—	—	
Alone	1.42	1.01 – 2.00	0.044	0.95	0.60 – 1.48	0.8
With family members	1.05	0.56 – 1.95	0.9	0.41	0.15 – 1.08	0.070
Housing						
Owner	—	—		—	—	
Tenant / Free accommodation / Other	0.94	0.66 – 1.35	0.7	1.13	0.72 – 1.78	0.6
Usufruct	1.84	1.19 – 2.85	0.006	0.93	0.48 – 1.81	0.8
Social activity						
Yes	—	—		—	—	
No	1.46	0.98 – 2.18	0.061	1.97	1.09 – 3.55	0.024
IADL						
8	—	—		—	—	
<8	1.64	1.20 – 2.24	0.002	0.95	0.63 – 1.44	0.8
MMSE						
> 24	—	—		—	—	
20-24	0.96	0.66 – 1.40	0.8	1.60	1.00 – 2.57	0.050
10-19	0.83	0.35 – 1.95	0.7	0.00	0.00 – 0.00	<0.001
<10	0.45	0.26 – 0.78	0.004	1.87	1.09 – 3.20	0.022
GDS						
≤ 9	—	—		—	—	
> 9	1.53	0.91 – 2.56	0.11	0.70	0.29 – 1.72	0.4

[1]OR = Odds Ratio. CI = Confidence Interval

5. CONCLUSION

Through the intervention of UPSAV, we have demonstrated the importance of closely monitoring older adults to prevent both initial and recurrent falls. This study allows us to identify fall risk factors within specific subgroups, categorized by their fall trajectories within the total study population. The predictive factors identified for each trajectory will provide specialists with a deeper understanding of faller profiles, enabling the development of more targeted and personalized prevention strategies. Furthermore, these findings emphasize the need for early identification of at-risk individuals to implement timely interventions that reduce fall incidence and associated complications.

By recognizing distinct trajectories, practitioners can better tailor rehabilitation programs, home modifications, and mobility assistance to meet the needs of each group. Incorporating these predictive factors into clinical decision-making may also enhance fall prevention guidelines and public health strategies. Future studies should explore how integrating real-time monitoring tools, such as wearable sensors or AI-driven assessments, could further improve risk stratification. Ultimately, a multidisciplinary approach, combining medical, environmental, and behavioral interventions, may offer the most effective long-term solutions for fall prevention in aging populations.

ACKNOWLEDGEMENT

The authors thank UPSAV, URCI, VieSanté, and the Chaire d'Excellence Intelligence Artificielle et Bien-Vieillir en Nouvelle-Aquitaine (IABV-NA) for their support. We also extend our gratitude to the study participants for their time, effort, and dedication.

REFERENCES

Grundstrom, Anna C., Clare E. Guse, and Peter M. Layde. (2012). "Risk Factors for Falls and Fall-Related Injuries in Adults 85 Years of Age and Older." *Archives of Gerontology and Geriatrics* 54 (3): 421–28. https://doi.org/10.1016/j.archger.2011.06.008.

"Health.Gouv." n.d. Accessed March 29, (2023). https://sante.gouv.fr/IMG/pdf/dp_plan-antichute-accessible28-02-2022.pdf.

Insee. (2016). "Population by Age - Tableaux de l'Économie Française | Insee." 2016. https://www.insee.fr/fr/statistiques/1906664?sommaire=1906743.

Moon, T.K. (1996). "The Expectation-Maximization Algorithm." *IEEE Signal Processing Magazine* 13 (6): 47–60. https://doi.org/10.1109/79.543975.

Moylan, Kyle C., and Ellen F. Binder. (2007). "Falls in Older Adults: Risk Assessment, Management and Prevention." *The American Journal of Medicine* 120 (6): 493.e1-6. https://doi.org/10.1016/j.amjmed.2006.07.022.

Nagin, Daniel S. (2014). "Group-Based Trajectory Modeling: An Overview." *Annals of Nutrition and Metabolism* 65 (2–3): 205–10. https://doi.org/10.1159/000360229.

Nagin, Daniel S., and Candice L. Odgers. (2010). "Group-Based Trajectory Modeling in Clinical Research." *Annual Review of Clinical Psychology* 6:109–38. https://doi.org/10.1146/annurev.clinpsy.121208.131413.

Park, Seong-Hi. (2018). "Tools for Assessing Fall Risk in the Elderly: A Systematic Review and Meta-Analysis." *Aging Clinical and Experimental Research* 30 (1): 1–16. https://doi.org/10.1007/s40520-017-0749-0.

Scikit-learn, developers. (n.d). "GaussianMixture." Scikit-Learn. Accessed February 23, 2025. https://scikit-learn.org/stable/modules/mixture.html.

Tchalla, Achille, Noëlle Cardinaud, Caroline Gayot, Nathalie Dumoitier, Michel Druet-Cabanac, Marie-Laure Laroche, Karen Rudelle, Cécile Laubarie Mouret, and Sophie Boyer. (2023). "Patterns, Predictors, and Outcomes of Frailty Trajectories in Community-Dwelling Older Adults: Results from the FREEDOM Cohort Study." *Archives of Gerontology and Geriatrics* 114 (November):105101. https://doi.org/10.1016/j.archger.2023.105101.

"The Prevention of Falls in Later Life. A Report of the Kellogg International Work Group on the Prevention of Falls by the Elderly." (1987). *Danish Medical Bulletin* 34 Suppl 4 (April):1–24.

Thierry, Xavier. (2010). "Accidents and physical assaults among the elderly: less frequent than among the young, but more severe." *Population & Sociétés* 468 (6): 1–4. https://doi.org/10.3917/popsoc.468.0001.

Welsh, T J, A L Gordon, and J R Gladman. (2014). "Comprehensive Geriatric Assessment – a Guide for the Non-Specialist." *International Journal of Clinical Practice* 68 (3): 290–93. https://doi.org/10.1111/ijcp.12313.

WHO. (2022). "Ageing and health." 2022. https://www.who.int/fr/news-room/fact-sheets/detail/ageing-and-health.

YOUDEN'S J STATISTIC IN CONSTRUCTING A DECISION TREE CLASSIFIER FOR MEDICAL DIAGNOSTIC REASONING

Irosh Fernando
Hunter New England Area Health, 72 Watt St, Newcastle, NSW, Australia.
School of Medicine and Public Health, Faculty of Health and Medicine, University of Newcastle, Callaghan, NSW, Australia

ABSTRACT

For clinical features with continuous data, a threshold (cutoff point) is necessary to indicate when it is pathological or benign in relation to a diagnosis. Also in diagnostic reasoning, it is important to rank clinical features in terms of their diagnostic importance in relation to a given diagnosis. The same is applicable in machine learning when it is necessary to binarise continuous data; also, when there are a large number of features related to an outcome variable, a dimensionality reduction is necessary by eliminating the features that are less relevant or important. There are already established techniques that are to be selected depending on the nature of the machine learning application. In this paper we present an approach for finding the best cutoff and an estimate for determining the importance of each feature in the context of medical diagnosis, using Youden's J statistic as a better metric for diagnostic reasoning. The proposed approach was implemented as a classification tree algorithm and was evaluated using the Pima Indians Diabetes Database and the results were compared with the standard Decision tree classifier algorithms.

KEYWORDS

Machine Learning, Decision Tree Classifier Algorithms, Medical Diagnostic Reasoning, Youden's J Statistic, Dimensionality Reduction, Binarisation of Continuous Data

1. INTRODUCTION

This paper addresses the following two problems in relation to the application of machine learning in medical diagnosis: 1) Ranking different clinical features according to their importance concerning a given diagnosis from a clinical reasoning perspective; 2) Estimating the best cutoff points to be used for binarisation of clinical features with continuous data in such a way that it maximises the accuracy of predicted diagnosis as the outcome.

There are well-established techniques that are effectively used in splitting continuous variables, particularly in constructing Decision tree classifiers (Breiman,1984). These include Genie impurity, Entropy, and log loss function. Similarly, there are well-established techniques for feature selections (Pudjihartono et al. 2022); these include: variance threshold (Fida et al. 2021); k score (Zhang et al. 2021); false positive rates, Geni impurity measures (Nembrini et al. 2018). The techniques that best suit may depend on the nature of the classification problem.

The purpose of this research work was to investigate the suitability of these techniques in relation to medical diagnosis using machine learning and improving the accuracy of the technique used. However, we have narrowed the scope of this paper to a comparison of Youden's J statistic (Youden, 1950) as a better metric according to clinical intuition, Gini impurity, entropy and log loss which are the standard metrics used in sklearn Decision tree classifiers (available via scikit-learn.org). We constructed a simpler decision tree classification algorithm using Youden's J statistic (referred to as J score in the rest of the paper), as the preferred metric, and the results were compared with what was obtained from the standard decision tree classification algorithms. Data from the Pima Indians Diabetes Database was used in the experiment. This dataset was originally from the National Institute of Diabetes and Digestive and Kidney Diseases and consists of data collected from 768 patients who were females of Pima Indian heritage. The dataset is made freely available via

Kaggle.com for use in machine learning to predict if a patient has diabetes or not, using a set of 8 clinical features which include the number of pregnancies each patient had, plasma glucose concentration, blood pressure, skin thickness, 2-hour serum insulin, body mass index (BMI), diabetes pedigree function, and age. As the outcome variable, the presence or absence of diabetes is recorded.

As the main contribution of this research work, we have demonstrated how the J score can be used to construct a simpler decision tree classifier with a reduced feature set and binarised data, that produced comparable results to sklearn decision tree classifier algorithms using Gini, entropy, and log-loss techniques, in predicting diabetes using Pima Indian diabetes dataset.

2. DIAGNOSTIC REASONING

Diagnostic reasoning involves an iterative process of generating hypotheses of differential diagnosis based on the symptoms reported by the patient and then searching for the other symptoms that are expected in each differential diagnosis (Ramoni et al. 1992); the final step of diagnostic reasoning involves matching all the clinical features that were elicited with what is expected in each differential diagnosis and determining the best fit which would be the final diagnosis. Thus, machine learning has been applied as a pattern recognition or classification technique (i.e., classifying clinical features into diagnoses) in arriving at a more definitive diagnosis (Ahsan et al. 2022). From a clinician perspective, the initial step of diagnostic reasoning involves guessing the likely diagnoses based on a given clinical feature. As more and more clinical features are elicited during the process of clinical assessment, the initial estimate of the differential diagnoses is refined by estimating its likelihood subjectively (i.e., how likely it is, not in the statistical sense, using clinical judgement) using the combination of more clinical features that are being elicited.

The process of estimating the likelihood of a diagnosis using a given set of clinical features can be formalised and estimated objectively if enough data are available from patients about the presence or absence of the combination of the different clinical features and the diagnosis, using sensitivity and specificity analysis process as described in Table 1. Sensitivity and specificity analysis are traditionally used in evaluating the accuracy of a diagnostic instrument or a test. This involves applying the diagnostic instrument or the test on a sample of patients with and without the diagnostic condition (Trevethan 2017). Using the results of the instrument or the test, the presence or absence of the diagnosis is predicted and the accuracy can be estimated using the following outcomes: 1) the number of patients with the condition that were correctly diagnosed (true positives); 2) the number of patients who did not have the conditions and were correctly diagnosed as not having the condition(true negatives); 3) the number of patients who did not have the conditions and were incorrectly diagnosed as having the condition (false positives); 4)the number of patients who did have the conditions and were incorrectly diagnosed as not having the condition (false negatives).

Table 1. Analysis of all possible outcomes of a diagnostic test or an instrument in predicting absence or presence of a diagnosis

	Predicted positive	**Predicted negative**
Diagnosis positive	True positive (TP)	False negative (FN)
Diagnosis negative	False positive (FP)	True negative (TN),

Using these outcome data from Table 1which is known as confusion matrix a number of metrics can be estimated as follows:

Sensitivity (True positive rate (TPR) or Recall) = $TP/(TP + FN)$;

Specificity (True negative rate (TNR)) = $TN/(TN + FP)$;

False positive rate (FPR)) = $FP/(TN + FP)$;

Positive Predictive Value (PPV) = $TP/(TP + FP)$;

Youden's J statistic (J) $= TPR + TNR - 1 = TPR - FPR$;

F1 score= $2(PPV \times TPR)/(PPV + TPR)$;

Accuracy= $(TP + TN)/(TP + FN + FP + TN))$.

3. APPLICATION OF YOUDEN'S J STATISTIC (J)

In the proposed approach we use the same sensitivity and specificity analysis that was described above, in predicting the presence or absence of the diagnosis using any given clinical feature as described in Table 2. For example, the feature glucose alone can be used to predict diabetes (see the last row of Table 3 corresponding to the model J Score (1) which is described in the next section).

Table 2. Predicting absence or presence of a diagnosis using a clinical feature

	Clinical feature present	**Clinical feature absent**
Diagnosis positive	True positive (TP)	False negative (FN)
Diagnosis negative	False positive (FP)	True negative (TN),

The diagnostic utility of any given clinical feature in diagnosing a given medical condition depends on the following two ratios: 1) the proportion of patients with the diagnosis having the clinical feature; and 2) the proportion of patients without the diagnosis having the clinical feature. In an ideal situation where a clinical feature has the highest diagnostic utility, the first proportion should be equal to one since all the patients with the diagnostic condition are supposed to have the clinical feature. Similarly, the second ratio should become zero since the patient without the diagnosis is not expected to have the clinical feature. The difference between the two ratios which is the same as the J score can be used as a clinically intuitive metric for diagnostic accuracy. With the clinical features having continuous data, the same analysis can be done for each value in the range.

3.1 Binarisation and Feature Selection

Thus, we used the following process for binarising the features with continuous data: For each feature calculate the J score for each value; use the value with the highest J score as the cutoff point. The higher the J score higher the importance of the feature. As an example, the J score for each value is plotted for Glucose and BMI. The highest J score was 0.433 which was achieved by Glucose followed by Age and BMI which achieved a score of 0.327 and 0.313 respectively. The corresponding cutoffs for Glucose, Age, and BMI were 124, 29 and 29.9 which were used in binarising the data and constructing of a decision tree classifier (Figure 2). The J score in relation to feature importance is visually compared with the feature importance measured by Gini impurity in the standard decision tree classifier (Figure 4) and the results are discussed in the next section.

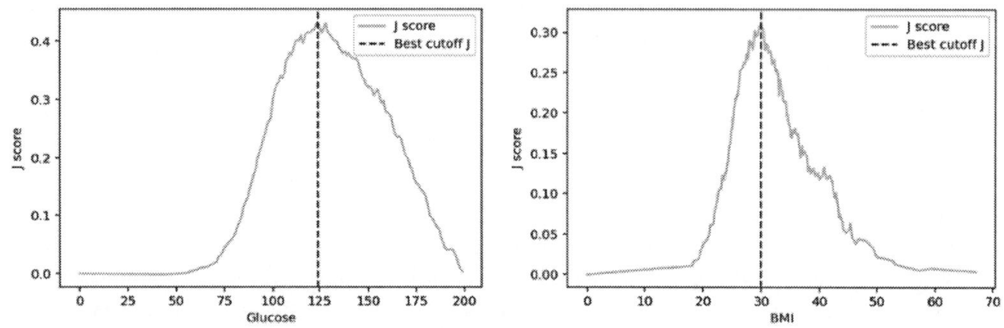

Figure 1. J score for each value in the feature (Glucose) and the maximum J score corresponding to the cutoff

3.1 A Decision Tree Classification Algorithm Using J Score

We constructed a probability decision tree using the cutoff scores to binarise the data and select the three clinical features having the highest score (Figure 2). Every permutation of the feature set is represented as branches in the tree in which 1 and 0 corresponds to 'Ture' and "False' respectively. For example, an entry with '1,0,1' in the binarised data set corresponds to the branch: Glucose≥ 124; Age <29; BMI ≤ 29.9. Training the algorithm involved populating the tree with the training data set and calculating the probabilities of the outcomes associated with each decision branch as a proportion. As we are using the presence or absence of diabetes as the outcome variable each decision branch is associated with the number of patients with diabetes and the number of patients without diabetes. In predicting the diagnosis for a given feature set, the predicted value can be either the probability of having diabetes (P_1) or the probability of not having proportions (P_0). We dichotomised the outcome using the following rule: 1 if $P_1 > P_0$, 0 otherwise. The model described above was implemented in Python and the source code will be made available via https://github.com/DA-Irosh-Fernando/Youden-s-J-statistic-in-constructing-a-decision-tree-classifier.

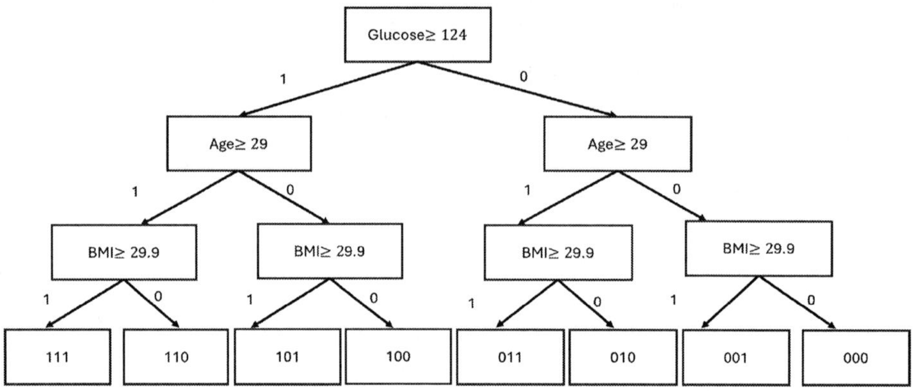

Figure 2. Probability decision three constructed using the three clinical features with the highest J score

We also explored how the model is improved progressively as more and more features are combined. In doing so, we first explored how the feature with the highest J score (Glucose) alone can predict diabetes (model J Score (1) in Table 3) followed by a decision tree classifer with the two features with the highest J scores (Glucose and Age) can predict diabetes (model J Score (2) in Table 3).

For comparison, we implemented a Decision tree classifier using sklearn DecisionTreeClassifier which uses either Gini, entropy or log-loss for splitting the data (nodes). The depth of the tree (parameter max_depth) was varied from "None" to 3 and 2 which have been denoted as w3 and w3 respectively in Table 3. By setting the maximum depth of the tree to 2, and using all the features in the dataset, the decision tree produced by the algorithm is shown in Figure 3. The diabetic dataset was split into two equal samples for training and predicting, which were used for both decision tree classifier algorithms. The findings were compared with what was obtained from Decision tree classification algorithms using Gini, entropy and log-loss for splitting the data. The outcome of predicting diabetes is summarised for comparison in Table 3, and the results are discussed in the following section.

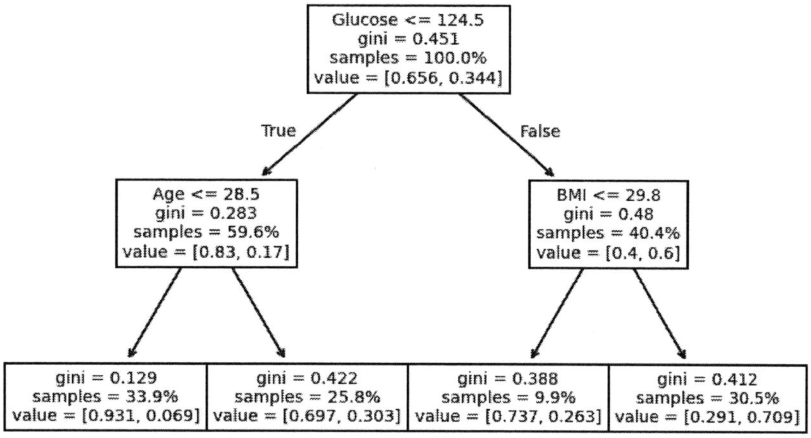

Figure 3. Decision tree classifier using sklearn DecisionTreeClassifier with all the features

4. RESULTS

Firstly, the top three features having the highest importance and their ordering according to the impurity measure (Glucose, BMI, Diabetes Pedegree Function) was different from the three features and their ordering that were selected according to J score (Glucose, Age, BMI), which can have an implication on the accuracy of the diagnostic prediction made using machine learning algorithms. Yet, interestingly, the decision tree produced by the sklearn DecisionTreeClassifier with a depth size of two using all the features, resulted in having the same three features that ranked with the highest J score. While the findings from a single dataset would not be adequate to make any firm conclusion about the correlation between the clinical importance and the feature importance selected according to these metrics, we believe that the J score is likely to give more accurate results than the impurity measure which is less intuitive clinically.

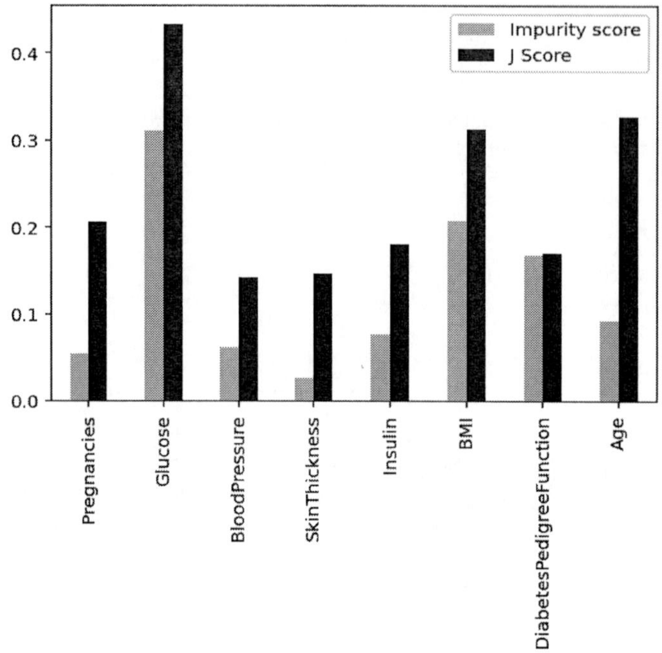

Figure 4. Visual comparison of feature importance derived from Gini impurity and J score

Secondly, as evident from the results summarised in Table 3 the accuracy of the prediction made using the decision tree constructed using J score, is comparable to the results from the standard decision tree classification algorithms using Gini, entropy and log-loss. The model J Score (2) which is the decision tree classifier we constructed using the two features with the highest scores (Glucose and Age), produced markedly improved results compared to the prediction made using the feature with the highest score (Glucose) alone. On the other hand, combining the next feature with the highest J score (model J Score (2)) did not improve the results, which is understandable given the relatively low values of the J scores. In relation to the results of the sklearn Decision tree classifiers the models using Gini, entropy and log loss produced similar results when the tree depth was fixed; the decision tree classifiers with a depth 3 produced slightly better results.

Table 3. Comparison of the performance of different models (J = Youden's J statistic. AU.= area under receiver operating characteristic (ROC) curve. Acc.=Accuracy)

DTC Model	**TP**	**TN**	**FP**	**FN**	**TPR**	**PPV**	**TNR**	**NPV**	**F1**	**J**	**AU.**	**Acc.**
Gini (None)	73	203	45	63	0.54	0.62	0.82	0.76	0.57	0.36	0.68	0.72
Entrop (None)	73	200	48	63	0.54	0.60	0.81	0.76	0.57	0.34	0.67	0.71
Log L.(None)	76	199	49	60	0.56	0.61	0.80	0.77	0.58	0.36	0.68	0.72
Gini (w3)	75	211	37	61	0.55	0.67	0.85	0.78	0.60	0.40	0.79	0.74
Entropy (w3)	74	212	36	62	0.54	0.67	0.85	0.77	0.60	0.40	0.81	0.74
Log-loss (w3)	74	212	36	62	0.54	0.67	0.85	0.77	0.60	0.40	0.81	0.74
Gini (w2)	74	212	36	62	0.54	0.67	0.85	0.77	0.60	0.40	0.77	0.74
Entropy (w2)	74	212	36	62	0.54	0.67	0.85	0.77	0.60	0.40	0.76	0.74
Log-loss (w2)	74	212	36	62	0.54	0.67	0.85	0.77	0.60	0.40	0.76	0.74
J Score (3)	75	213	35	61	0.55	0.68	0.86	0.78	0.61	0.41	0.78	0.75
J Score (2)	74	213	35	62	0.54	0.68	0.86	0.77	0.60	0.40	0.77	0.75
J Score (1)	55	143	105	81	0.40	0.34	0.58	0.64	0.37	-.02	0.49	0.51

In order to examine if there was a statistically significant difference in performance between the proposed models using the J score verses the other models, a receiver operating characteristic (ROC) analyses (DeLong test) was performed using the roccomp command of Stata 19 statistical software, by comparing the respective models with the highest scores (Figure 5) in Table 3. The null hypothesis that the areas under the respective ROC curves were the same, was not rejected as the resulting chi-square (2) score of 4.77 yielded a p-value of 0.09.

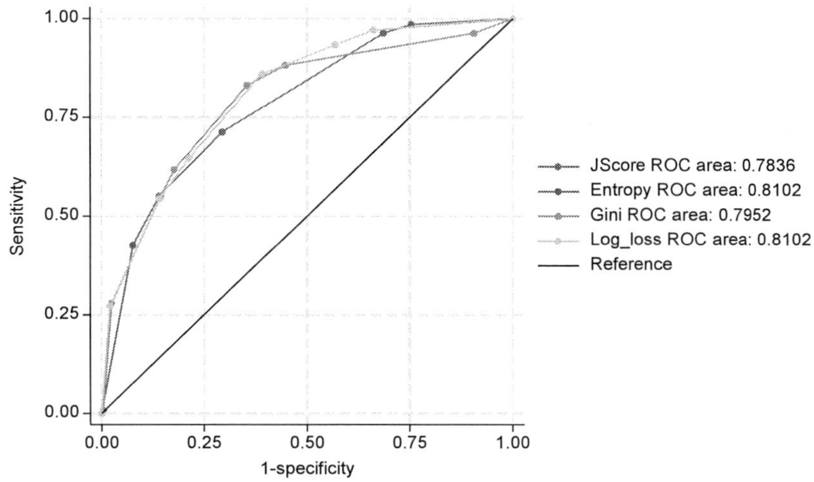

Figure 5. Comparison of the receiver operating characteristic (ROC) of JScore (3), Gini(w3), Entropy(w3), and Log-loss(w3) models

5. CONCLUSION

The approach presented in this paper is relatively naive and simpler than the standard decision tree classification algorithms. Yet, it performed equally well compared to the standard decision tree classification algorithm using the diabetic data set. Also, we have demonstrated that binarisation of clinical features using the J score enables fitting the probability decision tree with a training dataset set so naturally that every possible permutation of different clinical features is fitted into a branch of the tree. There are three main advantages of the proposed approach. Firstly, the feature importance derived using the J score can be used as a dimensionality reduction technique in machine learning by eliminating the features that are less important or less relevant. Secondly, the J score can be used as a clinically intuitive and accurate metric for splitting the nodes in constructing decision tree classifiers, as an alternative to gini, entropy and log loss measures. Thirdly, the proposed approach can be more appealing to clinicians as a clinically intuitive and easy-to-understand machine learning algorithm; furthermore, besides its use in machine learning, the J score can also assist as an educational tool to improve medical diagnostic reasoning and define clinical diagnoses using the most salient clinical feature set. One of the main limitations of the decision tree constructed using J score is the exponential growth of the tree as the number of features increases and the resulting need for a larger and balanced data set for training the tree. Thus, dimensionality reduction is necessary, which can be achieved by eliminating the features having low feature importance. Also, it is likely that in medical diagnostic reasoning, clinical diagnoses are often associated with only a relatively smaller set of salient clinical features. However, further research work is necessary to explore its utility in diagnostic prediction across a broader range of medical diagnostic conditions.

ACKNOWLEDGEMENT

Thanks to kaggle, UC Irvine Machine Learning Repository, and the National Institute of Diabetes and Digestive and Kidney Diseases for the dataset. Thanks to scikit-learn (Machine Learning in Python) for the open-source resources.

REFERENCES

Ahsan, M.M., Luna, S.A. and Siddique Z, (2022). 'Machine-Learning-Based Disease Diagnosis: A Comprehensive Review', *Healthcare (Basel)*, 10(3), doi:10.3390/healthcare10030541.

Fida, M.A.F.A, Ahmad, T. and Ntahobari, M. (2021). *Variance Threshold as Early Screening to Boruta Feature Selection for Intrusion Detection System.*

Nembrini, S., König, I.R. and Wright, M.N .(2018). 'The revival of the Gini importance?', *Bioinformatics*, 34(21): 3711-3718, doi:10.1093/bioinformatics/bty373.

Pudjihartono, N., Fadason, T., Kempa-Liehr, A.W. and O'Sullivan, J.M. (2022). 'A Review of Feature Selection Methods for Machine Learning-Based Disease Risk Prediction', *Front Bioinform*, 2:927312, doi:10.3389/fbinf.2022.927312.

Ramoni, M., Stefanelli, M., Magnani, L. and Barosi, G. (1992). 'An epistemological framework for medical knowledge-based systems', *IEEE Trans. Syst. Man Cybern.*, 22:1361-1375,

Trevethan, R. (2017). 'Sensitivity, Specificity, and Predictive Values: Foundations, Pliabilities, and Pitfalls in Research and Practice', *Frontiers in Public Health*, 5, doi:10.3389/fpubh.2017.00307.

Zhang, X., Fan, M., Wang, D., Zhou, P. and Tao, D. (2021). 'Top-k Feature Selection Framework Using Robust 0–1 Integer Programming', *IEEE Transactions on Neural Networks and Learning Systems*, 32(7):3005-3019, doi:10.1109/TNNLS.2020.3009209.

DIABETICA: A PERSONALIZED AI-POWERED PLATFORM FOR DIABETES MANAGEMENT

Mahmoud El Hamlaoui, Rdouan Faizi, Bouchaib Bounabat, Haitam El Hayani and Afraa Bouzgoura
ENSIAS, Mohammed V University in Rabat
Rabat, Morocco

ABSTRACT

Diabetes management is a critical global health challenge affecting millions of individuals worldwide. This paper presents Diabetica, an AI-driven web platform designed to assist diabetic patients in tracking key health metrics, managing appointments, and receiving personalized recommendations. Diabetica integrates machine learning models for diabetes prediction, ingredient safety assessment, and dish safety classification, along with an intelligent Chatbot for user engagement. This research paper introduces the platform's architecture, the AI models employed, and the potential impact of such a system on diabetes care. Our findings do confirm that the proposed AI platform can enhance patient self-management and improve doctor-patient collaboration, reducing risks and optimizing healthcare resources.

KEYWORDS

Digital Health Platforms, Diabetes Management, Artificial Intelligence

1. INTRODUCTION

With over 537 million adults living with diabetes globally as of 2021 (International Diabetes Federation, 2021) there is an increasing need for technological solutions that can facilitate disease management. The complexity of diabetes management lies in its requirement for constant monitoring of various health parameters and lifestyle factors, which is a challenging task for both patients and healthcare providers. Traditional management approaches, while foundational, often struggle to provide the immediate feedback and personalized guidance that modern technology can offer.

The emergence of artificial intelligence and machine learning technologies present unprecedented opportunities to transform diabetes care. These technologies enable the development of sophisticated systems that are capable of providing personalized recommendations, real-time monitoring, and predictive insights. However, existing solutions often focus on isolated aspects of diabetes management, failing to provide the comprehensive support that patients require. In response to these challenges, we present DIABETICA, an integrated web-based platform that harnesses the power of artificial intelligence to deliver personalized diabetes management support. Our platform represents a significant advancement in digital health technology, combining sophisticated AI algorithms with user-friendly interfaces to create a comprehensive diabetes management solution. The system not only monitors crucial health parameters but also provides intelligent insights and recommendations based on individual patient data. The significance of our work lies in its holistic approach to diabetes management. Rather than focusing solely on blood glucose monitoring or dietary tracking, DIABETICA integrates multiple aspects of diabetes care into a cohesive system. The platform incorporates advanced machine learning models for food safety analysis, predictive health monitoring, and personalized risk assessment, while maintaining the essential connection between patients and their healthcare providers.

The rest of the paper is structured as follows: Section 2 presents the system architecture of DIABETICA, detailing its key components and functionalities. It describes the AI models used for diabetes risk prediction, ingredient safety assessment, and dish classification. It also explains the Chatbot integration and the experimental results. Section 3 discusses the findings. Finally, Section 4 concludes the paper and outlines future research directions.

2. SYSTEM ARCHITECTURE

DIABETICA leverages Information and Communication Technologies (ICTs) to enhance diabetes management through an intelligent, integrated platform (Figure 1). The goal of our proposed solution is to provide diabetic people/potential users with a set of ready services to make their lives as simple as possible.

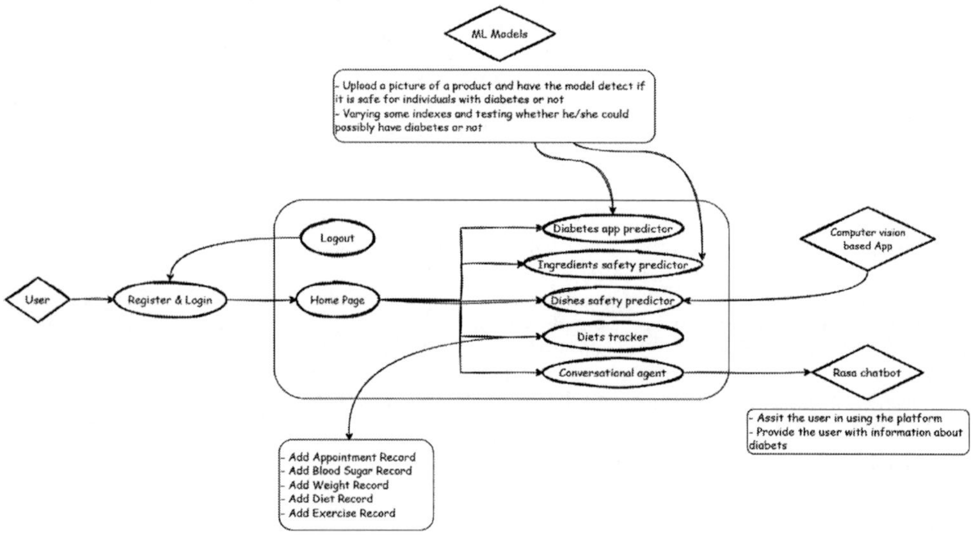

Figure 1. Global architecture of Diabetica

Once a user is registered, they can perform different actions:

- Interact with a Chatbot for more details about diabetes and to assist them in interacting with the platform.
- Check food safety by uploading dishes' pictures.
- Check product safety by uploading the ingredients' picture.
- Predict their possibility of being diabetic by changing some indexes.
- Track their habits by adding their blood sugar, weight, diet, exercise, and appointment records.

Diabetica leverages artificial intelligence to enhance diabetes management through predictive analytics, food safety assessment, and dietary recommendations. These models work together within the platform to provide an AI-powered decision-support system, empowering users to better manage their diabetes through predictive insights and dietary recommendations.

2.1 Diabets Risk Predictor

Diabetes Risk Predictor is based on a Random Forest classifier. It analyses various health parameters, including blood pressure, cholesterol levels, and BMI (Body Mass Index), to estimate the risk of developing diabetes. The model was trained on a merged dataset that combined two separate data collections while eliminating duplicates to improve its predictive capabilities.

Random Forest was chosen specifically because it excels at handling the complex interrelationships between multiple health variables involved in diabetes detection. The algorithm can effectively process both numerical data (like blood pressure readings) and categorical information (such as lifestyle factors), making it particularly suitable for this medical application. By learning these patterns, the system aims to identify potential diabetes risk early enough for preventive intervention.

The interactive interface shown in Figure 2 enhances the user experience by providing a user-friendly and intuitive platform for interaction. It features interactive form fields where users can input their personal data and relevant variables, such as age, gender, BMI, blood pressure, cholesterol levels, and lifestyle factors. Once the necessary information is submitted, the UI triggers the Random Forest model to process the inputs and generate a prediction regarding the user's diabetes status. The prediction is then displayed in a clear and comprehensible format. With its visually appealing layout and informative design, the UI empowers users to gain valuable insights into their diabetes risk, enabling them to make informed decisions about their health and lifestyle choices.

Figure 2. Diabetes risk predictor

2.2 Ingredient and Dish Safety Predictions

To support dietary decisions, the platform includes an ingredient safety predictor that employs Optical Character Recognition (OCR) to extract textual information from ingredient lists on food packaging. By comparing these ingredients against a curated database of items deemed high-risk for diabetics, the system provides instant feedback on whether a food product is safe for consumption. This capability reduces the complexity of meal planning and helps users make informed dietary choices.

Concerning the dish safety predictor, it utilizes a deep learning model based on Convolutional Neural Networks (CNN) to assess the suitability of prepared meals. By analyzing images of various dishes from the Food-101 dataset and supplementing them with nutritional information obtained via web scraping, the model determines whether a dish aligns with recommended dietary guidelines for diabetic individuals. This automated analysis ensures that users receive reliable nutritional insights without the need for manual ingredient verification.

The Food-101 dataset comprises 101,000 images evenly distributed among 101 diverse food categories, each category contains 1,000 images depicting dishes from a wide range of cuisines. For effective model training and evaluation, the dataset is divided into training and testing subsets, with approximately 75% (750 images per category) used for training and the remaining 25% (250 images per category) reserved for testing to assess generalization capabilities.

The base model used for dish recognition is InceptionV3, a pre-trained CNN model trained on the ImageNet dataset, which contains millions of images across 1,000 different classes. InceptionV3 is widely utilized in image classification tasks due to its ability to extract complex features from images. It consists of multiple convolutional layers, pooling layers, and fully connected layers, making it highly effective for detailed feature extraction and object recognition.

In addition to the image data, nutritional information, such as protein, calcium, fat, carbohydrates, vitamins, and sugar is gathered via web scraping of each dish of Food-101 dataset (Figure 3).

	name	protein	calcium	fat	carbohydrates	vitamins	sugar
0	apple pie	3.25	0.0	8.13	46.3	0.0	23.6
1	baby back ribs	18.8	0.026	9.82	0.0	0.0	0.0
2	baklava	6.6	0.041	29.0	37.8	0.0013	12.8
3	beef carpaccio	6.6	0.041	8.42	3.42	0.0019	12.8
4	beef tartare	17.4	0.028	15.6	0.34	0.0002	0.08
5	beet salad	1.32	0.018	5.29	11.0	0.4542	8.81

Figure 3. Sample of extracted nutritional information

The results of the food classification model training are presented in the following figures. In Figure 4, the x-axis represents the number of epochs, ranging from 0 to 30, while the y-axis represents accuracy, ranging from 0.1 to 0.8. The graph contains two lines: one representing training accuracy and the other representing validation accuracy. The training accuracy increases up to approximately 0.8, while the validation accuracy also rises to around 0.7 by epoch 30. The training and validation accuracies remain very close to each other throughout the epochs. The model's results are promising, as both accuracies increase consistently over the epochs and exhibit minimal divergence. This suggests that the model generalizes well to the validation data.

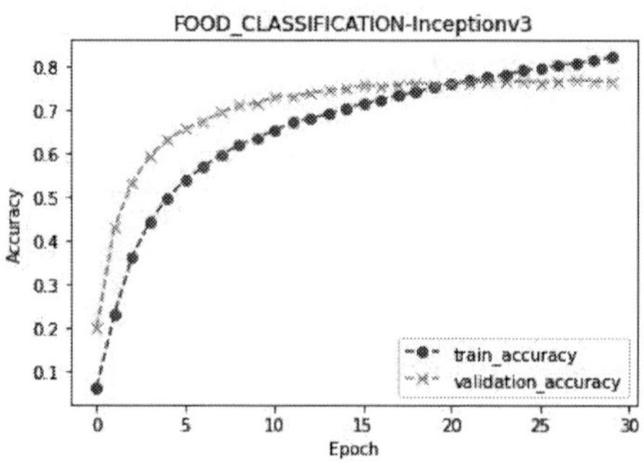

Figure 4. Training and validation accuracy of the food classification model

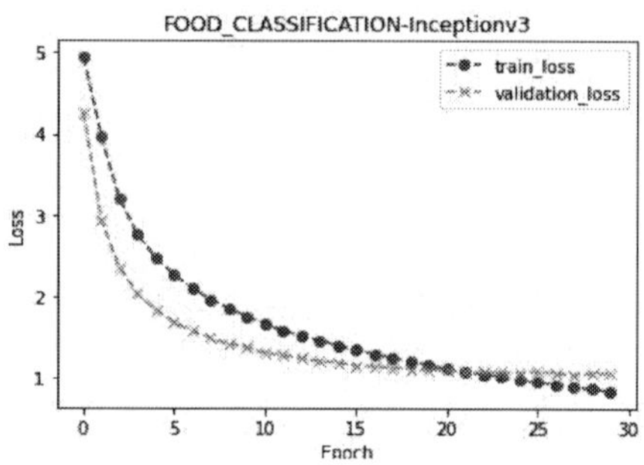

Figure 5. Training and validation loss of the food classification model

For the second graph, the x-axis represents the number of epochs, while the y-axis represents the error. Similar to the previous graph, there are two lines: one representing the training error and the other representing the validation loss. Both errors decrease steadily until reaching approximately. The model's results are promising, as both training and validation loss decrease consistently over the epochs and remain closely aligned. This suggests that the model is capable of correctly generalizing to the validation data.

From the user perspective, Figure 6 shows an example of the ingredient safety predictor notifying instantly that the ingredients of that product aren't safe to consume whereas Figure 7 demonstrates the model's real-time analysis of an uploaded dish image, providing immediate nutritional insights to guide user decisions.

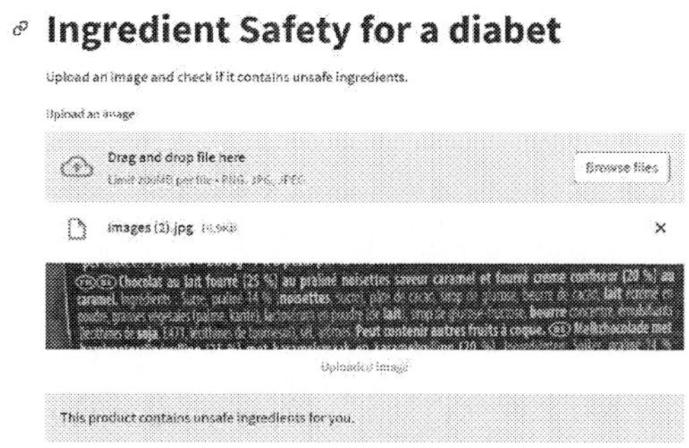

Figure 6. Ingredient safety predictor

Figure 7. Dish safety predictor

2.3 Chatbot Integration

The RASA-based conversational agent represents one of DIABETICA's most interesting components. This AI-powered Chatbot utilizes advanced natural language understanding capabilities to process and respond to user queries effectively. The implementation incorporates multiple layers of language processing, including intent classification, entity recognition, and dialogue management. The Chatbot's architecture enables it to maintain context throughout conversations, creating more natural and meaningful interactions. Through continuous learning from user interactions, the system improves its response accuracy and relevance over time. The integration with other platform features allows the Chatbot to provide comprehensive assistance, from answering general queries about diabetes management to offering specific guidance based on user health data.

The data collection process was conducted in multiple distinct phases. The project starting point is a dataset that comprises 325 rows and 26 columns (Crangle, 2018). The columns contain various types of information, including *'gender', 'person_ID', 'qx_ID', 'Qx', 'CAUSE', 'RISK', 'PREVENTION', 'DIAGNOSIS', 'MANIFESTATIONS', 'TREATMENT', 'ANATOMY', 'CURE', and 'DIET'*, among others.

The question filtering phase involved a meeting with Dr. blind1 (General Practitioner), who helped refine the dataset using the following criteria:

- Removing statements about diabetes, as the focus was solely on questions.
- Eliminating incorrect questions unrelated to diabetes.
- Reformulating unclear questions to ensure clarity.
- Generating additional questions from existing long sub-questions in the dataset.

Data validation was the most critical step in this process, as it established the credibility of the dataset. This phase required collaboration with multiple medical professionals, who were consulted to review and validate specific question-answer pairs and/or to remove or modify questions deemed inaccurate or irrelevant.

The validation of the dataset was conducted with the assistance of several medical professionals. Each expert reviewed and validated a set of question-answer (Q&A) pairs, ensuring the dataset's credibility and relevance. The table below summarizes the validation statistics:

Table 1. Summary of Data Validation Statistics

Medical Expert	*Q&A Sent*	*Q&A Validated*
Dr. Amer – General Practitioner	72	67
Dr. Loukach – General Practitioner	30	28
Dr. Saoud – Radiologist	52	45
Dr. Raissouni – General Practitioner	40	37
Dr. Khairoun – Internal Medicine (Endocrinology)	29	24
Anonymous Physician	17	15

This validation process played a crucial role in refining the dataset by ensuring that the retained Q&A pairs were accurate, relevant, and medically reliable.

However, to ensure that the model understands the various ways in which a question could be formulated, we generated multiple reworded versions of each question. Using Microsoft's Bing AI Chatbot, we produced between 30 and 50 paraphrased variations per question, thereby enriching the dataset with diverse linguistic expressions.

Next, configuring the NLP (Natural Language Processing) model was a critical step, carried out in the config.yml file. This configuration defines an NLP model for a virtual assistant capable of understanding and responding to queries in English. It employs a series of text-processing steps and machine learning components to perform tasks such as user intent classification, entity extraction, and selection of appropriate responses.

The final model's performance is summarized as follows (Figure 8):

- Total Loss (t_loss) = 1.46: This represents the total error at the end of training. A lower loss value indicates an improvement in model performance.
- Validation Loss (loss) = 0.928: This is the model's loss on the validation dataset. A lower validation loss suggests that the model generalizes well to unseen data.
- Accuracy (acc) = 0.98: This measures the proportion of correct predictions made by the model. A value of 0.98 indicates that the model achieves a high accuracy of approximately 98%, correctly classifying most examples.

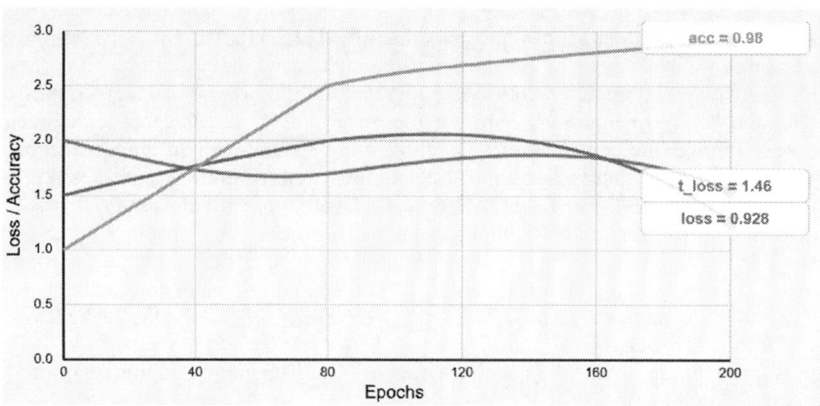

Figure 8. Model performance evaluation

These results demonstrate that the Chatbot trained with RASA performed well, with a low total loss, a low validation loss, and a high accuracy of 98%. This suggests that the model effectively understands user inputs and can make precise predictions when tested on validation data.

3. DISCUSSION

The development of digital health solutions for diabetes management has gained significant attention in recent years, particularly with the integration of artificial intelligence (AI) and web-based platforms. Various research efforts have focused on leveraging AI to provide personalized support, enhance patient monitoring, and improve health outcomes. Several studies have explored the role of AI in diabetes management, focusing on real-time monitoring, personalized recommendations, and predictive analytics. Platforms such as BlueLoop and MySugr have been developed to assist diabetic patients by tracking blood glucose levels, diet, and medication adherence. These platforms emphasize data-driven decision-making, enabling users to better manage their condition. Similar to Diabetica, they incorporate user-friendly interfaces and data visualization tools to enhance patient engagement (Contreras and Vehi, 2018). Furthermore, AI models have been employed for early diabetes detection. Research has demonstrated that machine learning algorithms, such as Random Forest and Support Vector Machines, can achieve high accuracy in predicting diabetes based on clinical data (Dagliati et al., 2023). The model used in the Diabetes Risk Predictor, integrated within Diabetica, aligns with such research efforts by utilizing AI for diabetes risk prediction.

Chatbots have emerged as effective tools in the healthcare domain, facilitating patient education and support. AI-driven Chatbots like Ada Health and Woebot have been designed to provide health advice, mental health support, and chronic disease management. Rasa-based Chatbots, in particular, have been widely adopted due to their ability to process natural language inputs and generate personalized responses (Tudor Car et al., 2024). The Gluco-Bot Chatbot in Diabetica shares similarities with these systems by providing real-time guidance to diabetic patients, enabling them to navigate their health concerns without immediate medical intervention.

To further enhance the adaptability and contextual responsiveness of DIABETICA's conversational agent, prompt engineering techniques offer significant potential. The work in (Tamenaoul et al., 2023) introduces a user prompt meta-model that formalizes the structure and semantics of prompts for GPT-based agents. Integrating such a meta-model into our system can guide future iterations of our Chatbot, enabling more precise, context-aware, and user-aligned interactions, especially when dealing with medical advice, symptom interpretation, or nutritional guidance.

Nutritional analysis and food safety evaluation have also been explored through AI and machine learning models. Optical Character Recognition (OCR) techniques have been leveraged to extract ingredient information from product labels, which is then analyzed against predefined dietary guidelines. Studies have highlighted the effectiveness of deep learning models, such as Convolutional Neural Networks (CNNs), in classifying food images and determining their nutritional content (Boucher et al., 2023). The ingredient and dish safety predictors within Diabetica employ similar techniques, offering a novel approach to assessing dietary risks for diabetic individuals.

While existing solutions provide essential support for diabetes management, Diabetica distinguishes itself by integrating multiple AI models within a unified platform. Unlike conventional diabetes tracking applications, Diabetica offers a holistic suite of functionalities, including appointment scheduling, diet tracking, and interactive Chatbot assistance. Additionally, the incorporation of food safety analysis sets it apart from traditional health monitoring applications.

4. CONCLUSION AND PERSEPECTIVES

The advancements in AI-driven healthcare solutions have paved the way for more efficient and personalized diabetes management. The related work discussed highlights the significant strides made in this domain, with Diabetica contributing a comprehensive and integrated approach to diabetes support. Future research may explore further enhancements, such as real-time physician collaboration and multilingual support, to expand the platform's usability and impact. The integration of multiple AI modules within DIABETICA, ranging from predictive analytics to conversational interfaces, necessitates seamless alignment of heterogeneous models. To address this complexity, collaborative decision-making approaches can be used. In (Bennani et al. 2019), the method used promotes consensus-building among diverse stakeholders and supports iterative alignment across technical layers, which can be particularly useful for refining DIABETICA's modular interactions between vision, prediction, and language processing components.

As DIABETICA evolves towards integration with Clinical Decision Support Systems (CDSS) and Electronic Health Records (EHR), ensuring coherent interaction between varied healthcare data models becomes essential. In this context, the work in (El Hamlaoui et al. 2021) emphasizes the alignment of system views through abstraction layers and transformation rules, which aligns well with our vision of a scalable and interoperable platform within complex healthcare ecosystems.

By integrating heterogeneous systems such as machine learning, computer vision, and conversational AI to enhance diabetes management. It ensures meaningful, accessible, and permanently available data by allowing patients to log and track their health records, thereby improving self-management and reducing clinical risk. The inclusion of AIDC systems, such as potential IoT-based glucose monitors and wearables, further automates health monitoring and data collection.

ACKNOWLEDGEMENT

This research was supported by AL-KHAWARIZMI Grant AÏDA-Artificial Intelligence for DiAbetes under N°12-2024

REFERENCES

Bennani, S., Ebersold, S., El Hamlaoui, M., Coulette, B., & Nassar, M. (2019). A collaborative decision approach for alignment of heterogeneous models. In the international conference on enabling technologies: Infrastructure for collaborative enterprises, 112-117. IEEE.

Boucher, E., Latour, B., & Desrosiers, J. (2020). AI-driven food recognition for personalized nutrition. *Frontiers in Public Health*, 8, 173. Available at: https://www.frontiersin.org/articles/10.3389/fpubh.2020.00173/full

Contreras, I., & Vehi, J. (2018). Artificial intelligence for diabetes management and decision support: Literature review. *Journal of Medical Internet Research*, 20(5), e10775. Available at: https://www.jmir.org/2018/5/e10775

Crangle, C. (2018). Questions on diabetes from patients and the public. Self-Published Report. Accessed on: 24/02/2025. Available at: https://figshare.com/articles/dataset/Questions_on_Diabetes_from_Patients_and_the_Public/7038584

Dagliati, A., Marini, S., Sacchi, L., et al. (2018). Machine learning methods to predict diabetes complications. *Journal of Diabetes Science and Technology*, 12(2), 295–302. Available at: https://journals.sagepub.com/doi/10.1177/1932296817706375

El Hamlaoui, M., Ebersold, S., Bennani, S., Anwar, A., Dkaki, T., Nassar, M., & Coulette, B. (2021). A Model-Driven Approach to align Heterogeneous Models of a Complex System. The Journal of Object Technology, 20(2), 1-24.

International Diabetes Federation. (2021). *IDF Diabetes Atlas* (10th ed.). Accessed on: 24/02/2025. Available at: from https://diabetesatlas.org/

Tamenaoul, H., El Hamaloui, M., Nassar, M. (2023). Prompt Engineering: User Prompt Meta Model for GPT Based Models. The International Conference on Artificial Intelligence and Smart Environment, 428-433, Springer.

Tudor Car, L., Dhinagaran, D. A., Kyaw, B. M., et al. (2020). Conversational agents in healthcare: Scoping review and conceptual analysis. *Journal of Medical Internet Research*, 22(8), e17158. Available at: https://www.jmir.org/2020/8/e17158/

WEARABLE SENSORS FOR STRESS MONITORING IN MENISCUS INJURY REHABILITATION

Catarina de Sá Baio Pires[1], Fernando José Velez[1], Paulo Jorge Coelho[2,3], António Oseas Pataca[4] and Ivan Miguel Pires[4]

[1]*Instituto de Telecomunicações and Universidade da Beira Interior DEM - Faculdade de Engenharia, Covilhã, Portugal*
[2]*School of Technology and Management, Polytechnic University of Leiria, Leiria, Portugal*
[3]*Institute for Systems Engineering and Computers at Coimbra, Coimbra, Portugal*
[4]*Instituto de Telecomunicações, Escola Superior de Tecnologia e Gestão de Águeda, Universidade de Aveiro, Águeda, Portugal*

ABSTRACT

Monitoring vital signs is essential in identifying physiological stress, particularly in rehabilitating meniscus injuries. Stress can compromise the effectiveness of physiotherapy, affecting recovery. This paper reviews monitoring technologies, such as those that consider heart rate and sweating sensors, and their application in remote patient monitoring. It highlights the importance of collecting real-time data to personalize treatments and optimize recovery and explores the challenges and benefits of this approach. Integrating these technologies can significantly improve therapeutic results, prevent stress, and improve patients' quality of life.

KEYWORDS

Meniscus Injuries, Stress, Wearable Sensors, Remote Monitoring, Physiotherapy, Personalized Medicine

1. INTRODUCTION

The menisci are two structures of fibrous cartilage in the knee that absorb shocks and contribute to its stability (Makris et al., 2011). There is a lateral meniscus on the outside and a medial meniscus on the inside. The femur is rounded, and the tibia is flatter, creating little congruence between them. The menisci increase congruence and provide a larger area for load transmission in the knee (Fox et al., 2012).

Sensors detect and respond to physical or chemical stimuli, converting them into electrical signals that can be measured and analyzed (Heikenfeld et al., 2018). In the context of stress monitoring, wearable sensors have been developed to continuously capture physiological signals, allowing for tracking in various areas of health, such as post-traumatic stress disorder (PTSD) (Zamkah et al., 2020). These devices offer continuous and real-time stress data collection, which can be used in clinical diagnostics or personal monitoring.

Meniscus injuries represent a significant challenge for patients undergoing rehabilitation, especially due to the associated biomechanical and psychological consequences (Rao et al., 2015). Several studies have shown that the knee's stability during the landing of a jump can be a crucial factor in predicting and monitoring these injuries, which underlines the importance of using sensors to analyze the body's movement and response. However, in addition to physical factors, some patients suffer a traumatic injury, such as a meniscus tear, and may develop symptoms of post-traumatic stress, impacting their recovery supported by physiotherapy (Thomson et al., 2002). The response to post-traumatic stress can be monitored through various vital signs, such as heart rate variation, electrodermal activity, and cortisol levels (Vavrinsky et al., 2021). Sensors can help analyze knee stability, providing a comprehensive approach to optimizing rehabilitation (Porciuncula et al., 2018).

As during meniscus rehabilitation, the emotional state of patients may fluctuate, it is worthwhile to facilitate monitoring tools for the rehabilitation team to prevent physiological stress and avoid its negative impact. Instead of only focusing on physical symptoms, the proposed holistic approach seeks to rehabilitate individuals by considering their physical, mental, and psychological dimensions.

In this paper, we explore the monitoring of vital signs and biomechanics in the identification of post-traumatic stress in patients with knee injuries, discussing the sensors used and the types of data collected. In addition, we propose a new test that combines the assessment of stability during the landing of a jump with the analysis of the physiological stress response to control recovery and prepare a personalized approach for each patient more precisely.

The remainder of this manuscript is organized as follows. In section 2, the methodological part of the paper is addressed. Next, the set of sensors considered in the study is described, and methodologies used to identify physiological indicators associated with PTSD are addressed, followed by a study of the implementation techniques identified in the literature. Section 3 proposes a novel approach for stress identification in the injured person and discusses its advantages and limitations. The conclusions are drawn in section 4, where suggestions for further research are also discussed.

2. METHODOLOGY

This paper is based on the analysis of studies published in 2019 and 2025, selected from recognized scientific databases such as PubMed and Web of Science. The research was conducted using keywords related to physiological monitoring, physiotherapy, meniscus, and PTSD, ensuring the relevance of the included studies. For the selection of articles, those that directly address the use of sensors in the assessment of vital signs and in the rehabilitation of patients with meniscus injuries have been considered, as well as their application in identifying physiological responses to stress. Studies that do not present empirical data, reviews without critical analysis of the methods used, and works that do not address the relationship between stress and recovery have been excluded.

In this way, we can analyze the diagnosis of injuries with sensors and how much stress they cause after they occur. This type of injury is common, especially among athletes, and can result in pain, swelling, and joint dysfunction (Buckwalter, 2003). Traditionally, the diagnosis is based on the patient's clinical history and detailed physical examinations that assess the integrity of the meniscus through specific knee movements. However, the accuracy of these tests can vary depending on the doctor's experience and the specific characteristics of each patient.

Sections 2.1 and 2.2, respectively entitled "Sensors" and "Implemented techniques", present a description of the main types of sensors analyzed in this work and the methodologies used to identify physiological indicators associated with PTSD. Figure 1 summarizes the relationship between the sensors mentioned, the techniques used, and the main results obtained from the analysis of the data collected.

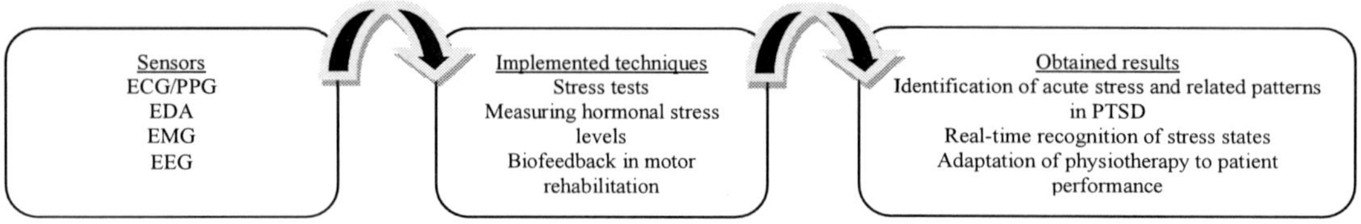

Figure 1. Link between sensors, implemented techniques, and results obtained

2.1 Sensors

With technological advancements, sensors are emerging as a promising tool in diagnosing these injuries, offering more objective and quantitative assessments. On the one hand, one of the most common is the heart rate variability (HRV) sensor, which measures the variation in the intervals between heartbeats (Kranjec et al., 2014). It can be based on the electrocardiogram (ECG), which uses electrodes on the skin to record the heart's electrical activity, or on photoplethysmography (PPG), which uses light to measure variations in blood flow (Affanni, 2020; Iqbal et al., 2022). HRV is an essential indicator because post-traumatic stress can cause autonomic dysfunction, altering the response of the nervous system (Kranjec et al., 2014).

On the other hand, electromyography (EMG) sensors measure muscle electrical activity, which can be very useful in detecting responses associated with conditions like PTSD (Morales-Sánchez et al., 2022). This examination captures muscle contractions and the pattern of muscle activity in real-time, which can reflect tensions in the body, such as those caused by stress.

The electroencephalogram (EEG) can also be used to identify brain wave patterns related to PTSD (Roohi-Azizi et al., 2017). Small electrodes on the head capture the brain's electrical activity, allowing the detection of hyperactivity in the amygdala, a region associated with fear, and dysfunction in the prefrontal cortex, which regulates emotions (Al-Ezzi et al., 2020; Roohi-Azizi et al., 2017).

Another important sensor is the Electrodermal Activity/Galvanic Skin Response (EDA/GSR), which measures the activity of the sympathetic nervous system (SNS) through the galvanic skin response (Affanni, 2020). Small electrodes detect variations in the skin's electrical conductance associated with involuntary sweating, reflecting increased anxiety and stress (Giannakakis et al., 2022). In addition, motion sensors, such as Inertial Measurement Units (IMUs), help identify symptoms like kinesiophobia, a condition where the individual feels excessive, irrational, and debilitating fear in performing movements (in this case, of the injured knee), common in patients with PTSD (Feldhege et al., 2015).

Elevated or dysregulated cortisol levels are often associated with stress, making it a relevant biomarker for diagnosis and monitoring (Dhama et al., 2019). Hence, sensors that measure cortisol levels, the stress hormone, can be applied through devices that analyze sweat, saliva, or blood in real-time (Wang et al., 2022).

The main advantages of using tiny sensors include early diagnosis, continuous monitoring, and greater accessibility compared to applied imaging equipment. However, challenges such as standardizing the protocols, interpreting the collected data, and patient adherence to these devices must be considered.

In Table 1, we can analyze the monitoring and treatment of post-traumatic stress, where studies are investigated that describe the use of physiological sensors to monitor the autonomic response to stress as a way to assess the psychological recovery of patients undergoing physiotherapy.

Table 1. Summary of the sensors considered in the literature

Name	Reference	Purpose
ECG	(Affanni, 2020), (Al-Ezzi et al., 2020)	Electrical activity of the heart
PPG	(Iqbal et al., 2022)	It uses light to measure variations in blood flow
EMG	(Morales-Sánchez et al., 2022)	Electrical activity of the muscles
EEG	(Roohi-Azizi et al., 2017)	Regulates emotions
EDA	(Affanni, 2020)	The activity of the SNP through the galvanic skin response
Motion sensor	(Feldhege et al., 2015)	Identification of stress-related symptoms
Cortisol level sensing	(Wang et al., 2022)	They analyze sweat, saliva, or blood in real-time.

2.2 Implemented Techniques

Based on the analysis, one of the analyzed studies used wearable sensors to monitor the physiological response to stress in a group of 35 healthy volunteers, who used devices equipped with PPG sensors to record variations in heart rate, providing relevant data for identifying patterns associated with PTSD in a recovery context (Iqbal et al., 2022). The participants were subjected to different stimuli to induce states of stress and relaxation. Among the identified protocols were the Stroop Test, used to assess the impact of cognitive effort on physiological response, the Trier Social Stress Test, which simulates social situations to trigger stress reactions, and the Hyperventilation Test, which induces respiratory changes to evaluate the response of the autonomic nervous system (Iqbal et al., 2022). The periods of exposure to stress were interspersed with moments of rest to allow for the comparison between distinct physiological states. The collected data included continuous HRV and skin conductance measurements, two indicators widely used to assess the autonomic stress response (Kranjec et al., 2014). The subsequent analysis of the signals allows for identifying activation patterns in the sympathetic and parasympathetic systems, providing insights into how the body reacts to this type of stimuli.

Another study aimed to evaluate the accuracy of a wearable sensor system developed to measure physical activity and knee angle in patients with neuromuscular disorders and gait alterations (Feldhege et al., 2015).

To this end, a device composed of accelerometers and flexion sensors was created, capable of recording the movement of the knee joint during different motor activities. The system was validated by comparing sensor data with reference measurements made by a digital goniometer and a laboratory motion capture system (Feldhege et al., 2015). The participants performed a series of controlled movements, including knee flexion and extension, walking at different speeds, and specific activities. This approach highlights the importance of wearable sensors in objective rehabilitation monitoring and facilitates continuous patient progress tracking. The integration of these devices with other vital sign monitoring tools can contribute to a broader analysis of the relationship between physiological stress and recovery, especially in patients with PTSD associated with meniscus injuries (Feldhege et al., 2015), (Kuczyński et al., 2025).

A paper that used electromyographic biofeedback (EMG-BF) to improve muscle activation in 10 football players who had undergone a meniscectomy has also been analyzed (Morales-Sánchez et al., 2022). During 6 to 10 sessions, sensors were used to measure the electrical activity of the quadriceps muscles. These data were used to help the participants improve muscle control and facilitate the recovery of motor function, promoting adequate muscle activation.

Although the study focused on recovery, it was possible to conclude that monitoring muscle activity and using biofeedback can directly impact the stress response during the recovery process (Morales-Sánchez et al., 2022). The approach could be applied to patients with PTSD, as emotional and psychological factors play a fundamental role in physical recovery (McNally et al., 2003).

We have paid attention to a study that monitors certain physiological signals, namely EDA and HRV, obtained through an ECG (Affanni, 2020). The methodology involves placing EDA sensors on the surface of the hands, which are responsible for capturing the galvanic skin response. In contrast, in (Affanni, 2020), an ECG sensor was positioned on the chest region to record cardiac activity. Measurements were taken under different experimental conditions to validate the system's effectiveness, comparing the sensors' responses to stimuli that induce stress (Affanni, 2020). The parameters analyzed included skin conductance and variations in the R-R intervals of the ECG, as both indicators are recognized by the autonomic nervous system (Affanni, 2020). This approach allowed for the real-time assessment of stress impact. It paved the way for future applications in clinical contexts, such as monitoring patients in motor recovery, where stress can directly influence rehabilitation.

In turn, we have found that using EEG to investigate the correlations between patterns of brain activity and the severity of PTSD symptoms may be of potential interest, as described by the authors from (Roohi-Azizi et al., 2017). The sample consisted of individuals diagnosed with this disorder, subjected to resting EEG sessions and during exposure to trauma-related stimuli (Meyer et al., 2018). For data collection, electrodes were positioned according to the international 10-20 system, recording neuronal activity in different brain regions (Roohi-Azizi et al., 2017). The methodology included the analysis of the spectrum of EEG signals, focusing on the alpha, beta, delta, and theta frequency bands, where associations between these and PTSD symptoms were sought (Roohi-Azizi et al., 2017).The relevance of this approach for the recovery of patients with meniscal injuries and PTSD lies in the possibility of controlling the impact of stress, contributing to the personalization of rehabilitation strategies.

Table 2 summarizes all the methods that can be used in tests using the sensors mentioned in Table 1.

Table 2. Summary of all the methods mentioned

Name	Reference	Method
PPG sensors	(Iqbal et al., 2022)	Different stimuli to induce states of stress and relaxation
Wearable sensor system	(Feldhege et al., 2015)	Measure physical activity and knee angle in patients with neuromuscular disorders and gait alterations.
EMG-BF	(Morales-Sánchez et al., 2022)	Measure the electrical activity of the quadriceps muscles
EDA and HRV through an ECG	(Affanni, 2020)	EDA sensors on the surface of the hands, while an ECG sensor was positioned on the chest
EEG	(Meyer et al., 2018)	Resting EEG sessions and during exposure to trauma-related stimuli

3. SOLUTION PROPOSAL

This study have revealed key findings on techniques that strongly support the proposed solution. The use of PPG and ECG sensors allowed for the real-time identification of stress states through changes in HRV, especially when patients were exposed to cognitive and social stressors (Giannakakis et al., 2022). This finding emphasizes the feasibility of integrating HRV-based metrics into rehabilitation sessions to adjust therapy intensity dynamically. Additionally, the application of EDA sensors revealed patterns of electrodermal activity consistent with sympathetic nervous system activation, which was shown to correspond to patient-reported anxiety during physiotherapy exercises (Fernandez Rojas et al., 2023). This insight helps clinicians identify stress peaks that may compromise performance or recovery.

EMG biofeedback methods demonstrated that targeted muscle activation can be improved with real-time monitoring, reducing involuntary tension and stress linked to poor motor control (Morales-Sánchez et al., 2022). These findings highlight the benefit of integrating EMG sensors into early rehabilitation stages, especially after procedures such as meniscectomy. Furthermore, EEG analysis revealed correlations between specific brain wave patterns (notably in the alpha and beta bands) and PTSD symptoms, reinforcing the importance of including neurophysiological stress markers in recovery programs for individuals with post-traumatic profiles (Micoulaud-Franchi et al., 2021). Collectively, these discoveries provide a robust foundation for developing a comprehensive, sensor-based monitoring system that detects physiological stress and contributes to the personalization of rehabilitation strategies. Combining these validated techniques makes it possible to reformulate rehabilitation plans in real-time, focusing on improving motor confidence and emotional regulation during recovery from meniscal injuries.

The present solution proposes the implementation of vital signs monitoring system based on the use of state-of-the-art wireless electrodes capable of accurately measuring heart rate fluctuations during the execution of specific exercises (Virginia Anikwe et al., 2022). Simultaneously, applying highly sensitive sensory devices on the patient's hands is employed, designed to detect changes in sweating, a physiological biomarker widely associated with activating the sympathetic nervous system and, consequently, stress (Zamkah et al., 2020). These two methods enable an exact monitoring of the patient's physiological response during exertion, especially in meniscal injury rehabilitation. The collected data would be transmitted directly to healthcare professionals responsible for therapeutic monitoring so that no time is lost in receiving the information. In this way, doctors and physiotherapists will be able to access an objective and continuous analysis of certain patient parameters, allowing for more informed clinical decision-making tailored to each case's individual needs. Based on the information received, it becomes possible to adjust the intensity and nature of therapeutic exercises, reducing the risk of overload and optimizing the recovery process. Table 3 compares the pros and cons of the solution we would like to implement in the diagnostics.

In addition to the limitations already mentioned, the proposed solution faces several technical and practical challenges worth highlighting. Firstly, sensors such as ECG and EDA are highly sensitive to movement artifacts, which can compromise data quality during intense physical exercise, requiring additional care in fixation and calibration. Inter-individual variability in the physiological stress response also requires an initial calibration period to establish personalized reference values, which can delay immediate practical application.

Another relevant obstacle is the complexity inherent in integrating multimodal data. Time synchronization between different sensors and the joint interpretation of physiological signals requires advanced data fusion systems, which are demanding to implement and maintain. In addition, the continuous collection of vital signs raises critical privacy and security issues, requiring strict compliance with the General Data Protection Regulation (GDPR) and the existence of robust technological infrastructures.

Table 3. Advantages and disadvantages of the proposed solution

Advantages	Disadvantages
Real-time monitoring	High cost
Personalization of treatment	Dependence on the internet connection
Reduction of injury risk	Noise and possible interferences
Remote monitoring	Synchronization between sensors
Integration with emerging technologies	Acceptance by the patient
	Need for technical training of healthcare professionals

When monitoring cardiac electrical activity using ECG, the electrodes are generally placed in the chest area, which allows direct capture of the electrical signals generated by the heart, particularly the QRS complex. In this way, a more accurate and stable reading of HRV is obtained.

To detect EDA or GSR, electrodes are usually placed on the palms of the hands or the tips of the fingers. These regions have a high density of sweat glands, which makes them ideal for detecting small variations in the electrical conductance of the skin caused by changes in sweating. During times of stress or emotional strain, the sympathetic nervous system is activated, increasing sweat secretion and, consequently, skin conductance. These devices work integrated into a WBAN (Wireless Body Area Network), allowing the real-time collection of the patient's responses without the wires or movement restrictions. The data is then sent to a sensor integrated into a watch, enabling both the patient and healthcare professionals to easily access the results and make informed decisions for the rehabilitation process. Figures 2 and 3.a) summarize the location of the ECG and EDA/GSR electrodes, respectively, and Figure 3.b) shows a smartwatch that is used to receive the data collected from the aforementioned electrodes via WBAN.

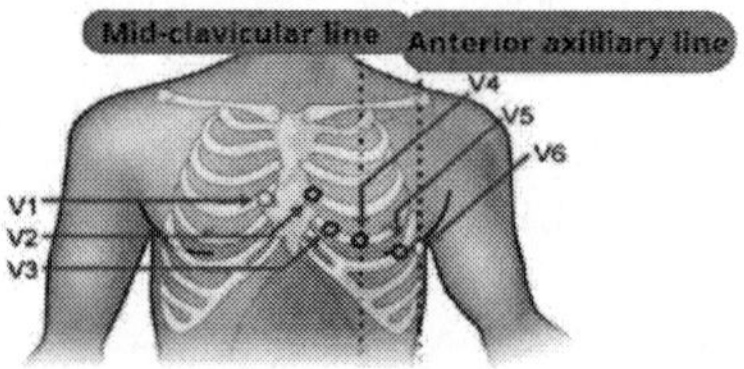

Figure 2. When monitoring cardiac electrical activity with an ECG, electrodes are usually placed on the chest area to directly capture the heart's electrical signals, especially the QRS complex

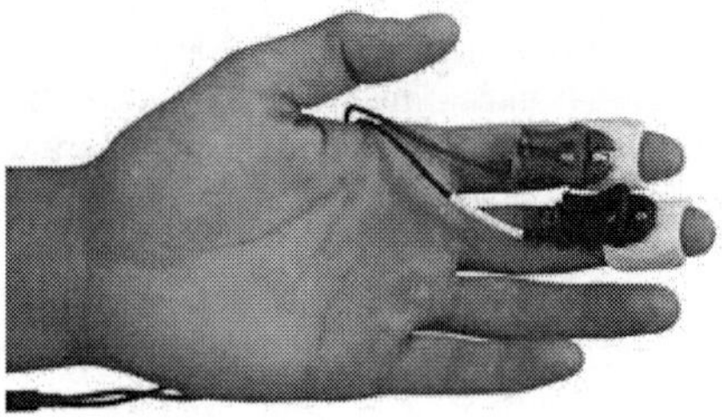

a)

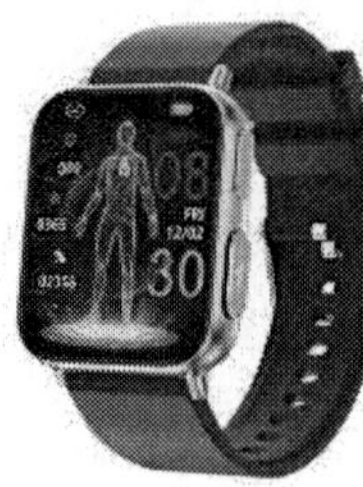

b)

Figure 3.a) To detect electrodermal activity, electrodes are usually placed on the palms of the hands or the tips of the fingers, where there is a higher concentration of sweat glands. b) Example of a smartwatch that will be used to receive the data collected

It should also be noted that, although this solution can identify physiological reactions associated with stress, it is not always possible to distinguish between emotional and physical stress, leading to incorrect interpretation without proper contextualization. Finally, the high cost of acquisition (before price reduction due to future mass-production) and the need for frequent maintenance and technological updates may represent additional barriers to its large-scale application.

4. CONCLUSION

The relationship between stress and physiotherapy is undeniably complex, as both factors directly influence the effectiveness and progress of injury recovery, including meniscal injuries. Stress, especially physical and emotional, can negatively affect the patient's performance during rehabilitation, potentially increasing the risk of complications. This factor can interfere with performing exercises correctly, hindering muscle regeneration.

Therefore, precise and real-time monitoring of vital signs, such as HRV and sweating, becomes essential to understand these variables during treatment.

The proposed approach represents a significant advancement in physiotherapy in the future. Combining these two methods makes obtaining a detailed and dynamic view of the patient's physiological response to stress possible. Technology offers the opportunity to carry out quick interventions tailored to each person's reality, according to their level of effort and stress tolerance. These systems can identify recovery patterns and predict possible complications, allowing physiotherapists to anticipate and further personalize treatment according to the patient's needs.

As a prospect for future work, it is planned to progressively implement the solution developed in hospitals and clinics specialized in physical rehabilitation to validate its applicability in real clinical scenarios. This phase will allow the system to be tested in a supervised environment, assessing its ability to monitor physiological indicators associated with stress during physiotherapy sessions accurately. In addition, the impact of therapeutic personalization based on biometric data will be analyzed, as well as the acceptance and usability of the system by health professionals and patients.

In the medium and long term, this approach aims to support the integration of continuous monitoring technologies into existing clinical protocols, contributing to the evolution of more effective, personalized, and patient-centered rehabilitation models.

ACKNOWLEDGEMENT

This work is funded by FCT/MECI through national funds and, when applicable co-funded EU funds under UID/50008: Instituto de Telecomunicações. This research is also supported by COST CA20120 INTERACT - Intelligence-Enabling Radio Communications for Seamless Inclusive Interactions, and by the Portuguese National Distributed Computing Infrastructure (INCD) through the FCT Advanced Computing Projects 2023.10865.CPCA.A1 and https://doi.org/10.54499/2023.10865.CPCA.A1.

REFERENCES

Affanni, A. (2020). Wireless Sensors System for Stress Detection using ECG and EDA Acquisition. *Sensors*, *20*(7), 2026. https://doi.org/10.3390/s20072026

Al-Ezzi, A., Kamel, N., Faye, I., & Gunaseli, E. (2020). Review of EEG, ERP, and Brain Connectivity Estimators as Predictive Biomarkers of Social Anxiety Disorder. *Frontiers in Psychology*, *11*, 730. https://doi.org/10.3389/fpsyg.2020.00730

Buckwalter, J. A. (2003). Sports, Joint Injury, and Posttraumatic Osteoarthritis. *Journal of Orthopaedic & Sports Physical Therapy*, *33*(10), 578–588. https://doi.org/10.2519/jospt.2003.33.10.578

Dhama, K., Latheef, S. K., Dadar, M., Samad, H. A., Munjal, A., Khandia, R., Karthik, K., Tiwari, R., Yatoo, Mohd. I., Bhatt, P., Chakraborty, S., Singh, K. P., Iqbal, H. M. N., Chaicumpa, W., & Joshi, S. K. (2019). Biomarkers in Stress-Related Diseases/Disorders: Diagnostic, Prognostic, and Therapeutic Values. *Frontiers in Molecular Biosciences*, *6*, 91. https://doi.org/10.3389/fmolb.2019.00091

Feldhege, F., Mau-Moeller, A., Lindner, T., Hein, A., Markschies, A., Zettl, U., & Bader, R. (2015). Accuracy of a Custom Physical Activity and Knee Angle Measurement Sensor System for Patients with Neuromuscular Disorders and Gait Abnormalities. *Sensors*, *15*(5), 10734–10752. https://doi.org/10.3390/s150510734

Fernandez Rojas, R., Brown, N., Waddington, G., & Goecke, R. (2023). A systematic review of neurophysiological sensing for the assessment of acute pain. *Npj Digital Medicine*, *6*(1), 76. https://doi.org/10.1038/s41746-023-00810-1

Fox, A. J. S., Bedi, A., & Rodeo, S. A. (2012). The Basic Science of Human Knee Menisci: Structure, Composition, and Function. *Sports Health: A Multidisciplinary Approach*, *4*(4), 340–351. https://doi.org/10.1177/1941738111429419

Giannakakis, G., Grigoriadis, D., Giannakaki, K., Simantiraki, O., Roniotis, A., & Tsiknakis, M. (2022). Review on Psychological Stress Detection Using Biosignals. *IEEE Transactions on Affective Computing*, *13*(1), 440–460. https://doi.org/10.1109/TAFFC.2019.2927337

Heikenfeld, J., Jajack, A., Rogers, J., Gutruf, P., Tian, L., Pan, T., Li, R., Khine, M., Kim, J., Wang, J., & Kim, J. (2018). Wearable sensors: Modalities, challenges, and prospects. *Lab on a Chip*, *18*(2), 217–248. https://doi.org/10.1039/C7LC00914C

Iqbal, T., Simpkin, A. J., Roshan, D., Glynn, N., Killilea, J., Walsh, J., Molloy, G., Ganly, S., Ryman, H., Coen, E., Elahi, A., Wijns, W., & Shahzad, A. (2022). Stress Monitoring Using Wearable Sensors: A Pilot Study and Stress-Predict Dataset. *Sensors*, *22*(21), 8135. https://doi.org/10.3390/s22218135

Kranjec, J., Beguš, S., Geršak, G., & Drnovšek, J. (2014). Non-contact heart rate and heart rate variability measurements: A review. *Biomedical Signal Processing and Control*, *13*, 102–112. https://doi.org/10.1016/j.bspc.2014.03.004

Kuczyński, N., Boś, J., Białoskórska, K., Aleksandrowicz, Z., Turoń, B., Zabrzyńska, M., Bonowicz, K., & Gagat, M. (2025). The Meniscus: Basic Science and Therapeutic Approaches. *Journal of Clinical Medicine*, *14*(6), 2020. https://doi.org/10.3390/jcm14062020

Makris, E. A., Hadidi, P., & Athanasiou, K. A. (2011). The knee meniscus: Structure-function, pathophysiology, current repair techniques, and prospects for regeneration. *Biomaterials*, *32*(30), 7411–7431. https://doi.org/10.1016/j.biomaterials.2011.06.037

McNally, R. J., Bryant, R. A., & Ehlers, A. (2003). Does Early Psychological Intervention Promote Recovery From Posttraumatic Stress? *Psychological Science in the Public Interest*, *4*(2), 45–79. https://doi.org/10.1111/1529-1006.01421

Meyer, T., Quaedflieg, C. W. E. M., Weijland, K., Schruers, K., Merckelbach, H., & Smeets, T. (2018). Frontal EEG asymmetry during symptom provocation predicts subjective responses to intrusions in survivors with and without PTSD. *Psychophysiology*, *55*(1), e12779. https://doi.org/10.1111/psyp.12779

Micoulaud-Franchi, J. A., Jeunet, C., Pelissolo, A., & Ros, T. (2021). EEG Neurofeedback for Anxiety Disorders and Post-Traumatic Stress Disorders: A Blueprint for a Promising Brain-Based Therapy. *Current Psychiatry Reports*, *23*(12), 84. https://doi.org/10.1007/s11920-021-01299-9

Morales-Sánchez, V., Falcó, C., Hernández-Mendo, A., & Reigal, R. E. (2022). Efficacy of Electromyographic Biofeedback in Muscle Recovery after Meniscectomy in Soccer Players. *Sensors*, *22*(11), 4024. https://doi.org/10.3390/s22114024

Porciuncula, F., Roto, A. V., Kumar, D., Davis, I., Roy, S., Walsh, C. J., & Awad, L. N. (2018). Wearable Movement Sensors for Rehabilitation: A Focused Review of Technological and Clinical Advances. *PM&R*, *10*(9S2). https://doi.org/10.1016/j.pmrj.2018.06.013

Rao, A. J., Erickson, B. J., Cvetanovich, G. L., Yanke, A. B., Bach, B. R., & Cole, B. J. (2015). The Meniscus-Deficient Knee: Biomechanics, Evaluation, and Treatment Options. *Orthopaedic Journal of Sports Medicine*, *3*(10), 2325967115611386. https://doi.org/10.1177/2325967115611386

Roohi-Azizi, M., Azimi, L., Heysieattalab, S., & Aamidfar, M. (2017). Changes in the brain's bioelectrical activity in cognition, consciousness, and some mental disorders. *Medical Journal of the Islamic Republic of Iran*, *31*(1), 307–312. https://doi.org/10.14196/mjiri.31.53

Thomson, L., Handoll, H. H., Cunningham, A. A., & Shaw, P. C. (2002). Physiotherapist-led programs and interventions for rehabilitation of anterior cruciate ligament, medial collateral ligament, and meniscal injuries of the knee in adults. *Cochrane Database of Systematic Reviews*, *2010*(1). https://doi.org/10.1002/14651858.CD001354.pub2

Vavrinsky, E., Stopjakova, V., Kopani, M., & Kosnacova, H. (2021). The Concept of Advanced Multi-Sensor Monitoring of Human Stress. *Sensors*, *21*(10), 3499. https://doi.org/10.3390/s21103499

Virginia Anikwe, C., Friday Nweke, H., Chukwu Ikegwu, A., Adolphus Egwuonwu, C., Uchenna Onu, F., Rita Alo, U., & Wah Teh, Y. (2022). Mobile and wearable sensors for data-driven health monitoring system: State-of-the-art and future prospect. *Expert Systems with Applications*, *202*, 117362. https://doi.org/10.1016/j.eswa.2022.117362

Wang, B., Zhao, C., Wang, Z., Yang, K.-A., Cheng, X., Liu, W., Yu, W., Lin, S., Zhao, Y., Cheung, K. M., Lin, H., Hojaiji, H., Weiss, P. S., Stojanović, M. N., Tomiyama, A. J., Andrews, A. M., & Emaminejad, S. (2022). Wearable aptamer-field-effect transistor sensing system for noninvasive cortisol monitoring. *Science Advances*, *8*(1), eabk0967. https://doi.org/10.1126/sciadv.abk0967

Zamkah, A., Hui, T., Andrews, S., Dey, N., Shi, F., & Sherratt, R. S. (2020). Identification of Suitable Biomarkers for Stress and Emotion Detection for Future Personal Affective Wearable Sensors. *Biosensors*, *10*(4), 40. https://doi.org/10.3390/bios10040040

MODELING VIOLENT DEATH RATES: AN EXPLAINABLE AI ANALYSIS OF THE ASSOCIATION BETWEEN SOCIAL AND HEALTHCARE SERVICES AND HOMICIDE, SUICIDE, AND SUBSTANCE CAUSED DEATH

Qing Zeng-Treitler[1,2,*], Yijun Shao[1,*], Ryan Wu[1,**], Adnan Lakdawala[1,****], Alicia Jones[2,***], Megan Koch[3,*****], Yingxuan Liu[2,**], Lori Post[2,*], and Maryann Mason[2,*]

[1]*George Washington University, Washington, DC, USA*
[2]*Northwestern University, Chicago, IL, USA*
[3]*Illinois Department of Public Health, Springfield, IL, USA*
**PhD, **MS, ***BS, ****MBBS, *****MPH*

ABSTRACT

We assembled a novel dataset by linking national data sources and applied three types of machine learning (ML) models, Random Forest, XGBoost, and Neural Networks, along with an explainable AI approach to analyze the association between the density of social and healthcare services and violent death rates across 53,680 census tracts in the United States. The ML models demonstrated substantially better performance, as measured by R^2, than traditional linear regression, achieving approximately 70% explained variance for homicide and substance-related deaths. Novel explainable AI analyses further revealed that the impact of service density on violent death rates varied not only by type of outcome but also across individual tracts, highlighting significant geographic and contextual heterogeneity. These findings suggest that a one-size-fits-all strategy may be ineffective and that tailored, locally informed approaches are essential for effective violence prevention efforts.

KEYWORDS

Violent Death, Social/Healthcare Service, Explainable AI

1. INTRODUCTION

Violent deaths, including homicide and suicide, claim tens of thousands of lives in the United States (US) each year.(Hatton et al., 2024, Sakran et al., 2024, Kleinman and Morris, 2023) The homicide rates in the US peaked during the COVID pandemic,(Degli Esposti et al., 2025) reaching 8.2 per 100,000 people in 2021 and has since decreased. The suicide rates, on the other hand, have seen a gradual increase over the past decades,(Garnett and Curtin, 2024) reaching 14.3 deaths per 100,000 people in 2022. The increase in suicide rates coincided with the opioid crisis in the US. Both intentional and unintentional overdoses are recognized as significant public health challenges.(Aronowitz and Reed, 2025)

It has been suggested that social and environmental factors play a significant role in the prevention of violent deaths.(Boeck et al., 2022) Social and healthcare services can have direct and indirect roles in mitigating the risks of violent deaths.(Choi et al., 2020, Sipsma et al., 2017, Tenso et al., 2024, Montgomery et al., 2024) Some community-based intervention programs, for example, provide recreational opportunities and positive role models to at-risk populations.(DeBlieux et al., 2024, Blackburn et al., 2023) Mental health services could help reduce homicide or suicide rates by treating pain, trauma, substance abuse, or behavioral issues.(Cantor et al., 2025, Doupnik et al., 2025, Englund, 2024, Cerel et al., 2016, Bishop et al., 2020)

A key challenge in the study of violence prevention is the lack of comprehensive data. The US Center for Disease Control (CDC)'s National Violent Death Reporting System (NVDRS) is a highly detailed dataset of violent deaths with over 600 variables.(Christensen, 2016) However, NVDRS does not contain extensive social and environmental contexts.

Another challenge when analyzing a dataset like NVDRS with statistical or machine learning (ML) methods to examine the potential contribution of social and healthcare services is the lack of controls since only individuals with violent deaths are included in the NVDRS. In this study, we measured mortality rates in different communities. We first linked NVDRS with several US national datasets to examine the association between social and healthcare services and violent death. We then experimented with statistical and AI (ML) approaches to model these relationships. Since AI models are often viewed as black boxes, we conducted explainable AI analyses to shed light on the relationships captured by the models (Shao et al., 2021).

2. DATA

2.1 NVDRS

NVDRS is the leading national dataset for violent deaths in the US. It is the only state-based surveillance system pooling data from multiple sources. We obtained NVDRS data from the CDC's Restricted Access Data program through a data use agreement. The dataset we received contains ZIP (Zone Improvement Plan) codes. A ZIP Code is used by the United States Postal Service to identify specific geographic regions for the efficient sorting and delivery of mail. For our analyses, we aggregated the death rates (number of homicides, suicides, and substance caused deaths per 100,000 individuals) from 2017 to 2021 and took the mean as an outcome variable.

NVDRS includes a "substance cause death" variable defined as "Substance n: Substance clearly identified as the case of death." If a death is due to poisoning, this variable will be endorsed. We selected this variable as a potential indicator of drug-related overdose death in NVDRS.

We mapped ZIP codes to census tracts using a crosswalk table created by the US Department of Housing and Urban Development (HUD). A Census Tract is a small geographic subdivision of a county or equivalent entity in the US, defined by the U.S. Census Bureau. These tracts are designed to provide a consistent framework for analyzing population and demographic data over time.

2.2 EJI

The Environmental Justice Index (EJI) dataset,(Services, 2024) released by the CDC in August 2022, is a significant step forward, aiding efforts to quantify and explore environmental health disparities in the US. The EJI groups environmental vulnerabilities into three distinct modules to provide a holistic framework for evaluating environmental justice issues: social vulnerability, environmental burden, and health vulnerability. This study included all social vulnerability factors, such as socioeconomic status, education levels, housing conditions, and racial or ethnic composition, and all environmental burdens, such as air pollution, water contamination, proximity to hazardous waste facilities, and extreme weather events, as predictors.

2.3 NaNDA Social Services

The National Neighborhood Data Archive (NaNDA) social services dataset was developed by the University of Michigan's Institute for Social Research. The dataset includes detailed information on the number and density of social service establishments such as community centers, child and youth services, senior services, meal delivery programs, substance abuse counseling, and daycare facilities across U.S. Census Tracts and ZIP Code Tabulation Areas (ZCTAs) from 1990 to 2021. In this analysis, we aggregated the density of social services (number of services per 1000 individuals) from 2017 to 2021 and took the mean as predictors.

2.4 NaNDA Healthcare Services

The NaNDA healthcare services dataset was also developed by the University of Michigan's Institute for Social Research. It offers data on the number and density of healthcare services (including doctors, dentists, mental health professionals, nursing homes, diagnostic laboratories, and ambulance services) across U.S. Census

Tracts and ZIP Code Tabulation Areas (ZCTAs) from 1990 to 2021. As with social services, we aggregated the densities from 2017 to 2021 and took the mean as predictors.

3. METHOD

We merged NVDRS with EJI and NaNDA Social and Healthcare Services datasets using census tract IDs (n = 53,680). The study focused on three outcomes: homicide, suicide, and substance-related death rates (per 100,000 individuals). A total of 48 predictors were used, including 7 NaNDA social services, 8 NaNDA healthcare services, 2 EJI demographic variables, 15 EJI social variables, and 16 EJI environmental variables. Missing values were imputed using the mean of each variable. These variables are distinct in their focus: NaNDA variables capture the availability of services within communities, such as the density of childcare services or mental health providers. In contrast, EJI variables reflect the broader social and environmental characteristics of communities, including the percentages of youth and elderly residents, the proportion of immigrants with limited English proficiency, air pollution levels, and walkability. Because the availability of services is often linked to a community's social and environmental context, we included EJI variables as covariates in the analysis.

We began by calculating Spearman correlations between predictors and outcomes. Spearman is a nonparametric test that assesses monotonic relationships based on ranked data.

We first fitted linear regression (LR) models to predict each of the three violent death outcomes using all EJI and NaNDA features. LR assumes a linear relationship between predictors and outcomes and is known for its simplicity and interpretability. The dataset was randomly split 80:20 for training and testing, and model performance was evaluated using mean squared error (MSE) and R-squared (R^2).

Next, we implemented random forest (RF) models—an ensemble method that builds multiple decision trees. For RF, data were split into 60% training, 20% validation, and 20% testing. A grid search with 3-fold cross-validation optimized hyperparameters including the number of trees (100, 300, 500), maximum depth (3, 6, 9, unlimited), feature sampling strategies ('auto', 'sqrt', 'log2'), and node split criteria (min samples per split: 2, 5, 10; min samples per leaf: 1, 2, 4). The best model was selected based on MSE.

We also trained XGBoost (Extreme Gradient Boosting) models, which build trees sequentially to correct previous errors. For each outcome, we tested combinations of learning rate (0.05, 0.1, 0.15), number of trees (250, 300, 350), max depth (5, 6, 7), row sampling (0.8, 1.0), and feature sampling (0.8, 1.0). Like RF, XGBoost used 3-fold cross-validation and optimized for negative MSE with a similar train/validation/test split.

Finally, after exploring several other architectures, we developed neural network (NN) models using a deep and wide architecture. The wide path consists of a dense layer with 64 units and linear activation to capture simple interactions. The deep path includes batch normalization followed by four dense layers (256, 128, 64, 32 units) with swish activation, L2 regularization, and dropout (rate = 0.4) after the first three layers to prevent overfitting. Outputs from both paths are concatenated and fed into a final dense layer with linear activation for regression. The model was compiled with the Adam optimizer (learning rate = 0.001) and trained with early stopping (patience = 10). Data partitioning and evaluation mirrored that of the RF and XGBoost models.

We compared the MSE and R^2 results from LR, RF, XGBoost, and NN. Because the explanation of a model can only be as good as a model's fitting, we explained the best-fitted models, which are XGBoost and RF models. Unlike LR, ML models have thousands to millions of hyperparameters, which are not suitable for human interpretation. We adapted an explainable AI approach called impact score for this analysis.(Shao et al., 2021) Essentially, the impact score quantifies the change in the outcome when the value of a predictive feature changes, offering a model-agnostic approach to interpretability. On a logistic regression model, the impact score corresponds to the log odds ratio for a variable. The revised impact score corresponds to the variable's coefficient on a linear regression model. The ML models, however, are often highly nonlinear. This makes it important to calculate the individual-level scores, which are expected to differ from the population mean in nonlinear models.

Let $y = F(x_1, \ldots, x_n)$ be the XGBoost model that predicts y, the outcome of interest. The impact score is defined as

$$\frac{f(\ldots, x_i^c, \ldots) - f(\ldots, x_i^r, \ldots)}{(x_i^c - x_i^r)}$$

where x_i^c and x_i^r are, respectively, the current and reference values of the variable x_i on the instance, and "..." represents the current values of all other variables than x_i. Note that the individual interaction score is only defined in those instances such that $x_i^c \neq x_i^r$. The data set-level score is the median of all instance-level scores.

4. RESULTS

According to conventional thresholds, (Shao et al., 2021) correlation coefficients between -0.3 and 0.3 indicate weak/no correlation. Based on the Spearman test results, all social and healthcare services have weak/no correlation with violent death rates.

The model fitting performance is summarized in Table 1 and Figure 1. For every outcome, all ML models achieved better fitting with lower MSE and higher R^2 when compared with LR, suggesting that the relationships between the predictors and violent deaths are not linear. The XGBoost models have the best fitting for two of the three outcomes. R^2 indicates the proportion of variance in the dependent variable that a model can explain. The XGBoost models can explain about 70% of variance in the homicide and substance caused death rates on the census tract level, but only about 27% variance in the suicide rates. The RF model reached 30% R^2 for suicide. All models achieved the highest R^2 on homicide and the lowest R^2 on suicide, suggesting that the associations between the predictors and outcomes are the strongest for homicide and weakest for suicide.

Table 1. R^2 and MSE of the four types of models for homicide, suicide, and substance caused death

	R2			
Outcome	**LR**	**RF**	**XGBoost**	**NN**
Homicide Per 100,000	0.42	0.66	0.72	0.66
Suicide Per 100,000	-0.28	0.31	0.27	0.17
Substance Caused Death Per 100,000	0.03	0.52	0.68	0.57
	MSE			
Outcome	**LR**	**RF**	**XGBoost**	**NN**
Homicide Per 100,000	860.83	515.74	421.22	505.84
Suicide Per 100,000	720.41	390.02	410.84	465.71
Substance Caused Death Per 100,000	177.02	88.39	57.66	78.98

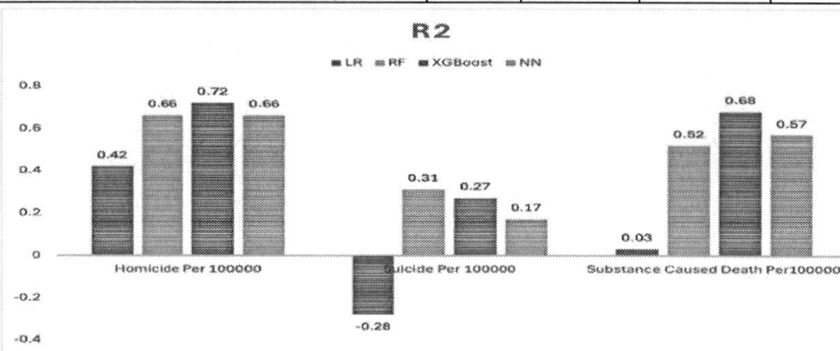

Figure 1. R^2 of the four types of models for homicide, suicide, and substance caused death. A negative R^2 value indicates that a model's predicted values are worse than using the average as a predicted value

Table 2 shows the median and standard deviations (STD) of the impact scores of each social/environmental factor based on the best-performing ML models. The impact score measures the impact of the social and healthcare services on the violent death rates, specifically the change in the number of deaths per 100,000 individuals. A negative score indicates reduced deaths, and a positive score indicates increased deaths. All services have both negative and positive impact scores on the individual census tract level.

Some social and healthcare services exhibit negative median impact scores, suggesting they are associated with reduced rates of homicide, suicide, or substance-related deaths across many census tracts. However, in nearly all cases, the standard deviations of these scores exceed the medians, reflecting substantial variability in individual-level impacts. In practical terms, this means increased service density may lead to improved outcomes in some communities but not in others.

This finding underscores a key difference between linear and nonlinear modeling approaches. In a linear model, the individual-level effect of a predictor is assumed to be constant across all observations. Therefore, the individual impact score for any given feature is always the same as the population-level effect, rendering the calculation of metrics such as the median and standard deviation of individual scores unnecessary. In contrast, machine learning models such as RF, XGBoost, and NN are able to model complex, context-specific interactions and nonlinearities. This enables them to reveal nuanced patterns. For instance, a service may reduce violent deaths in some areas while having no effect or even a counterintuitive association in others.

Some may assume that increasing the density of social or healthcare services is universally beneficial, reasoning that services are designed to address needs rather than contribute to harm. However, the presence of services in a community does not guarantee their accessibility, utilization, or benefit for residents. For example, a large urban hospital may be physically located in a high-poverty neighborhood with elevated crime rates. However, its services may primarily serve patients from more affluent or insured populations living elsewhere. In such cases, the facility's presence does not necessarily translate into improved outcomes for the surrounding community. Additionally, resources are finite and must be effectively matched to local needs.

The differing prevalence and risk profiles of various types of violent deaths further illustrate this complexity. Suicide is generally more prevalent than homicide, and homicide is more prevalent than substance-related deaths. As a result, the same type of service may have a different impact depending on the dominant public health challenge in a given area. Moreover, the effectiveness of services is likely shaped by the community's characteristics, including demographic structure, socioeconomic status, rurality or urbanicity, built environment, and availability of complementary services. As shown in Table 3, the California community might benefit from more childcare services to reduce homicide, while the Virginia community would not. The former has low daycare density, high homicide rates, more poverty and unemployment, and lower levels of education and insurance.

Table 2. Median impact scores and standard deviations of social/healthcare services for three types of violent deaths

	Homicide Per 100,000	Suicide Per 100,000	Substance Caused Death Per 100,000
Density of elder/disability services	-1.92 (5.63)	3.40 (1.64)	0.72 (1.35)
Density of child/youth services	3.16 (3.20)	0.36 (0.56)	-0.09 (0.60)
Density of emergency disaster services	1.35 (2.63)	0.13 (0.25)	0.09 (0.13)
Density of meal delivery programs	0.00 (1.83)	0.00 (0.03)	0.13 (0.63)
Density of general counseling services	-3.26 (7.03)	1.05 (1.51)	-0.38 (5.65)
Density of substance abuse counseling	-1.18 (2.15)	0.26 (0.35)	0.61 (2.33)
Density of child daycare services	-1.32 (9.91)	2.79 (10.76)	-7.21 (30.60)
Density of physical health physicians	-0.45 (2.56)	0.64 (0.89)	-0.21 (1.69)
Density of mental health physicians	1.51 (1.42)	0.66 (0.43)	-0.03 (0.36)
Density of mental health practitioners	0.21 (0.94)	0.07 (0.46)	0.45 (1.18)
Density of rehab practitioners	0.68 (3.83)	-0.30 (0.73)	0.44 (2.86)
Density of alt. medicine practitioners	0.00 (0.66)	0.03 (0.38)	-0.08 (0.23)
Density of chiropractors	0.30 (2.83)	0.21 (0.67)	1.12 (6.04)
Density of optometrists	0.91 (2.88)	1.57 (1.50)	0.64 (0.71)
Density of dentists	-1.23 (2.85)	0.18 (0.42)	0.01 0.63)

Taken together, these findings highlight the critical importance of tailoring service provision and public health strategies to the specific contexts of individual communities. Aggregate measures can obscure important differences in how services function across populations, and one-size-fits-all approaches may fail to address the nuanced realities on the ground. By leveraging explainable AI to examine both median and individual impact scores, this study reveals the diverse ways social and healthcare services intersect with community health outcomes.

Table 3. Selected characteristics of two communities, one in California and one in Virginia. They have different densities of child daycare services, impact scores for homicide, and socioenvironmental features

	A Community in California	A Community in Virginia
Impact Score of Density of Child Daycare Services	-9.6	2.6
Density of Child Daycare Services	0.1	0.7
Population	4165	6106
Day Population	1702	6142
Ozone	0.7	0.3
PM2.5	12.7	8.8
Diesel Particulate Matter	0.9	0.4
National Priority List Sites	45.3	31.5
Toxic Release Inventory Sites	66.7	0.0
Recreational Parks	100.0	49.1
Houses Built Pre-1980	91.0	3.5
Walkability	15.1	7.5
Railways	100.0	0.0
Impaired Surface Water	93.4	48.8
Minority Status	99.4	52.9
Poverty	84.7	15.8
No High School Diploma	54.6	4.0
Unemployment	8.8	2.8
Housing Tenure	90.6	28.5
Housing Burdened Lower- Income Households	66.6	19.3
Lack of Health Insurance	19.7	5.6
Lack of Broadband Access	35.1	0.6
Age 65 and Older	6.3	6.7
Age 17 and Younger	28.5	29.6
Civilian with a Disability	11.0	5.3
Speaks English "Less than Well"	32.1	0.6
Mobile Homes	1.5	0.0
Group Quarters	0.7	0.0
Homicide Per 100,000	48.8	9.4

5. DISCUSSION AND CONCLUSIONS

5.1 Significance

This study assembled a novel dataset by linking NVDRS with EJI and NaNDA. *To our knowledge, this is the first study to apply explainable AI techniques to the field of violence death prevention.* Adopting a model-agnostic approach to explainable AI not only represents a methodological innovation but also leads to valuable empirical insights. Prior research on risk factors has typically focused on average effects, which are appropriate within the context of linear models. However, the superior performance of machine learning methods such as RF, XGBoost, and NN compared to LR suggests that the relationship between the density of social and healthcare services and violent deaths is highly nonlinear. Given the limited number of features and our sample size, we did not implement deeper and more complex models. Nonetheless, the presence of strong nonlinear associations underscores the need for analytic strategies that go beyond population averages.

Indeed, the explainable AI analyses revealed the complexity of these relationships. Depending on the specific characteristics of a census tract, higher service density could be associated with either higher or lower rates of a given type of violent death. While the median impact scores may trend positive or negative, the wide distribution of individual scores indicates substantial heterogeneity across communities. These findings emphasize that the effect of service density is not uniform and highlight the importance of localized, context-sensitive approaches in public health interventions.

5.2 Implication

Because social and healthcare services require substantial resource allocation from government agencies, nonprofit organizations, and private-sector partners, strategic and evidence-based planning is essential to maximize impact. The use of explainable AI models offers a promising pathway toward this goal. Rather than applying a one-size-fits-all approach, policymakers and service providers can use these insights to tailor interventions sensitive to local demographic, socioeconomic, and geographic characteristics. For instance, in some areas, increasing the density of services may be strongly associated with a decrease in violent deaths, while in others, it may have little effect due to factors such as service accessibility, community demographics, or pre-existing infrastructure.

By leveraging explainable AI, decision-makers can better understand not only where to invest but also how to design interventions that align with the complex realities on the ground. This approach enables more targeted and effective violence prevention strategies, ultimately enhancing public health outcomes. For example, as illustrated in Table 3, investing in childcare services in the California community could potentially help reduce the high homicide rate.

5.3 Limitations and Future Work

This study's cross-sectional design limits causal interpretation. While the ML models revealed associations between service density and violent death rates, these should not be seen as evidence of direct causation. For instance, it is unlikely that child daycare services increase suicide risk; observed associations may reflect unmeasured confounders. Suicide rates vary by age and racial/ethnic groups, but the EJI's broad categories (e.g., under 18, over 65, and aggregated minority status) may overlook important distinctions. For example, suicide is more common among White and American Indian populations but less common among Black Americans—details obscured by the dataset. Additionally, using current residence as a proxy for environmental exposure is not optimal, as relocation can disconnect individuals from long-term service impacts.

The low R^2 values observed for suicide outcomes (ranging from 27 to 31 percent) suggest limited predictive power, indicating that suicide is inherently more challenging to model compared to other outcomes. This aligns with existing literature that emphasizes the complexity of suicidal behavior and the limitations of predictive models in capturing its nuances. Suicide, particularly completed suicide, results from an interplay of diverse factors, including psychiatric illness, psychosocial stressors, genetic vulnerability, access to means, and often unpredictable acute triggers.(Turecki et al., 2019) As a result, the prediction accuracies of models are often extremely poor.(Franklin et al., 2017, Belsher et al., 2019)

Future work will address these limitations by incorporating longitudinal data, exploring interactions with environmental factors, and integrating electronic health records to examine individual-level outcomes.

6. CONCLUSION

This study created a novel dataset linking NVDRS with EJI and NaNDA and is the first to apply explainable AI to violence death prevention. Machine learning models (RF, XGBoost, NN) outperformed linear regression, highlighting the complex, nonlinear relationship between service density and violent deaths. Explainable AI revealed that the impact of service density varies widely across communities, depending on local characteristics. These insights underscore the need for strategic, evidence-based planning rather than one-size-fits-all approaches. By leveraging explainable AI, policymakers and service providers can design more targeted and effective violence prevention strategies tailored to specific community needs.

ACKNOWLEDGEMENT

The authors on this study have been funded by CDC/APHA data demonstration project 2024-0008 and NIH AIM AHEAD Pilot 1OT2OD032581-02-388.

REFERENCES

Aronowitz, S. V. & Reed, M. K. (2025). The Evolving Overdose Crisis: United States Needs Robust Policy Change to Address the Fourth Wave. *Am J Public Health,* 115, 457-459.

Belsher, B. E., Smolenski, D. J., Pruitt, L. D., Bush, N. E., Beech, E. H., Workman, D. E., Morgan, R. L., Evatt, D. P., Tucker, J. & Skopp, N. A. (2019). Prediction Models for Suicide Attempts and Deaths: A Systematic Review and Simulation. *JAMA Psychiatry,* 76, 642-651.

Bishop, T. M., Walsh, P. G., Ashrafioun, L., Lavigne, J. E. & Pigeon, W. R. (2020). Sleep, suicide behaviors, and the protective role of sleep medicine. *Sleep Med,* 66, 264-270.

Blackburn, N. A., Ramos, S., Dorsainvil, M., Wooten, C., Ridenour, T. A., Yaros, A., Johnson-Lawrence, V., Fields-Johnson, D., Khalid, N. & Graham, P. (2023). Resilience-Informed Community Violence Prevention and Community Organizing Strategies for Implementation: Protocol for a Hybrid Type 1 Implementation-Effectiveness Trial. *JMIR Res Protoc,* 12, e50444.

Boeck, M. A., Wei, W., Robles, A. J., Nwabuo, A. I., Plevin, R. E., Juillard, C. J., Bibbins-Domingo, K., Hubbard, A. & Dicker, R. A. (2022). The Structural Violence Trap: Disparities in Homicide, Chronic Disease Death, and Social Factors Across San Francisco Neighborhoods. *J Am Coll Surg,* 234, 32-46.

Cantor, J., Schuler, M. S., Kerber, R., Purtle, J. & Mcbain, R. K. (2025). Changes in Specialty Crisis Services Offered Before and After the Launch of the 988 Suicide and Crisis Lifeline. *JAMA Psychiatry.*

Cerel, J., Singleton, M. D., Brown, M. M., Brown, S. V., Bush, H. M. & Brancado, C. J. (2016). Emergency Department Visits Prior to Suicide and Homicide: Linking Statewide Surveillance Systems. *Crisis,* 37, 5-12.

Choi, K. R., Saadi, A., Takada, S., Easterlin, M. C., Buchbinder, L. S., Johnson, D. C. & Zimmerman, F. J. (2020). Longitudinal Associations Between Healthcare Resources, Policy, and Firearm-Related Suicide and Homicide from 2012 to 2016. *J Gen Intern Med,* 35, 2043-2049.

Christensen, M. (2016). National Violent Death Reporting System (NVDRS). *Dela J Public Health,* 2, 34.

Deblieux, P. J., Alexander, L. F., Nookala, N. & Nereim, C. (2024). A Review of Community-Based Gun Violence Prevention Programs and the Physician's Role. *Adv Pediatr,* 71, 41-54.

Degli Esposti, M., Schell, T. L. & Smart, R. (2025). The Recent Rise in Homicide: An Analysis of Weekly Mortality Data, United States, 2018-2022. *Epidemiology,* 36, 174-182.

Doupnik, S. K., Bowden, C. F., Worsley, D., Keating, C., Cassidy, K., Foster, A. A., Quarshie, W., Min, J., Meisel, Z. & Marcus, S. C. (2025). Suicide Prevention and Telehealth in Children's Hospital Emergency Departments. *Pediatr Emerg Care,* 41, e10-e17.

Englund, H. (2024). Improving suicide risk screening in the emergency department. *Emerg Nurse,* 32, 21-25.

Franklin, J. C., Ribeiro, J. D., Fox, K. R., Bentley, K. H., Kleiman, E. M., Huang, X., Musacchio, K. M., Jaroszewski, A. C., Chang, B. P. & Nock, M. K. (2017). Risk factors for suicidal thoughts and behaviors: A meta-analysis of 50 years of research. *Psychol Bull,* 143, 187-232.

Garnett, M. F. & Curtin, S. C. (2024). Suicide Mortality in the United States, 2002-2022. *NCHS Data Brief.*

Hatton, V. R., Clark, E. & Huber, R. S. (2024). Temporal Patterns in Youth Suicide Deaths Reported in the National Violent Death Reporting System. *J Adolesc Health,* 74, 1049-1052.

Kleinman, R. A. & Morris, N. P. (2023). Suicide, Homicide, and Other Violent Deaths Among People Experiencing Homelessness in the United States: A Cross-sectional Study. *Public Health Rep,* 138, 309-314.

Montgomery, A. E., Blosnich, J. R., Derussy, A., Richman, J. S., Dichter, M. E. & True, G. (2024). Association between Services to Address Adverse Social Determinants of Health and Suicide Mortality among Veterans with Indicators of Housing Instability, Unemployment, and Justice Involvement. *Arch Suicide Res,* 28, 860-876.

National Neighborhood Data Archive (NaNDA): Health Care Services by Census Tract and ZCTA, United States, 1990-2021 [Online]. University of Michigan. Available: https://nanda.isr.umich.edu/project/health-care-services/ [Accessed December 8, 2024].

National Neighborhood Data Archive (NaNDA): Social Services by Census Tract and ZCTA, United States, 1990-2021 [Online]. University of Michigan. Available: https://nanda.isr.umich.edu/project/social-services/ [Accessed December 28, 2024].

Sakran, J. V., Lunardi, N., Mehta, A., Ezzeddine, H. M., Chammas, M., Fransman, R., Byrne, J. P., Stevens, K. & Efron, D. (2024). Increasing Injury Intensity among 6,500 Violent Deaths in the State of Maryland. *J Am Coll Surg,* 238, 710-717.

Services, U. S. D. O. H. A. H. (2024). *Environmental Justice Index* [Online]. U.S. Department of Health and Human Services. Available: https://www.hhs.gov/climate-change-health-equity-environmental-justice/environmental-justice/index/index.html [Accessed 10/29/2024 2024].

Shao, Y., Cheng, Y., Shah, R. U., Weir, C. R., Bray, B. E. & Zeng-Treitler, Q. (2021). Shedding Light on the Black Box: Explaining Deep Neural Network Prediction of Clinical Outcomes. *J Med Syst,* 45, 5.

Sipsma, H. L., Canavan, M. E., Rogan, E., Taylor, L. A., Talbert-Slagle, K. M. & Bradley, E. H. (2017). Spending on social and public health services and its association with homicide in the USA: an ecological study. *BMJ Open,* 7, e016379.

Tenso, K., Strombotne, K., Garrido, M. M., Lum, J. & Pizer, S. (2024). Virtual Mental Health Care and Suicide-Related Events. *JAMA Netw Open,* 7, e2443054.

Turecki, G., Brent, D. A., Gunnell, D., O'connor, R. C., Oquendo, M. A., Pirkis, J. & Stanley, B. H. (2019). Suicide and suicide risk. *Nat Rev Dis Primers,* 5, 74.

VALIDATING MAINTAINABILITY OF OPEN-SOURCE EHR SYSTEMS USING ISO STANDARDS: AN OPENMRS CASE STUDY

Marianne Lodvir Hemsing
Department of Business, Strategy and Political Science, University of South-Eastern Norway, Kongsberg, Norway

ABSTRACT

Ensuring the quality and maintainability of open-source software is critical, especially in healthcare. This study evaluates OpenMRS against ISO/IEC/IEEE 14764:2022 (software maintenance) and ISO/IEC 25010:2023 (software quality). Using Static Source Code Analysis, we extract key metrics (Cyclomatic Complexity, Cognitive Complexity, Maintainability Index) and Design Visualisations and map them to ISO-defined attributes. The results show that OpenMRS aligns well with ISO-defined maintainability attributes, positioning it as a robust and scalable open-source healthcare system. This research also provides support for a framework for ISO-based evaluation of maintainability of open-source software.

KEYWORDS

Software Maintainability, Software Quality, Software Quality Metrics, Software Architecture and Design

1. INTRODUCTION

In healthcare, unreliable or difficult-to-maintain software can lead to critical failures, affecting patient care and system sustainability. Open-source healthcare software, like OpenMRS, must balance rapid innovation with long-term maintainability. This study evaluates OpenMRS against ISO standards to assess its long-term viability and sustainability as a trusted healthcare solution.

OpenMRS has emerged as a widely adopted open-source electronic health record (EHR) system, particularly in resource-constrained settings. With over two decades of development since its launch in 2004, OpenMRS has accumulated a substantial codebase that includes both new and older programming practices (OpenMRS, 2024). The coexistence of new and legacy code is known to potentially increase system complexity and maintainability trough architectural drift, design decay and increased complexity (Lehman, 1980) (Le, et al., 2018) (Lanza, 2003). The OpenMRS community actively works on modernizing the codebase by refactoring legacy code, updating libraries, and adopting newer technologies where feasible (OpenMRS, 2024).

Maintenance typically represents 40-80% of the total software cost, e.g. (Buchmann, I. et al, 2011) (Robert L. Glass, 2001) (Vention Developers, 2024). Making software maintainability a real concern for long-term sustainability. As an open-source alternative in the healthcare sector, validating OpenMRS's maintainability is crucial for trust and adoption.

ISO standards provide a structured framework for quality and maintainability, and alignment facilitates broader acceptance within the healthcare industry (Villa-Gallón, J. E. et al, 2024). Previous research highlights the need for standardized quality assessments in open-source electronic health record (EHR) systems (Shaikh, M. et al., 2022); A adherence to international standards is essential for ensuring the reliability, safety, and interoperability of healthcare software and implementing ISO-based evaluations can enhance the credibility and adoption of open-source EHR systems in clinical settings.

In this research paper we aim to evaluate the maintainability of OpenMRS and to assess compliance with best practices in software engineering.

By conducting this evaluation, we aim to:

1. Demonstrate the effectiveness of free tools in software maintainability assessment using ISO standards.
2. Validate OpenMRS as a maintainable, well-structured healthcare system.

3. Provide a structured methodology for evaluating open-source software maintainability using ISO standards, that can be extended to other open-source projects.

1.1 ISO Software Quality Standards

In this paper we evaluate OpenMRS against the maintainability criteria outlined in two internationally recognized standards that define software maintenance processes and software quality models.

ISO/IEC/IEEE 14764:2022 "Software engineering — Software life cycle processes — Maintenance" (ISO, 2022) provides a framework for managing software maintenance activities, including corrective, adaptive, perfective, and preventive maintenance, ensuring that software remains functional and relevant over time.

The standard ISO/IEC 25010:2023 (System and Software Quality Models) (ISO, 2023) defines quality characteristics (e.g., maintainability, modularity, and testability) that align with reverse engineering goals, ensuring systems meet modern software quality expectations essential for assessing the robustness of a software system.

Other standards are also relevant but not analyzed further in this paper. Software lifecycle standards, such as ISO/IEC 12207, emphasize maintainability as a key factor in ensuring long-term software sustainability and evolution. Additionally, ISO 82304-1, which focuses on the lifecycle of health software, reinforces the importance of maintaining software quality and safety throughout its operational life. ISO 13408, which governs lifecycle considerations in healthcare systems, highlights the necessity of structured maintenance practices to ensure reliability, security, and compliance in critical environments. Furthermore, IEC 62304, the international standard for the lifecycle of medical software, explicitly defines software maintenance as an essential process, ensuring that safety-critical healthcare applications remain functional, secure, and adaptable to changing requirements.

2. METHOD

To evaluate the quality of OpenMRS the two against international software standards, we employed a structured comparison methodology that maps empirical software metrics and design artifacts extracted using free Static Source Code Analysis tools, to the quality attributes defined in ISO/IEC 25010:2023 and ISO/IEC/IEEE 14764:2022.

2.1 Static Source Code Analysis

The Static Source Code Analysis was performed with the free tools Moose version 11 with VerveineJ for Java version v3.2.1 (Moose Technology, 2024), and IntelliJ IDEA version 2024.2.4 with MetricsReloaded version 1.12 (JetBrains, 2024).

The Moose analysis platform was used to extract source code metrics and Visual representations (dependency structures, and architectural diagrams) from the OpenMRS codebase. The selected limits are listed in Table 1., and are based on previous research on maintainability and complexity in Java applications (Yamashita et al., 2016) (Mishra, A. et al, 2021) (Allan, M. et al, 2018) (Microsoft, 2024).

Table 1. Metrics and threshold values for maintainability

Metric	Definition	High	Medium	Low
Cyclomatic Complexity (CC)	Measures the number of independent execution paths in a method.	**1 - 10**	**11 - 20**	**> 20**
Essential Cyclomatic Complexity (ECC)	Evaluates complexity by removing structured control constructs.	**1 - 3**	**4 - 6**	**> 6**
Cognitive Complexity (CoC)	Measures how difficult the code is to read and understand.	**1 - 10**	**11 - 15**	**> 15**
Design Complexity (DC)	Assesses the complexity of module interactions and architecture.	**1 - 3**	**4 - 7**	**> 7**
Simplified Maintainability Index (sMI)	Quantitative measure used to assess how easily a software system can be maintained.	**>85**	**65-85**	**<65**
C-Range Simplified Maintainability Index (CsMI)	Simplified Maintainability index set to lie between 0 and 100.	**20-100**	**10-19**	**0-9**

The two Maintainability Indexes are calculated as shown in Figure 1.:

$$sMI = 171 - 0.23 \times CC - 16.2 \times \ln(LOC)$$

$$CsMI = (171 - 0.23 \times CC - 16.2 \times \ln(LOC)) \times \frac{100}{171}$$

Figure 1. Simplified MI Calculations

Interpretation of Limits:

- High Maintainability (Good): The code is simple, easy to maintain, and testable.
- Medium Maintainability (Acceptable): The code may require some monitoring but is still manageable.
- Low Maintainability (Needs Refactoring): The code is hard to maintain, understand, and test, making it prone to errors and technical debt. Results

2.2 Static Source Code Analysis Alignment with ISO Standards

Data retrieved proved adequate to assess software maintainability in alignment with ISO principles.

Table 2 and 3 summarize how the design diagrams and metrics retrieved from Moose and MetricsReloaded align with the ISO/IEC/IEEE 14764:2022 and ISO/IEC 25010:2023 maintenance attributes, providing actionable insights for maintaining and improving OpenMRS.

Table 2. Alignment with ISO/IEC/IEEE 14764:2022

ISO/IEC/IEEE 14764:2022 Maintenance Activity	Static Source Code Analysis Metrics and Diagrams
1. Corrective Maintenance. Identifying and fixing defects to restore functionality	**Cyclomatic and Cognitive Complexity**: Highlights high-complexity methods prone to defects; focus on outliers like methods with Cyclomatic Complexity > 20.
	Dependency and Call Graphs: Pinpoint fault propagation and isolate root causes.
	Class and Package Diagrams: Aid in understanding module relationships for targeted defect correction.
	Simplified Maintainability Index: Evaluate if software is well-structured and easier to debug, reducing the effort required for defect correction.
2. Adaptive Maintenance Modifying software to adapt to changing environments or requirements.	**Dependency Graphs**: Assess impact of changes by showing interactions between modules.
	Design Complexity: Identifies tightly coupled areas that may need decoupling to enhance adaptability.
	Heat Maps: Focus adaptation efforts on areas of high complexity or technical debt.
	Simplified Maintainability Index: Evaluate if software is adaptable to evolving requirements, allowing for smoother modifications without introducing instability.
3. Perfective Maintenance Improving performance, maintainability, or usability based on feedback.	**Class Diagrams and Heat Map Visualizations**: Provide a structural overview to optimize or enhance specific areas.
	Cognitive Complexity: Focus on higher values to improve readability and usability.
	Heat Maps: Pinpoint high-complexity areas for refactoring.
4. Preventive Maintenance Reducing the risk of future failures and improving system stability.	**Structural Complexity**: Use Design Complexity to identify and reduce coupling.
	Dependency Graphs: Highlight interdependent modules for decoupling.
	Cyclomatic Complexity: Monitor and simplify high-complexity methods to minimize future risks.
	Simplified Maintainability Index: helping identify areas that require proactive refactoring to prevent technical debt accumulation.

Table 3. Alignment with ISO/IEC 25010:2023

ISO/IEC 25010:2023 Maintainabiltiy Attribute	**Static Source Code Analysis Metrics and Diagrams**
Maintainability *(Modularity, Reusability, Analyzability, Modifiability, Testability)*	- **Cyclomatic Complexity** Identifies complex, high-risk, hard-to-maintain areas in the code.
	- **Cognitive Complexity** Assesses code readability and maintainability, helping reduce developer effort in comprehension.
	- **Class and Dependency Diagrams** Relationships and Dependencies to assess change impact and detect tight coupling, and highlighting complex or redundant code.
	Heat Map Highlights complexity hot spots, aiding in refactoring, risk detection, and test prioritization. **Butterfly** Map Identifies core components with many dependencies, helping visualize architectural structure.
	Design Complexity: Evaluates architectural complexity, ensuring a well-structured, maintainable design.
	Simplified Maintainability Index: Reflects how easily a system can be understood, modified, and tested. A high MI' suggests modular, well-structured code, while a lower MI' indicates higher complexity and potential maintenance challenges.

This alignment validates free tools such as Moose and MetricsReloaded capable of delivering industry-standard analyses, enabling assessment of maintainability and architectural health in line with established quality models. The use of standards ensures that the methods and tools are not only technically sound but also provide a framework for systematic and repeatable software maintenance, elevating their practical relevance and value.

In summary, the use of Static Source Code Analysis on OpenMRS directly supports the goals outlined in ISO/IEC/IEEE 14764:2022, offering a structured, standards-aligned approach to managing the maintenance lifecycle of a complex software system.

2.3 Open MRS Quality Metrics

The data collected is available on FigShare (Figshare Dataset, 2024). Table 4 lists the key software quality metrics that were retrieved:

Table 4. OpenMRS Software Quality Metrics

Metric	**Purpose**	**Value**
Average Cyclomatic Complexity (CC)	Number of paths through a function	1,69
Average Essential Cyclomatic Complexity	Complexity of core logic, without structured constructs	1.19
Average Cognitive Complexity	Effort required to understand code	1.06
Average Design Complexity	Structural dependencies between modules	1.52
Number of Methods	Quantitative measure	13 049
Average Lines of Code (LOC) per Method	Quantitative measure for size	12,15
Average C-range Simplified Maintainability Index (CsMI)	Quantitative measure for maintainability	76
	Total Maintainability level	**High**

19 methods had a Cyclomatic complexity above 20.
7 methods had a Cognitive Complexity above 100.

An average cyclomatic complexity of 1.69 is relatively low, suggesting that most methods are straightforward and have simple control flows. This low average could indicate that the majority of methods in OpenMRS perform single, linear tasks, or they are broken down into small, manageable functions, which is generally positive for maintainability and readability.

Average Essential Cyclomatic Complexity: An average of 1.19 suggests that, even after reducing structured complexity, the code retains minimal intrinsic complexity. This low value is positive, indicating that most methods are straightforward and do not rely on complex branching logic. This contributes to readability and further supports maintainability.

Average Cognitive Complexity: 1.06 An average of 1.06 is very low, suggesting that the codebase is easy to comprehend on a cognitive level. This low cognitive complexity means that OpenMRS code is accessible to developers, which is advantageous for onboarding new contributors, enhancing readability, and minimizing misunderstandings in code comprehension. Cognitive complexity measures the human effort required to understand a software program, considering factors like the amount of information, control structures, and the spatial distribution of code elements (Wijendra & Hewagamage, 2021).

Average Design Complexity An average of 1.52 indicates relatively simple design structures. Low design complexity suggests a modular, loosely coupled design, which enhances scalability, flexibility, and ease of modification. **Cyclomatic Complexity above 20** is generally considered high and often signals methods that are harder to test and maintain. These complex methods may have many branching conditions (e.g., multiple if statements, loops, and switch cases). But **only a very small fraction** of the code is particularly complex. This concentration of complexity suggests "hot spots" in the codebase. Focusing on these methods could improve overall system quality by reducing potential sources of bugs and enhancing maintainability.

Cognitive complexity scores above 100 are extremely high and indicate that these methods are likely difficult to understand even for experienced developers. High cognitive complexity implies that these methods are not only logically complex but also contain deep nesting, recursion, or intricate logic that challenges readability and comprehension.

CsMI value falls into the High Maintainability range. This confirms that individual methods in OpenMRS are easy to maintain, indicates well-structured code, making it easier to identify and fix defects efficiently. Systems with a high MI are easier to modify and extend without introducing errors, and High maintainability correlates with better testability.

2.4 Open MRS Design Diagrams

We generated 10 sets of design diagrams with Moose, each providing valuable insights into the OpenMRS codebase.

The Butterfly Map clearly highlighted the core components and their dependencies, making it easy to identify critical modules. Components with many connections appeared centrally, visually reinforcing their importance in the system.

The Class Diagram offered a clear and structured representation of class hierarchies, attributes, and relationships. It was intuitive and effective in communicating OpenMRS's object-oriented design.

The Dependency Graph provided a concise overview of module dependencies.

The Blueprint/Heat Map, exemplified in Figure 2 visually represented software complexity, using color coding to highlight hot spots—areas with high coupling, complexity, or size. These diagrams were clear and well-structured, enabling quick identification of areas needing optimization or refactoring.

Overall, the diagrams provided a clear overview of relationships and hierarchies, confirming the metrics in that the software followed structured software practices.

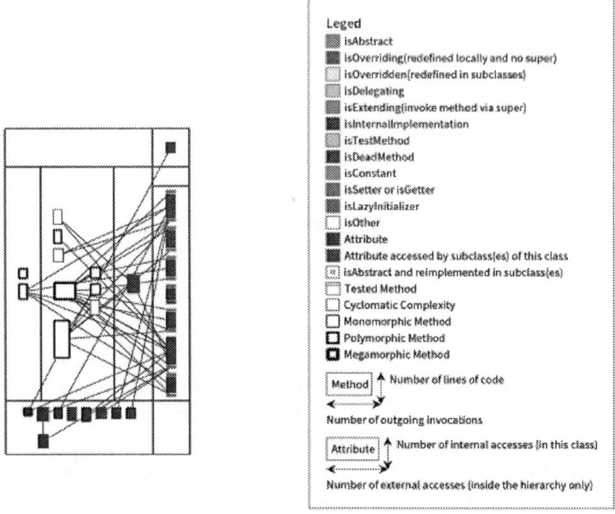

Figure 2. Example of OpenMRS Heat Map

2.5 Implications for OpenMRS

The recovered metrics and diagrams indicate that OpenMRS has a Manageable Complexity Level and exhibits a well-maintained and modular architecture, adhering to strong software engineering principles that contribute to its long-term sustainability. The system's low complexity metrics and structured modularity reinforce its maintainability and adaptability, making it a reliable foundation for future development.

Several key factors highlight OpenMRS's maintainability strengths:

- Easier Feature Expansion and Updates: The modular architecture enables scalable enhancements without introducing excessive complexity.
- Lower Technical Debt and Maintenance Costs: With complexity metrics below critical thresholds, OpenMRS requires less refactoring, reducing long-term maintenance effort.
- Adaptability to Evolving Healthcare Needs; The well-structured design ensures OpenMRS remains flexible for future improvements and integrations.
- Potential for Regulatory Compliance: Adhering to ISO-based quality benchmarks could improve OpenMRS's adoption in hospitals, research institutions, and organisations.
- Need for Ongoing Complexity Monitoring: As new features are introduced, continued evaluation of complexity and modularity will be essential to maintain these strengths.

This evaluation confirms that OpenMRS aligns well with ISO-defined maintainability attributes, positioning it as a robust and scalable open-source healthcare system.

This study focuses on static analysis and does not account for runtime behaviours or real-world performance. Additionally, while OpenMRS is a significant case study, results may not generalize to other open-source healthcare software.

2.6 Future Work

Future work should expand this approach to other open-source systems, applying similar ISO-based validation frameworks to further enhance software credibility and adoption in regulated industries.

While this study evaluated OpenMRS against ISO-defined maintainability attributes, future research could compare OpenMRS with other open-source EHR systems to establish broader benchmarks for software maintainability in healthcare IT. Analyzing how different EHR projects handle modularity, complexity, and long-term sustainability could provide valuable insights for standardizing quality assessments across open-source medical software.

Further research should explore how AI-powered static analysis tools can automate software maintainability assessments.

3. CONCLUSIONS

This study demonstrates the feasibility of using free tools for static source code analysis and systematic maintainability analysis in accordance with ISO/IEC/IEEE 14764:2022 and ISO/IEC 25010:2023. The extracted design diagrams and quality metrics effectively support key software maintenance practices, bridging the gap between open-source development and structured quality evaluation. The results confirm that automated tools like Moose and MetricsReloaded can provide reliable insights into software maintainability, making them valuable for benchmarking open-source projects against international standards.

The evaluation of OpenMRS revealed that maintainability metrics were within acceptable thresholds, and the visualized architecture was clear, structured, and readable. These findings suggest that OpenMRS is a well-structured, high-quality open-source system, demonstrating low complexity, modular design, and strong maintainability attributes.

This study reinforces the value of ISO standards as a validation tool for open-source software. The findings validate the relevance of standardized metrics and methodologies in a real-world context, demonstrating how internationally recognized quality models can provide structured, objective, and comparable assessments of software maintainability. By aligning open-source projects with ISO guidelines, developers can enhance credibility, foster adoption in regulated industries, and ensure long-term sustainability.

ACKNOWLEDGEMENTS

We would like to acknowledge the work performed by the communities of the OpenMRS Project published under the Mozilla Public License 2.0 (OpenMRS, 2024), the IntelliJ IDEA with MetricsReloaded published under the Apache License 2.0 (JetBrains, 2024), and the Moose Project with VerveineJ licensed under the MIT License (Moose Technology, 2024).

REFERENCES

Allan, M. et al, (2018). Evaluating domain-specific metric thresholds: an empirical study. *Proceedings of the 2018 International Conference on Technical Debt (TechDebt '18). Association for Computing Machinery, New York, NY, USA, 41–50.*

Buchmann, I. et al, (2011). Towards an Estimation Model for Software. *European Conference on Software Maintenance and Reengineering.*

Figshare Dataset, 2024. *10.6084/m9.figshare.28477178.* s.l.:s.n.

ISO, (2022). ISO/IEC/IEEE 14764:2022 Software engineering — Software life cycle processes — Maintenance.

ISO, (2023). ISO/IEC 25010:2023 Systems and software engineering — Systems and software Quality Requirements and Evaluation (SQuaRE) — Product quality model.

JetBrains, (2024). *IntelliJ IDEA Community Edition.* [Online] Available at: https://www.jetbrains.com/idea/

Lanza, M., (2003). *Object-Oriented Reverse Engineering.* s.l.:PhD University of Bern.

Le, D. M., Link, D., Shahbazian, A. & Medvidovic, N., (2018). An Empirical Study of Architectural Decay in Open-Source Software. *IEEE International Conference on Software Architecture (ICSA).*

Lehman, M., (1980). Programs, life cycles and the laws of software evolution. *Proc. IEEE, 15 (3).*

Microsoft, (2024). *Code metrics - Maintainability index range and meaning.* [Online] Available at: https://learn.microsoft.com/en-us/visualstudio/code-quality/code-metrics-maintainability-index-range-and-meaning?view=vs-2022

Mishra, A. et al, (2021). echniques for Calculating Software Product Metrics Threshold Values: A Systematic Mapping Study. *Applied Sciences, 11(23), 11377.*

Moose Technology, (2024). *Moose.* [Online] Available at: https://moosetechnology.org

OpenMRS, (2024). *A Visual History of OpenMRS Software.* [Online] Available at: https://openmrs.org/

Robert L. Glass, (2001). Frequently Forgotten Fundamental Facts about Software Engineering. *IEEE SOFTWARE.*

Shaikh, M. et al., (2022). Open-source electronic health record systems: A systematic review of most recent advances. *Health Informatics Journal 28(2).*

Vention Developers, (2024). *Software Maintenance Costs - 2024 Benchmark Overview.* [Online] Available at: https://ventionteams.com/enterprise/software-maintenance-costs

Villa-Gallón, J. E. et al, (2024). ISO Standards in Healthcare Organizations: Research Evolution and Trends from a Bibliometric Analysis. *Publications, 12(3), 27.*

Wijendra, D. R. & Hewagamage, K., (2021). Analysis of Cognitive Complexity with Cyclomatic Complexity Metric of Software. *International Journal of Computer Applications* .

Yamashita et al., (2016). Thresholds for Size and Complexity Metrics: A Case Study from the Perspective of Defect Density. *2016 IEEE International Conference on Software Quality, Reliability and Security (QRS).*

PREDICTING SMALL DATA SEQUENCES WITH MISSING VALUES IN HEALTHCARE: FREQUENTIST VS BAYESIAN APPROACH

Alexandra Vatian, Ivan Tomilov, Keram Goguev and Natalia Gusarova
ITMO University, 49 Kronverksky av., St. Petersburg, Russia

ABSTRACT

Predicting health trajectories with small and incomplete datasets remains a critical challenge in modern healthcare. This study comparatively evaluates frequentist and Bayesian approaches for forecasting small sequences with missing values across diverse medical datasets. We systematically examined four prominent models (ARIMA, Gradient Boosting, Fully Connected Neural Networks, and Transformers) in both classical and Bayesian implementations, using three representative healthcare datasets characterized by different sequence properties: extremely short sequences, large sequences with missing values, and multimodal sequences with irregular patterns. By employing a comprehensive experimental design that varied training set sizes and missing value proportions, we investigated how different model configurations address prediction tasks under constrained data conditions. The study utilized the Symmetric Mean Absolute Percentage Error (SMAPE) metric to assess predictive performance. Notably, Bayesianization consistently improved forecasting precision across most models and datasets, with particularly significant improvements observed in complex sequence scenarios. Our findings demonstrate that Bayesianization effectively balances model complexity, especially for neural network architectures, and provides a promising approach for handling small, incomplete healthcare datasets. The results offer valuable insights for developing more robust predictive models in medical domains characterized by limited and sparse data, potentially enhancing personalized patient monitoring and preventive healthcare strategies.

KEYWORDS

Bayesian Forecasting, Healthcare Prediction, Small Data, Missing Values, Time Series

1. INTRODUCTION AND MOTIVATION

Predicting health status is essential both for personalized patient treatment and monitoring and for preventive medicine, improving quality of life, and reducing healthcare costs. Today, this forecasting is largely based on the trajectory of previous changes, including the medical care received by individual patients as well as on the dynamics of the population as a whole (Pinaire, 2021; Patharkar, 2024; Flothow, 2023; Kaushik, 2020). Addressing the challenges of prediction in the domain of medicine and healthcare has recently been shown to involve several issues.

Firstly, forecasting based on small and missing data is becoming increasingly important for healthcare. In the perspective of emerging and rapidly developing epidemics (as clearly demonstrated by COVID-19), a forecast has to be obtained as early as possible, without waiting for the data to be sufficient for classical statistics.

Secondly, the data processed has been significantly diversified. Today, effective forecasting does not only require the use of clinical and laboratory data in tabular form, but also high-tech medical images, linguistic data (including clinical reports and judgement of individual physician), as well as administrative, genomic, geospatial, and other data (Avinash, 2025; Morid, 2021; Hong, 2023; Tesema, 2020). All the above has to be used together, in a multimodal mode. In addition, the scenarios of sequence formation in healthcare have recently become increasingly volatile, which results in the changes in their statistical properties. Alongside with the conventional missingness mechanisms (Ren, 2024), namely missing completely at random (MCAR), missing at random (MAR), and missing not at random (MNAR), sets of extremely short sequences have to be processed, which places particularly high demands on the processor in terms of identifying the autocorrelation properties of the sequence.

The above issues place special demands on time series forecasting techniques, which today in most cases use machine learning. Among them Autoregressive Integrated Moving Average (ARIMA) models (Box, 1970) remain one of the most commonly used time series forecasting techniques, followed by other regression methods (Krone, 2017) and exponential smoothing (Gardner, 2006). To combat overfitting in small data conditions, the number of tunable ML parameters is reduced using various types of regularization (Tumpach, 2021; Zhang, 2018). Tree-based models handle missing values by their very nature by means of Missingness Incorporated in Attributes (MIA) strategy (Josse, 2019), and the desired balance is achieved through ensembling. More sophisticated models include the Kalman filter (Kone, 2023), which reduces the number of model parameters and thereby ensures accurate short-term predictions for short time series. In (Cui, 2021) the gray system theory-based model (Kayacan, 2010) is used, which exploits a combination of discrete integration and differential equations and, accordingly, provides few model parameters.

The above algorithms implement a two-stage prediction procedure: filling in (imputation) the missing values followed by the analysis of the complete sequence. Numerous literature reviews show that multiple imputations remain the most common missing data handling method used in medical data forecasting (Patharkar, 2024; Ren, 2024). On the other hand, instead of using a separate imputation of missing values followed by a conventional classification, recurrent architecture aggregates both operations based on the internal state vector. To a certain extent, this alleviates the issue of the balance between the amount of data and the model parameters. Several solutions have been proposed in this respect (Che, 2018; Tan, 2020), however, all of them have been tested on fairly long sequences.

The emergence of transformers (Vaswani, 2017) with their native ability to aggregate the operations of imputation and classification, could not but affect the area of sequence prediction (Wen, 2023). Although transformers are primarily used to impute the missing values into a sequence (Ding, 2024; Liu, 2025; Casella, 2024), in (Tipirneni, 2022) clinical outcomes from irregularly sampled multivariate clinical time-series are predicted without any imputation required. However, the estimates of the effectiveness of transformers in time-series prediction still remain controversial [compare (Peng, 2024) and (Zeng, 2023)]. The only work detected containing experimental comparative assessments of ARIMA-, LSTM- and transformer-based algorithms for short-term forecasting on small samples (Ammann, 2022) focuses on energy consumption, therefore, its results cannot be directly transferred to data in the healthcare domain.

Despite the diversity of the predictive models described above, they all rely on frequency, which limits their capabilities in the perspective of small data. Additional enrichment of the available data can be achieved by moving from point estimates of parameters to probabilistic estimates using Bayes' theorem. Healthcare domain involves predictive models based on Gaussian processes, kriging and splines (Hong, 2023; Tesema, 2020; Hsu, 2023; Enjilela, 2019) are presented. Initially built as Bayesian, these models are nonparametric, and their quality largely depends on the successful choice of the kernel, which requires lengthy and costly heuristic.

The so-called Bayesianization procedure (Shi, 2024; Tan, 2019) eliminates the above issues. It consists in transforming a classical model already having been trained into a Bayesian one by replacing the values of its parameters with some probability distributions. Noteworthily, the new model is in a certain sense equivalent to those discussed above. For example, the average value of the parameter distributions might be chosen to be equal to their original values. However, until recently this model has not been widely used for predicting short and sparse sequences [only (de Alba, 2007; Hryniewicz, 2015) can be mentioned as references before 2017]. A revival of interest in Bayesianization has recently been due to the advent of transformers (Muller, 2022; Sankararaman, 2022; LaBarr, 2019; Sinha, 2020; Peng, 2024; Oliveira, 2024). Meanwhile, the authors (Smid, 2019) warn against the danger of overbias when predicting short sequences using Bayesianization.

To summarize, the discussed theoretical and experimental studies on forecasting small sequences with missing values present a mixed and largely controversial picture, which s to provide an informed choice of an IT solution for a specific situation. In an effort to fill this gap, we set the goal to compare the effectiveness of the most popular approaches to forecasting small sequences with missing values in healthcare domain in both their frequentist and Bayesian implementations. Our experimental approach can be described as specific in the following three directions:

Models. In our experiments we aimed to demonstrate the effects of Bayesianization in their purest form. In doing so, we selected the models typical for benchmarking the efficiency of sequences prediction (Oliveira, 2024; Perez-Lebel, 2022), namely, models from the families of ARIMA, Gradient boosting with trees (GB), fully connected neural networks (FCNN), and transformers (the latter - without any imputation). All model implementations were examined both in their classical and in Bayesian forms.

Datasets. While remaining within the framework of the generally accepted classification of the mechanisms of missing values formation (Ren, 2024), we introduce its further refinement. Namely, we use the MCAR model, however, we apply the removal procedure either to single samples with extremely short sequences or to the entire sequences when it represents the whole dataset. This distinction is made to preserve the semantic coherence of the dataset components. To include the scenarios with small data in all cases, we similarly perform a subsequent reduction in the training data size. Besides, following (Perez-Lebel, 2022), we deliberately use real rather than synthetic datasets, thereby preserving the statistical properties of datasets that are characteristic of real healthcare practice.

Task. We have chosen 1-step forecasting as the basic task in the time series forecasting. It features fewer subtle nuances than multi-step forecasting. Due to its simplicity, it identifies the main substantive details of models behavior in aforementioned conditions of small data and missing values. We have chosen forecasting in a regression form, as categorical sequential data can be forecasted likewise by choosing a link function (Fokianos, 2003).

The rest of the article is organized as follows. Section 2.1 presents the structure of the models used in their frequentist and Bayesian implementations. In section 2.2, we describe the experimental scenario, whereas Section 2.3 discusses the experimental data and settings as well as the accepted metrics for evaluating the model's effectiveness. In Section 3, we present and discuss the experimental evaluation of the analyzed models effectiveness and their implementations depending on the input dataset size and model complexity. Section 4 concludes the work and presents some prospects for further research.

2. METHOD AND MATERIALS

2.1 Models and their Bayesianization

We have notably neglected grey system theory-based models (Kayacan, 2010) since their fundamental limitations are their processing only positive input data and featuring a basically exponential form of prediction.

In implementing a GB regression, we followed (Zhang, 2019) to handle a multidimensional sequence forecasting task. To handle missing values in sequences, we adopted the approach from (Khosravi, 2020) as one of the simplest. The model's complexity was regulated by the variable depth of trees. The number of trees was fixed at a standard value of 100 trees.

To handle missing values with the ARIMA model, the maximum likelihood approach from (Velicer, 2005) was used as an example of an approach adopted to MCAR type missing values.

A FCNN was implemented and applied in a standard way for sequence forecasting as described in (Borovykh, 2019). The number of layers was chosen as the complexity parameter for this model.

A transformer model (Figure 1) was built in the simplest form (Wen, 2023) as a composition of multi-head attention layers with standard feed forward network and normalization between them to elicit the effect of a small training dataset in its pure form.

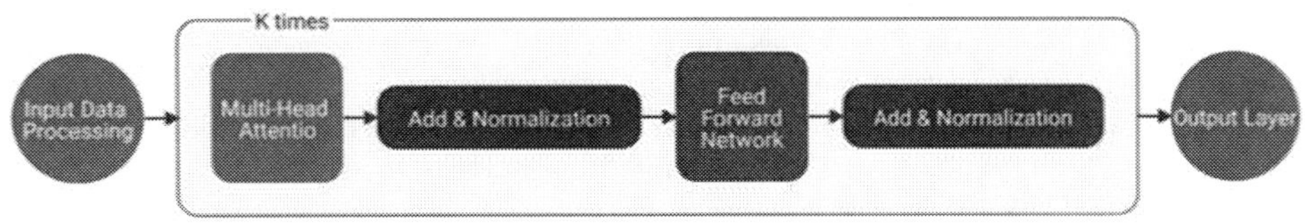

Figure 1. Transformers scheme in the frequentist variant

To convert models to Bayesian form we used a scheme (Figure 2) with the parameters estimated by fully factorized gaussians (Kochurov, 2018). This scheme was chosen as one of the simplest, the fastest and the most stable among Bayesian inference for deep learning (DL) models to increase the reliability of results.

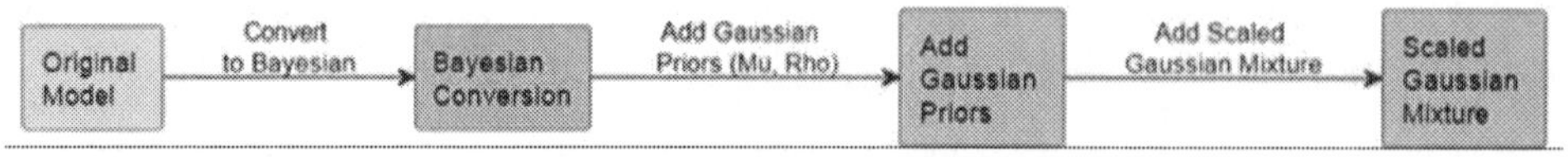

Figure 2. Bayesianization scheme

Scheme (Figure 2) was applied to ARIMA to obtain its analogue in the form of Bayesian structural mode as in (Zhang, 2024). Bayesianization was only applied to the last fully connected layers of the transformer model (Figure 3) to increase stability and convergence (Okolie, 2024).

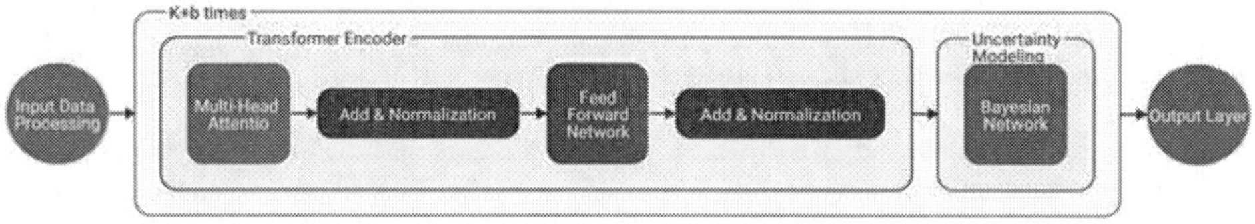

Figure 3. Transformers scheme: Bayesian variant

2.2 Datasets Used

We used three datasets which correspond to the following types of sequencing: (i) a set of extremely short sequences; (ii) a long sequence with regular patterns; (iii) a sequence with irregular patterns.

Type (i) is represented by the dataset (Tsanas, 2009), referred to as the Parkinson Telemonitoring. It contains the results of a six-month study on remote monitoring of patients with early-stage Parkinson's disease and includes 19 features.

Type (ii) is represented by the dataset (COVID-19, 2023), referred to as the COVID-19 dataset.

It contains some global time series on the spread of COVID-19 collected by Johns Hopkins University from January 2020 to March 2023. We selected the one including daily statistics on confirmed cases, deaths, and recoveries for Mexico.

Type (iii) is represented by the dataset (Cajas, 2024), referred to as A Multi-Modal Satellite Imagery Dataset. It brings together time series and metadata for 81 municipalities in Colombia for the period 2007-2019. The core is epidemiological indicators (dengue fever cases), supplemented by three key categories of related data: Sociodemographic, Climate and Epidemiological records.

2.3 Experimental Scenario and Metrics

Our experiment pursued two objectives: to investigate how different model configurations would address the prediction task with a variable training set size and the volume of missing values, and to compare the quality of Bayesian alternatives with classical frequentist models. We aimed to arrange all scenarios from the same methodological standpoint, generally following the methodology of the benchmarking (Nejad, 2024).

For this purpose, the fragments of various lengths (100%, 75%, 50%, 25%) were randomly selected from the training part of each experimental dataset. Data fragmentation for the training set was performed so as to preserve the semantic coherence of the original data set. Namely, sequences of adjacent data were cut out from the data sets consisting of a single time series; individual time series were selected without disturbing them from the data sets representing a set of short time series.

To simulate the missing values, following MCAR model, we randomly replaced varied (0%, 25%, 50%, 75%) proportions of values with the missing ones.

We repeated the above procedures 30 times with each dataset, thereby forming sub-datasets for training the models, namely, 480 sub-datasets for each original dataset. All the analyzed models were trained independently and from scratch on each such sub-dataset. Their efficiency was measured using the SMAPE metric:

$$\mathrm{SMAPE} = \frac{100}{\mathrm{n}} \sum_{i=1}^{n} \frac{|y_i - \bar{y}_i|}{(|y_i| + |\bar{y}_i|)/2},$$

where y_i is the target value of i-th example, $\bar{y}_i$ is the prediction for i-th example.

3. RESULTS AND DISCUSSION

As shown in Section 2.3, the full experimental results are too extensive to be included in the paper. They can be provided upon request. To demonstrate the identified patterns, the most representative cases were selected for each of the dataset groups described in Section 2.2. Namely, the following groups of cases were formed:

1) Full dataset without missing values, referred to below as "Full dense";
2) 10% of the original dataset, but without missing values, referred to below as "Small dense";
3) Full dataset, but with 75% missing values, referred to below as "Full missing";
4) 10% of the original dataset with 75% missing values, referred to below as "Small missing".

Figure 4 demonstrates the behavior of model's SMAPE for the dataset of type (i).

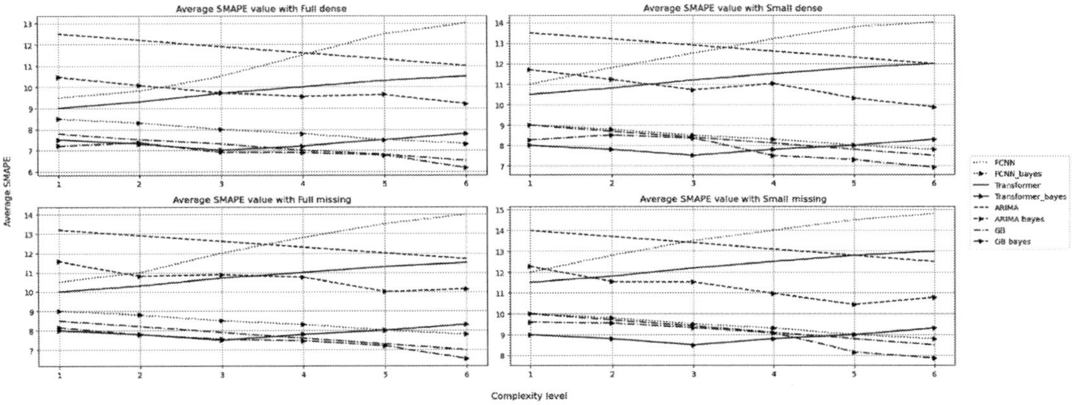

Figure 4. SMAPE of models for the dataset of type (i) – a set of extremely short sequences

Figure 4 shows that in all cases Bayesianization significantly improves the efficiency of all models regardless of their complexity. It reduces SMAPE in each case by the value from 0.5 to 2. Notably, in the cases of FCNN and transformers SMAPE monotonically increases by 1 and 3 correspondingly, but for their Bayesianized versions SMAPE monotonically drops by 1. This suggests that Bayesianization balances model's complexity according to the dataset.

Bayesianized GB and Bayesianized transformer have approximately the same quality in all cases, while classical GB outperforms classical transformer. This is notable because GB is one of the best models for tabular datasets, and the case of extremely short sequences partially reduces to tabular form. Meanwhile, Bayesianization results in the transformer achieving the same performance, even though its ability of the effective processing of autocorrelations ceases to provide advantages against its high complexity.

Figure 5 demonstrates the behavior of a model's SMAPE for the dataset of type (ii).

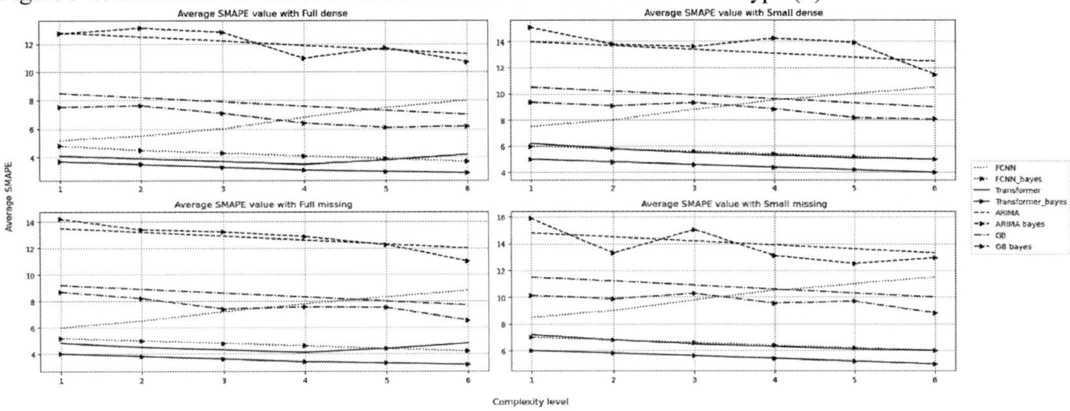

Figure 5. SMAPE of models for the dataset of type (ii) – large sequence with a large proportion of missing values

Figure 5 shows that in all cases Bayesianization improves the efficiency of all models, except ARIMA, regardless of their complexity. It reduces SMAPE by the value from 0.3 to 1. The contrast with the results on the dataset (i) is noticeable in transformers and FCNN cases, where classical FCNN's SMAPE still increase, but classical transformers SMAPE decreases.

In thy discussed case Bayesianization also improves transformers' quality less than on previous dataset (0.6% against 1.2%), however, for all other models improvement is comparable. This suggests that in the case of large sequences, transformers do not require balance of the complexity provided by Bayesianization as much as other models.

Figure 6 demonstrates the behavior of SMAPE models for datasets of type (iii).

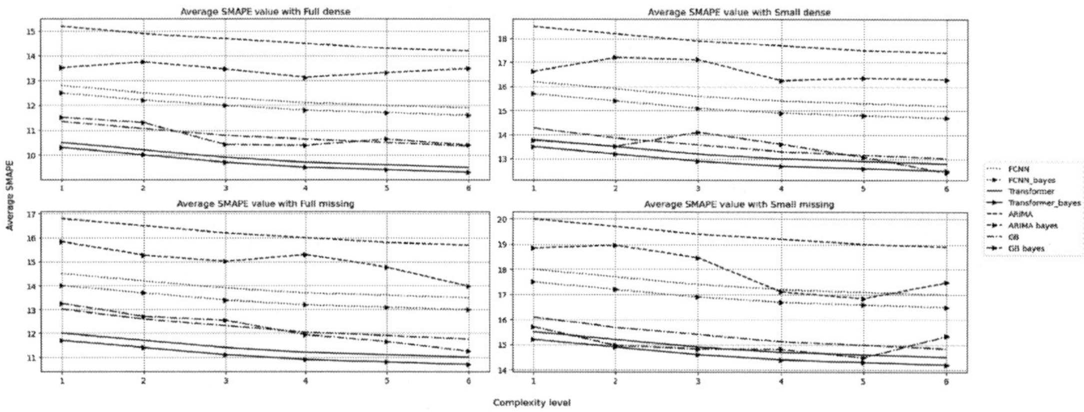

Figure 6. SMAPE of models for the dataset of type (iii) – a multimodal sequence with irregular patterns

Figure 6 shows that in all cases Bayesianization improves the efficiency of all models. Namely, their SMAPE reduces approximately by the value of 0.3 or larger. For all models, with complexity increasing their SMAPE reduces, which is consistent with the dataset being complex due to irregular patterns contained. Notably, in this case transformers' performance still increases with Bayesianization, however, the increase is much smaller than with the other two datasets.

The GB and its Bayesian version achieve approximately the same SMAPE in the case of small dense data, however, they perform worse in the case of small missing data. For both cases of full data the gap in performance is larger. This indicates the ability of transformer to use autocorrelations in the data flexibly, which helps when the sequence is long or sparse.

To summarize, in all cases Bayesianization improves quality, except for ARIMA, for which there is no stable improvement on large sequence dataset. For the case of transformers and FCNN models Bayesianization appears to not only improve the quality, but also to balance their complexity. This is due to the fact that on simple datasets (type (i) and (ii)) their SMAPE grows with increasing complexity of architecture, while after Bayesianization it only decreases. Notably, on short sequences, GB and its Bayesianized version shows the highest quality, and Bayesianized (but not classical) transformers reach that quality as well. On other datasets, which are less reducible to tabular form, transformers manifest the best quality.

4. CONCLUSION

In this paper, we compared the effectiveness of various approaches to forecasting small sequences with missing values in their frequentist and Bayesian implementations in medical domain. For comparison, we used three types of datasets characteristic for the healthcare domain, namely: (i) a set of extremely short sequences; (ii) a large sequence with a large proportion of missing values; (iii) a small sequence with a large proportion of missing values. We considered the most commonly used approaches, namely, models from the families of ARIMA, GB, FCNN, and transformers, with varying architectural complexity. We used the SMAPE metric to assess the quality of prediction. In all cases Bayesianization was observed to provide an improvement of regression forecasting precision.

Our approach has the advantage of a unified methodological standpoint. Besides, using real rather than synthetic datasets includes addressing common practical challenges of healthcare domain preserving the statistical properties of datasets that are characteristic of real practice.

The main limitation of our work may be considered its focus on one-step forecasting. However, the latter is a typical task in medical domain. We would leave anomaly detection, classification and other types of tasks for further research. The second limitation may be an insufficient coverage of practical cases by the datasets chosen, and we envisage addressing it in further research.

The obtained results show that Bayesianization is a promising way to improve the quality of forecasting in the case of small data sequences, in the case of missing values and in the especially difficult case of a combination of the above conditions. These results can be amply used in the processing of small amounts of data in today's healthcare practice including newly discovered diseases and patients with comorbidities.

Further work will be aimed at studying the effectiveness of various Bayesianization techniques addressing sequence forecasting in the most complicated cases.

ACKNOWLEDGEMENT

This work was supported by Russian Science Foundation, Grant № 23-11-00346.

REFERENCES

Ammann G. (2022) Using LSTMs And Transformers to Forecast Short-term Residential Energy Consumption. Tilburg University June 2022. https://arno.uvt.nl/show.cgi?fid=160767

Avinash, G., Pachori, H., Sharma, A. *et al.* (2025) Time series forecasting of bed occupancy in mental health facilities in India using machine learning. *Sci Rep* 15, 2686. https://doi.org/10.1038/s41598-025-86418-9

Borovykh A., Oosterlee C.W., Bohte S.M. (2019) Generalization in fully-connected neural net-works for time series forecasting. arXiv:1902.05312v2 (stat.ML) 27 Jul 2019.

Box, G.E.P. and Jenkins, G.M. (1970). Time Series Analysis: Forecasting and Con-trol. San Francisco: Holden-Day.

Cajas, S. A., et al. (2024). A Multi-Modal Satellite Imagery Dataset for Public Health Analysis in Colombia (version 1.0.0). PhysioNet. https://doi.org/10.13026/xr5s-xe24.

Casella M., et al. (2024) Transformers deep learning models for missing data imputation: an application of the ReMasker model on a psychometric scale. *Front. Psychol., Sec. Quantitative Psychology and Measurement.* V.15. 2024. https://doi.org/10.3389/fpsyg.2024.1449272

Che, Z., et al. (2018) Recurrent Neural Networks for Multivariate Time Series with Missing Values. Sci Rep 8, 6085 (2018). https://doi.org/10.1038/s41598-018-24271-9

COVID-19 (2023) Dataset provided by the Mexican government: https://datos.gob.mx/busca/dataset/informacion-referente-a-casos-covid-19-en-mexico

Cui, Z., et al. (2021). A hybrid rolling grey framework for short time series modelling. Neural Comput & Applic 33, 11339–11353. https://doi.org/10.1007/s00521-020-05658-0

de Alba E., Mendoza M. (2007) Bayesian Forecasting Methods for Short Time Series. FORESIGHT, January 2007. https://gente.itam.mx/mendoza/Foresight.pdf

Ding S. et al. (2024) TIformer: A Transformer-Based Framework for Time-Series Forecasting with Missing Data. In: Chen, T., Cao, Y., Nguyen, Q.V.H., Nguyen, T.T. (eds) Data-bases Theory and Applications. ADC 2024. Lecture Notes in Computer Science, vol 15449. Springer, Singapore. https://doi.org/10.1007/978-981-96-1242-0_6

Enjilela, E., et al. (2019). Cubic-Spline Interpolation for Sparse-View CT Image Recon-struction With Filtered Backprojection in Dynamic Myocardial Perfusion Imaging. *Tomography,* 5(3), 300-307. https://doi.org/10.18383/j.tom.2019.00013

Flothow, A., Novelli, A. & Sundmacher, L. (2023) Analytical methods for identifying sequences of utilization in health data: a scoping review. *BMC Med Res Methodol,* 23, 212. https://doi.org/10.1186/s12874-023-02019-y

Fokianos K., Kedem B. (2003) Regression Theory for Categorical Time Series. *Statistical Science*, Vol. 18, No. 3, 357–376.

Gardner E.S. (2006) Exponential smoothing: The state of the art—Part II, International Journal of Forecasting, Volume 22, Issue 4, 2006, Pages 637-666, https://doi.org/10.1016/j.ijforecast.2006.03.005.

Hong J., Chun H. (2023) A prediction model for healthcare time-series data with a mixture of deep mixed effect models using Gaussian processes. *Biomedical Signal Processing and Control*, Volume 84, July 2023, 104753.

Hryniewicz O., Kaczmarek K. (2015). Forecasting Short Time Series with the Bayes-ian Autoregression and the Soft Computing Prior Information. In: Grzegorzewski, P., Gagolewski, M., Hryniewicz, O., Gil, M. (eds) Strengthening Links Between Data Anal-ysis and Soft Computing. Advances in Intelligent Systems and Computing, vol 315. Springer, Cham. https://doi.org/10.1007/978-3-319-10765-3_10

Hsu CC, et al. (2023). A Stabi-lized Kriging Method for Mapping Disease Rates. *J Epidemiol.* 2023 Apr 5;33(4):201-208. doi: 10.2188/jea.JE20210276.

Josse J, et al. (2019) On the consistency of supervised learning with miss-ing values. 2019; arXiv:1902.06931.

Kaushik, S., et al. (2020) AI in Healthcare: Time-Series Forecasting Using Statistical, Neural, and Ensemble Architectures. *Front. Big Data*, 19 March 2020. *Sec. Medicine and Public Health.* Volume 3 – 2020.

Kayacan E., (2010) Grey system theory-based models in time series pre-diction. Expert Systems with Applications, Volume 37, Issue 2, 2010, pp 1784-1789, https://doi.org/10.1016/j.eswa.2009.07.064.

Khosravi P., Vergari A., Choi Y. (2020) Handling Missing Data in Decision Trees: A Probabilistic Approach. arXiv:2006.16341v1 (cs.LG) 29 Jun 2020 .

Kochurov M., et al. (2018) Bayesian Incremental Learning for Deep Neural Networks. arXiv:1802.07329v3 (stat.ML) 27 Mar 2018.

Kone L.A., et al. (2023). Short Time Series Forecasting Method Based on Genetic Programming and Kalman Filter. In: Collet, P., Gardashova, L., El Zant, S., Abdulkarimova, U. (eds) Complex Computational Ecosystems. CCE 2023. Lecture Notes in Computer Science, vol 13927. Springer, Cham. https://doi.org/10.1007/978-3-031-44355-8_6

Krone, T., Albers, C. J., & Timmerman, M. E. (2017). A comparative simulation study of AR (1) estimators in short time series. Quality & Quantity, 51, 1–21.

LaBarr A. (2019) The Bayesians are Coming! The Bayesians are Coming! The Bayesians are Coming to Time Series! Paper 3188-2019. https://support.sas.com/resources/papers/proceedings19/3188-2019.pdf

Liu C., et al. (2025) Heterogeneous multivariate time series imputation by transformer model with missing position encoding. Expert Systems with Applications. Available online 16 January 2025, 126435

Morid M.A., Sheng O.R. L., Dunbar J. (2021) Time Series Prediction Using Deep Learning Methods in Healthcare. *ACM Tr. on Management Information Systems*, V.14, Issue 1. Article No.: 2, Pages 1 – 29. https://doi.org/10.1145/3531326

Muller S., et al. (2022) Transformers Can Do Bayesian Inference. ICLR 2022. https://arxiv.org/pdf/2112.10510

Nejad A. S. et al. (2024) SERT: A transformer-based model for multivariate temporal sensor data with missing values for environmental monitoring. Computers & Geosciences, Volume 188, June 2024, 105601.

Okolie C.J., et al. (2024) Performance analysis of Bayesian optimised gradient-boosted decision trees for digital elevation model (DEM) error correction: interim results. June 2024. ISPRS Annals of the Photogrammetry Remote Sensing and Spatial In-formation Sciences X-2-2024:179-183. DOI:10.5194/isprs-annals-X-2-2024-179-2024

Oliveira JM, Ramos P. (2024) Evaluating the Effectiveness of Time Series Transformers for Demand Forecasting in Retail. *Mathematics*. 2024; 12(17):2728. https://doi.org/10.3390/math12172728

Patharkar A., et al. (2024) Predictive modeling of biomedical temporal data in healthcare applications: review and future directions. *Front Physiol.* 2024 Oct 15; 15:1386760. doi: 10.3389/fphys.2024.1386760

Peng S., et al. (2024) Mamba or Transformer for Time Series Forecasting? Mixture of Universals (MoU) Is All You Need. 28 Aug 2024. https://arxiv.org/html/2408.15997v1

Perez-Lebel A., et al. (2022) Benchmarking missing-values ap-proaches for predictive models on health databases. *Gigascience.* 2022 Apr 15;11:giac013. doi: 10.1093/gigascience/giac013

Pinaire J., et al. (2021). Sequential Pattern Mining to Predict Medical In-Hospital Mortality from Administrative Data: Application to Acute Coronary Syndrome. *Journal of Healthcare Engineering*, Volume 2021, Article ID 5531807.

Ren W, et al. (2024) Moving Beyond Medical Statistics: A Systematic Review on Missing Data Handling in Electronic Health Records. Health Data Sci. 2024 Dec 4;4:0176. doi: 10.34133/hds.0176.

Sankararaman K.A., Wang S., Fang H. (2022) BayesFormer: Transformer with Uncertainty Es-timation. arXiv:2206.00826v1 (cs.CL) 2 Jun 2022.

Shi H., et al. (2024) Training-Free Bayesianization for Low-Rank Adapters of Large Language Models. arXiv:2412.05723v1 (stat.ML) 7 Dec 2024.

Sinha K., Sahu P.K. (2020) Forecasting Short Time Series using Rolling Grey Bayesian Frame-work. International Journal of Statistical Sciences ISSN 1683-5603 Vol. 20(2), 2020, pp 207-224.

Smid, S. C., et al. (2019). Bayesian Versus Frequentist Estimation for Structural Equation Models in Small Sample Contexts: A Systematic Review. *Structural Equation Modeling: A Multidisciplinary Journal*, 2019, *27*(1), 131–161. https://doi.org/10.1080/10705511.2019.1577140

Tan M.T., Tian G.-L., Ng K.W. (2019) *Bayesian Missing Data Problems: EM, Data Augmentation and Noniterative Computation.* Chapman & Hall, 2019. 346 p. ISBN 9780367385309

Tan Q., Ye M., Yang B., et al. (2020) DATA-GRU: Dual-attention time-aware gated re-current unit for irregular multivariate time series. The Thirty-Fourth AAAI Conference on Artificial Intelligence (AAAI-20), 930–937.

Tesema G.A., Tadesse Z., Tamirat K.S. (2020) Decomposition and Spatio-temporal analysis of health care access challenges among reproductive age women in Ethiopia, 2005–2016. *BMC Health Services Research* 20:760 https://doi.org/10.1186/s12913-020-05639-y

Tesema G.A., Tadesse Z., Tamirat K.S. (2020) Decomposition and Spatio-temporal analysis of health care access challenges among reproductive age wom-en in Ethiopia, 2005–2016. BMC Health Services Research 20:760 https://doi.org/10.1186/s12913-020-05639-y

Tipirneni S., Reddy C.K. (2022) Self-supervised transformer for sparse and irregularly sampled multivariate clinical time-series. *ACM Trans. Knowl. Discov. Data*, 16 (6), pp. 1-17.

Tsanas, A. & Little, M. (2009). Parkinsons Telemonitoring (Dataset). UCI Machine Learning Repository. https://doi.org/10.24432/C5ZS3N.

Tumpach J., Kalina J., Holena M. (2021) A Comparison of Regularization Techniques for Shallow Neural Networks Trained on Small Datasets. 2021. https://ceur-ws.org/Vol-2962/paper38.pdf

Vaswani A., et al. (2017) Attention is all you need. In Advances in Neural Information Processing Systems 30, 2017, pages 5998–6008. Curran Associates, Inc.

Velicer W.F., Colby S.M. (2005) A Comparison of Missing-Data Procedures for Arima Time-Series Analysis. August 2005 *Educational and Psychological Measurement* 65(4):596-615. DOI:10.1177/0013164404272502

Wen Q., et al. (2023) Transformers in Time Series: A Survey. arXiv:2202.07125v5 (cs.LG) 11 May 2023.

Zeng A., et al. (2023) Are Transformers Effective for Time Series Forecasting? *The Thirty-Seventh AAAI Conference on Artificial Intelligence* (AAAI-23).

Zhang F., Li Y., Li X. (2024) Comparison of ARIMA and Bayesian Structural Time Series Models for Predicting the Trend of Syphilis Epidemic in Jiangsu Province. Infection and Drug Resistance. December 2024. 17:5745-5754. DOI:10.2147/IDR.S462998

Zhang Z., Jung C. (2019) GBDT-MO: Gradient Boosted Decision Trees for Multiple Outputs. arXiv:1909.04373v2 (cs.CV) 28 Dec 2019.

Zhang, Y., Ling, C. (2018) A strategy to apply machine learning to small datasets in materials science. npj Comput Mater 4, 25 (2018). https://doi.org/10.1038/s41524-018-0081-z

SMART TEXTILES FOR HEALTH MONITORING IN INDUSTRIAL ENVIRONMENTS: A FRAMEWORK FOR WEARABLE SENSOR INTEGRATION

Bruna Abreu Vilarinho[1], André Ferreira Costa Vieira[2], João Nunes-Pereira[2], João Miguel Nunes Parente[2], Ivan Miguel Pires[3] and Fernando José Velez[1]

[1]*Instituto de Telecomunicações and Universidade da Beira Interior, DEM - Faculdade de Engenharia, Covilhã, Portugal*

[2]*Universidade da Beira Interior, Centre for Mechanical and Aerospace Science and Technologies (C-MAST-UBI), Covilhã, Portugal*

[3]*Instituto de Telecomunicações, Escola Superior de Tecnologia e Gestão de Águeda, Universidade de Aveiro, Águeda, Portugal*

ABSTRACT

The occupational health and safety of workers in industrial plants are increasingly compromised through exposure to thermal, physical, and psychological stressors. Wearable sensors embedded in garments have been introduced as a promising technology for real-time, continuous health monitoring without compromising workers' mobility or comfort. This paper reviews recent literature on wearable sensor technologies embedded in textiles, with a focus on their applicability in industrial settings. The paper identifies the key physiological parameters commonly monitored (e.g., heart rate, skin temperature, respiratory rate, and skin conductance), the materials and sensor types used, as well as the methods of integration into garments. Based on the findings, a conceptual model for a smart textile monitoring system tailored to industrial workers is proposed. The aim is to support the development of ergonomic, accessible, unobtrusive, and effective solutions that promote occupational health and prevent work-related illnesses.

KEYWORDS

Wearable Sensors, Smart Textiles, Physiological Monitoring, Industrial Environments, Stress Detection, Respiratory Monitoring

1. INTRODUCTION

The development of wearable devices for monitoring the health of factory workers, integrating sensors, communication devices, processing units, memory, and a battery (or other energy supply device) contributes to creating a work environment that avoids health risks and prevents occupational accidents that may arise due to working conditions or the worker's health status. Using biomedical sensors integrated into clothing, along with data acquisition and associated signal processing, the device will allow the evaluation of the comfort conditions of factory workers during their daily work activities.

Occupational health issues are a significant concern, particularly in industrial environments where workers are subjected to constant exposure to high temperatures, prolonged physical strain, repetitive strain, and unfavorable conditions. Exposure may lead to physiological stress, fatigue, and long-term conditions. Therefore, there is a growing need for real-time physiological monitoring technologies to anticipate and prevent work-related health risks (Sharma, Suri et al. 2022).

Recent developments in wearable technologies have paved the way for the integration of physiological sensors directly into textiles, creating a new category of "smart garments". These systems enable the continuous and unobtrusive collection of vital health data such as heart rate, respiratory rate, skin temperature, and even electrodermal activity without restricting the wearer's mobility or comfort (Pinnelli, Lo Presti et al. 2024).

Textile-integrated sensors have been effectively applied in several contexts. For example, wearable systems have been deployed to monitor heat stress in Swiss railway workers, offering insight into the physiological impact of thermal exposure under controlled conditions (Vázquez, Fujs et al. 2024). In another study, conducted in India, a combination of environmental sensors and physiological data collection was used to predict worker stress using machine learning algorithms (Sharma, Suri et al. 2022). Furthermore, a large-scale study in Egypt demonstrated the feasibility of using wearable biosignal sensors for stress detection in diverse real-world populations (Abd Al-Alim, Mubarak et al. 2024).

These promising results underscore the growing relevance of smart textiles in occupational health. Despite their potential, challenges such as durability during washing, sensor calibration, energy consumption, and data privacy must be addressed. This paper aims to review current advancements in textile-based sensor systems applied to health monitoring of factory workers (mainly based on their respiratory performance data) and propose a conceptual monitoring model tailored to industrial plant work environments.

The remainder of this manuscript is organized as follows. In section 2 the methodological part of the paper is addressed. In section 3, the wearable sensor technologies and implemented techniques to be embedded into the garment are described. Section 4 proposes a novel approach for industrial workers' health monitoring through a conceptual model for the proposed wearable monitoring system. The conclusions are drawn in section 5 where suggestions for further research are also discussed.

2. METHODOLOGY

This study is based on a literature review focused on scientific publications between 2018 and 2024. Articles were retrieved from databases such as PubMed, IEEE Xplore, and ScienceDirect using keywords including "wearable sensors", "smart textiles", "occupational health", and "physiological monitoring". The inclusion criteria required that the selected studies: (i) employ wearable sensors for physiological monitoring; (ii) involve integration of sensors into textile or flexible substrates; and (iii) be applied in industrial environments or controlled laboratory simulations replicating workplace conditions. The primary parameters investigated in these studies were heart rate, respiratory rate, skin temperature, and skin conductance, the known indicators of physiological and thermal stress.

The study further detailed the use of different types of sensors, including piezoresistive, capacitive, thermal, and optical sensors, which have different advantages in the case of monitoring physiological parameters as shown in Figure 1. Piezoresistive sensors, for example, are highly effective in sensing strain and breathing motion, while capacitive sensors respond greatly to pressure variations. Thermal and optical sensors are useful in sensing skin temperature and blood flow, respectively, in contributing to overall health monitoring.

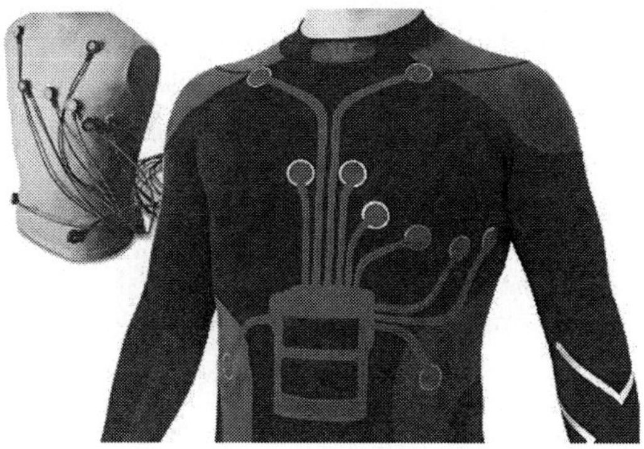

Figure 1. Wearable sensors for monitoring health physiological parameters

From an integration perspective, the techniques were widely distributed: textile lamination allows multilayering; embroidery with conductive yarns assures durability and washability; inkjet and screen printing allow precise positioning of conductive materials; and weaving of fibers provides seamless integration of sensors. These processes must be chosen based on the foreseen application of the garment, the degree of durability, and comfort.

Elastomers, carbon nanotubes, conductive polymers, and silver nanowires were selected on the basis of their high electrical conductivity, stretchability, and biocompatibility. For instance, carbon nanotubes in elastomeric matrices yield high sensitivity with mechanical robustness, while silver nanowires yield high thermal and electrical conductivity with minimal rigidity. An ideal approach to future applications can be the merging of silver nanowire meshes with breathable polymer substrates for the realization of both signal fidelity and comfort during long-term industrial applications.

Furthermore, the review also looked at the facilitating technological infrastructure surrounding wearable sensors. These are power supply subsystems such as flexible batteries and energy harvesting modules (e.g., triboelectric nanogenerators or thermoelectric generators) that provide smooth operation without compromising mobility. Data acquisition platforms usually rest on low-power microcontrollers (e.g., ESP32, STM32), enabling analog-to-digital conversion, pre-processing, and data compression at the edge to prevent transmission overload.

Wireless communication protocols such as Bluetooth Low Energy (BLE) and LoRa were utilized the most since they are appropriate for different industrial environments—BLE for short-range, low-latency applications and LoRa for long-distance, low-power transmission in large facilities.

On the software side, data interpretation is heavily reliant on native machine learning algorithms. These can run locally or in the cloud and are themselves trained to recognize abnormal patterns within physiological signals. Decision trees, support vector machines, and deep learning algorithms (e.g., LSTM networks for time-series physiological data) are typical examples. Incorporation of these systems not only facilitates real-time feedback on health but also predictive alerts, making the monitoring exercise proactive rather than reactive.

3. WEARABLE SENSOR TECHNOLOGIES IN TEXTILES

Wearable sensors integrated into textiles can monitor multiple physiological parameters. Piezoresistive sensors, for example, have shown high sensitivity in detecting respiratory movements and skin pressure, especially when combined with flexible substrates such as liquid elastomers and carbon nanotubes (Selvan T, Sharma et al. 2022). Authors of the study presented the development of a stencil-printable multifunctional adhesive sensor based on carboxylated multiwalled carbon nanotubes (c-CNTs) and functionalized liquid isoprene rubber (LIR). The methodology involved solvent-free fabrication using planetary centrifugal mixing and stencil printing techniques, followed by heating for characterization. Rheological studies identified the percolation threshold at 0.42 vol% c-CNT, correlating filler concentration with polymer chain mobility. Electrical properties were evaluated experimentally and through finite element simulations, demonstrating strong agreement and confirming a 3D percolation network. The sensor exhibited high sensitivity to temperature (0.347%/°C) and strain (gauge factor of 62.4 up to 40% strain), alongside excellent adhesion to various substrates. Its multifunctionality was validated through physiological monitoring, including pulse, respiratory rate, and phonetics detection. The robust methodology highlights the potential for scalable, cost-effective fabrication of wearable sensors for healthcare applications.

Capacitive and thermal sensors are also employed to detect temperature alteration and body heat distribution and play an important role in the identification of thermal stress. Their functionality has been established under environmental stressors and mechanical deformation by numerous studies. (Sharma, Thapa et al. 2024)works are particularly remarkable for having developed a stencil-printable graphene–liquid elastomer-based strain sensor for unobtrusive respiration monitoring that was also effective under dynamic thermal conditions.

Furthermore, authors from (Cui, Poblete et al. 2019) showed a breathable, stretchable temperature sensor utilizing silver nanowires specifically for thermoresistive sensing and with no effect of strain on sensing. Their method allowed accurate measurement of skin temperature under motion and is consequently most suitable for wearable technology in the industrial environment where thermal stress is the top priority.

The study presented the development of a stencil-printable graphene–liquid elastomer-based strain sensor for unobtrusive respiratory monitoring. The sensor addresses challenges like crosstalk, noise, and environmental interference (temperature and humidity) that limit the accuracy of flexible strain sensors. It is designed for mass utility using frugal engineering and avoids complex fabrication processes. The sensor was mounted on an N95 mask and demonstrated the ability to detect respiratory profiles at room and sub-ambient temperatures (−26 °C) with high sensitivity and stability over 500 cycles. It can accurately determine respiratory conditions like eupnea and tachypnea and transmit data wirelessly via Bluetooth. In terms of materials and composition, liquid polyisoprene rubber (LIR) was used as the matrix, and graphene served as the functional filler. Dicumyl peroxide (DCP) was employed as a crosslinker. The composition was mixed using hand blending and a planetary centrifugal mixer to create bubble-free ink. Sensor fabrication encompassed using stencil printing to print sensors on polyethylene terephthalate (PET) substrates. The sensors were cured at 185 °C for 8 minutes, and silver paste was applied as electrodes. Encapsulation with SYLGARD™ 517 gel was performed to eliminate humidity interference. Electrical conductivity has been assessed, showing a percolation threshold at 0.56 vol% of graphene. Dynamic mechanical analysis revealed changes in elastic modulus upon strain application. The sensors demonstrated gauge factors of −196.56 and 117.49 for strain ranges of 0–0.17% and 0.17–0.35%, respectively. Environmental Testing has also been a concern. The sensors were tested for low-temperature performance using liquid nitrogen and humidity sensitivity using a humidifier. Encapsulation successfully eliminated humidity interference, while a Wheatstone bridge configuration compensated for temperature effects. The study also encompassed breath monitoring. Sensors were attached to an N95 mask and tested on a volunteer. The breathing waveform was accurately detected, including respiratory rate, depth, inhalation/exhalation times, and total breathing time. FFT analysis was used to calculate respiratory rates for different breathing patterns. As simple wireless monitoring setup was explored. Using Arduino UNO and a Bluetooth module, the sensors transmitted respiratory data wirelessly to a smartphone, showcasing their applicability in real-time healthcare monitoring.

Smart textiles often employ printing techniques to embed these sensors directly into garments, ensuring minimal interference with workers' movement. Materials such as silver nanowires and graphene composites have demonstrated excellent conductivity and stretchability, ideal for demanding industrial plant settings (Cui, Poblete et al. 2019, You, Lu et al. 2025).

The reviewed literature reveals a growing trend toward using smart textiles for health monitoring in the workplace. For instance, research involving Swiss railway workers under controlled conditions demonstrated successful detection of perceived heat stress and physiological strain through wearable devices (Vázquez, Fujs et al. 2024).

In India, environmental and physiological data collected through embedded sensors enabled machine learning algorithms to predict stress conditions and adjust workloads accordingly (Sharma, Suri et al. 2022). Additionally, large-cohort studies in Egypt employed biosignal sensors to evaluate stress patterns in diverse populations, confirming their scalability and reliability (Abd Al-Alim, Mubarak et al. 2024).

Despite these successes, challenges remain, including the durability of textile-integrated sensors during washing, power management, data privacy, and user acceptability. Addressing these issues is essential for transitioning from experimental setups to real-world deployment.

When comparing the proposed conceptual model with the systems analyzed in the literature, its modularity, adaptability to different work functions and simultaneous integration of physiological and environmental sensors stand out. Unlike previous approaches focusing on controlled laboratory environments or specific applications (such as N95 masks or single sensors), this model aims for uniform integration into conventional workwear, with an emphasis on ergonomics, low cost and simplified maintenance. The combination of BLE and LoRa also allows for a hybrid solution adaptable to different scales of industrial space.

Several studies have investigated not only the technical feasibility but also the real-world adoption of wearable health-monitoring systems. (Sharma, Suri et al. 2022) reported positive reception among factory workers in India, highlighting improved stress management and workload regulation. Similarly, (Vázquez, Fujs et al. 2024) documented effective use of wearables for heat stress detection in Swiss railway workers, with high compliance and minimal discomfort reported. These examples demonstrate growing worker acceptance when systems are integrated seamlessly and with proper training. The proposed model addresses similar needs by emphasizing ergonomic design, real-time feedback, and minimal user intervention, aiming to facilitate successful adoption.

4. PROPOSED TEXTILE-BASED MONITORING SYSTEM

Based on the findings, a conceptual model for the wearable monitoring system is proposed. The system includes:

- **Sensor modules** embedded in key garment areas (chest, back, limbs), designed to monitor heart rate, temperature, respiratory rate, and skin conductance;
- **Flexible interconnections** using conductive threads or printed circuitry to connect the sensors to the processing unit without compromising comfort;
- **A processing unit** composed of a lightweight microcontroller capable of analog-to-digital conversion, basic signal processing, and wireless transmission (e.g., via Bluetooth or LoRa);
- **A mobile or web-based interface** that receives and visualizes data in real time, with functionalities for trend tracking, alerts, and personalized dashboards for occupational health professionals;
- **Machine learning algorithms** embedded in the application or server backend to learn individual baseline physiological profiles and detect anomalies indicative of thermal stress, fatigue, or overexertion.

The suggested system architecture is modular, scalable, and adaptable to different job profiles and environmental situations. It prioritizes low-cost production, extended wearability over time, and simple integration into existing workwear. Battery optimization and washability of sensorized textiles are also addressed in the design for longer lifetimes of the devices and greater user acceptance.

This comprehensive solution combines physiological sensing, real-time analytics, and human factors design to offer an active health monitoring system for workers in factories. The long-term objective of the solution is to help detect noxious conditions at an early stage and allow timely interventions that reduce occupational risk and enhance worker safety.

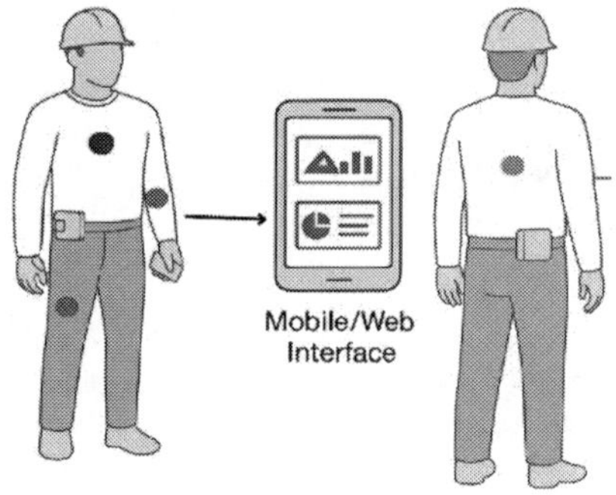

Figure 2. Proposal for the monitoring system with intelligent textiles for occupational health

Figure 2 shows the conceptual design of an occupational health monitoring system using smart textiles. Wearable sensors are incorporated in individual zones of the body, i.e., the chest, back, arms, and legs, to monitor vital physiological parameters like heart rate, body temperature, respiration activity, and joint movement in real time. Data collected is transmitted via Bluetooth or LoRa to a waist-mounted central processing unit, which forwards the information to a digital interface (web or mobile). The interface provides real-time visualization of the vital signs, detection of risk conditions at early stages, and alert generation. The system is modular, ergonomic, and compatible with professional workwear to provide user comfort, mobility, and unobtrusive health monitoring during the working day.

The long-term success of the proposed system depends significantly on its usability. Factors such as the comfort of wear, the ease of donning and doffing the smart garment, and the intuitiveness of the data interface are critical to ensure worker adoption. Preliminary usability guidelines were defined considering ISO 9241 ergonomic standards and ISO 62366 for medical systems usability. We propose involving both workers and occupational health professionals in co-design sessions to identify barriers and expectations. Usability studies will include both qualitative interviews and System Usability Scale (SUS) assessments during pilot testing.

Applying the model in a real environment faces significant challenges. The size of the central processing unit, as well as the weight and positioning of the power source, must ensure worker comfort and mobility.

Integration in the waist area is proposed with a lightweight belt, detachable for washing. The durability of the sensors, resistance to washing and thermal comfort are essential criteria to be validated experimentally. Acceptability of the system by workers also depends on co-design processes, training and continuous involvement of end users.

5. ONGOING EXPERIMENTAL VALIDATION

As part of the ongoing development, preliminary fabrication and integration attempts have been carried out, as shown in Figure 3. One of the first prototypes showed significant limitations in sensor adhesion and signal stability, especially under repeated strain cycles and washing conditions. This sample, although not fully functional, provided critical insights into material compatibility and integration techniques. Current empirical efforts are focused on optimizing the interface between the sensor and textile substrate using thermoplastic adhesives and breathable encapsulants. Further iterations of the sensor will be subjected to electrical characterization, washability tests, and user trials.

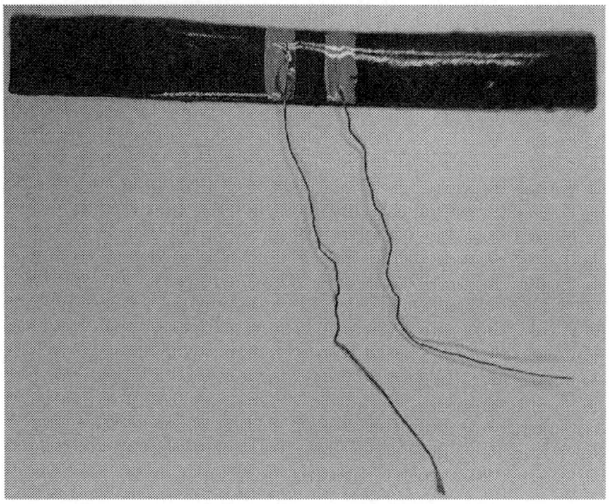

Figure 3. Initial attempt of producing a sensor by mixing carbon nanotubes and reduced graphene oxide in a polydimethylsiloxane matrix

A further step will be to provide a system vision toward the production of an integrated solution. These studies will help us figure out how well the system works technically and how well it can be used in diverse work situations. Usability studies with workers and occupational health professionals are also foreseen to find out how much they like the product, how comfortable it is, and how it affects their work routine. It will be important to set up common evaluation procedures to make sure that the transition to large-scale deployment goes smoothly.

6. CONCLUSION

The integration of sensors in wearables into textiles is a paradigm change in occupational health monitoring. Not only do these technologies enable constant, real-time measurement of physiological parameters but also offer a user-friendly, non-invasive solution for application in numerous diverse industrial workplaces. The application of such systems can significantly reduce the incidence of occupational diseases by identifying early signs of fatigue, heat stress, or physical overloading.

In short, wearable sensors improve worker safety by enabling real-time, continuous monitoring of physiological parameters such as heart rate, respiratory rate, skin temperature, and skin conductance. These sensors, integrated into textiles, provide unobtrusive data collection without compromising mobility or comfort. Key benefits include the following ones:

- Early Detection of Health Risks: Sensors can identify signs of fatigue, heat stress, or physical overexertion, allowing timely interventions to prevent occupational illnesses or accidents.
- Stress Monitoring: By tracking physiological and environmental data, wearable systems can predict stress levels and adjust workloads, accordingly, reducing psychological strain.
- Thermal Stress Management: Sensors detect temperature variations and heat distribution, helping workers avoid heat-related health issues.
- Enhanced Decision-Making: Data visualization and machine learning algorithms provide insights into worker health trends, enabling occupational health professionals to make informed decisions.

The theoretical model outlined in this paper synthesizes the key conclusions of contemporary literature and plots them into an operational system for use. By combining sensor technology, wireless data transfer, and smart algorithms, the proposed solution addresses both employee comfort and technical viability.

But their achievement depends on the solution of a series of technical and human problems. These are ensuring the durability and washability of sensorized clothing, determining the ideal energy consumption for extended use, protecting privacy and confidentiality of collected data, and creating user acceptance among occupational health professionals and workers.

The continuous collection of physiological data in work environments raises relevant ethical issues. It is essential to ensure that the data belongs to the workers and is only used for the purposes communicated in advance. Compliance with the General Data Protection Regulation (GDPR) must be ensured right from the design of the system, with anonymization and access control measures. Transparency and ethical acceptance will also depend on the involvement of ethics committees and the existence of clear information management policies.

Although prototype tests have not yet been carried out, functional tests in simulated industrial environments are planned, followed by validation in partnership with companies in the textile and metallurgical sectors. These studies will make it possible to assess the technical robustness and adaptability of the system to different work contexts. In addition, we intend to carry out usability studies with workers and occupational health professionals, in order to understand the degree of acceptance, comfort and impact on the work routine. To ensure an effective transition to large-scale implementation, it will be essential to define standardized evaluation guidelines. Future research should also be based on interdisciplinary collaborations between engineers, health professionals and industry representatives, to refine these technologies and ensure their responsible adoption in real industrial environments.

The continuous collection and analysis of workers' physiological data introduces significant ethical and legal challenges. Ensuring compliance with the General Data Protection Regulation (GDPR) is a fundamental requirement. This includes obtaining informed consent, ensuring data anonymization, providing transparency on how the data will be used, and implementing strict access control mechanisms. Additionally, data ownership should reside with the individual, and employers must commit to using the data solely for health and safety purposes, not for productivity surveillance or punitive action. Ethics committee involvement, clear governance policies, and ongoing worker engagement will be integrated into the deployment strategy to uphold ethical standards and build trust in the technology.

ACKNOWLEDGEMENT

This work is funded by FCT/MECI through national funds and, when applicable co-funded EU funds under UID/50008: Instituto de Telecomunicações. This research is also supported by COST CA20120 INTERACT - Intelligence-Enabling Radio Communications for Seamless Inclusive Interactions, and by the Portuguese National Distributed Computing Infrastructure (INCD) through the FCT Advanced Computing Projects 2023.10865.CPCA.A1 and https://doi.org/10.54499/2023.10865.CPCA.A1. The authors are also very grateful for the support granted by the Centre for Mechanical and Aerospace Science and Technologies (C-MAST-UBI), through the Project reference UIDB/00151/2020 (DOI: 10.54499/UIDB/00151/2020) and UIDP/00151/2020 (DOI: 10.54499/UIDP/00151/2020), funded by the Fundação para a Ciência e a Tecnologia, IP/MECI through national funds (PIDDAC). João Miguel Nunes Parente thanks FCT/MECI for the grant UI/BD/151477/2021 (DOI: 10.54499/UI/BD/151477/2021), and João Nunes-Pereira for the contract under the Stimulus of Scientific Employment, Individual Support: 2022.05613.CEECIND (DOI: 10.54499/2022.05613.CEECIND/CP1746/CT0001).

REFERENCES

Abd Al-Alim, M., R. Mubarak, N. M. Salem and I. Sadek (2024). "A machine-learning approach for stress detection using wearable sensors in free-living environments." Computers in Biology and Medicine 179: 108918.

Cui, Z., F. R. Poblete and Y. Zhu (2019). "Tailoring the Temperature Coefficient of Resistance of Silver Nanowire Nanocomposites and their Application as Stretchable Temperature Sensors." ACS Applied Materials & Interfaces 11(19): 17836-17842.

Pinnelli, M., D. Lo Presti, S. Silvestri, R. Setola, E. Schena and C. Massaroni (2024). "Towards the Instrumentation of Facemasks Used as Personal Protective Equipment for Unobtrusive Breathing Monitoring of Workers." Sensors (Basel) 24(17).

Selvan T, M., S. Sharma, S. Naskar, S. Mondal, M. Kaushal and T. Mondal (2022). "Printable Carbon Nanotube-Liquid Elastomer-Based Multifunctional Adhesive Sensors for Monitoring Physiological Parameters." ACS Applied Materials & Interfaces 14(40): 45921-45933.

Sharma, M., N. M. Suri and S. Kant (2022). "Analyzing occupational heat stress using sensor-based monitoring: a wearable approach with environmental ergonomics perspective." Int J Environ Sci Technol (Tehran) 19(11): 11421-11434.

Sharma, S., A. Thapa, S. Singh and T. Mondal (2024). "Crosstalk-free graphene–liquid elastomer based printed sensors for unobtrusive respiratory monitoring." Nanoscale 16(7): 3498-3509.

Vázquez, C. G., M. Fujs, M. F. Koller, P. Wolf and G. Da Poian (2024). "Beat the heat: wearable-based study of perceived heat stress and physiological strain in swiss track workers in a controlled climate chamber." Physiological Measurement 45(12): 125001.

You, J., M. Lu, L. Dazhen, M. Gao, R. Zhang, W. Li, F. Lei, W. Ren, G. Li and J. Yang (2025). "Anti-Motion Artifacts Iontronic Sensor for Long-Term Accurate Fingertip Pulse Monitoring." Adv Sci (Weinh) 12(15): e2414425.

DEVELOPMENT OF A DIGITAL TOOL FOR SHOWING TEST RESULTS FROM CERVICAL CANCER SCREENING

Jiaxin Li, Ophelia Prillard, Sandra Klonteig and Volker Hoffmann
SINTEF
Forskningsveien 1, 0373, Oslo, Norway

ABSTRACT

Cervical cancer screening is an effective way of detecting and preventing precancer from further developing. Especially when it´s done in the recommended frequency. However, the constant evolving screening guidelines make it difficult for both the doctors and screening participants to interpret test results and decide on the follow-up procedures. Not to mention that screening participants have limited access to their test results. Therefore, built upon the user requirements we collected from co-design workshops earlier, we have developed a digital solution aiming for enhancing transparency and accessibility. The prototype has 4 main features: (i) Latest screening result; (ii) Screening over-view; (iii) My data; and (iv) Information page. The wireframe was programmed into web application with assistance from AI tools.
We discussed the design challenges while developing the digital solution and concluded with a few design recommendations: (i) Balanced amount of information; (ii) Personalized risk communication; (iii) Provide reliable and accessible information. We have also identified some advantages and disadvantages of using AI tools to assist the coding process. Future work should involve a diverse sample of end users for testing the effectiveness and usability of the prototype as well as further investigating the use of AI tools to enhance code development.

KEYWORDS

Cervical Cancer Screening, Digital Tool Development, User Interface Design, Health Information Visualization, AI-Assisted Web Application

1. INTRODUCTION

Almost all cervical cancer cases are caused by persistent infection with high-risk human papillomavirus (HPV) subtypes (Cohen *et al.*, 2019). Screening is an effective way to detect and prevent precancerous changes. The Cancer Registry of Norway (CRN) runs CervicalScreen Norway (referred to as "the screening program" hereafter), targeting women aged 25–69 (Bjørge *et al.*, 2022). The main screening methods are the Pap test and HPV test, with practices updated based on research. Since July 2023, HPV testing has become the primary method for all women in this age group (*Primær HPV-test*, no date). However, recommendations can vary based on test history and new findings, making it challenging for participants and doctors to determine the optimal testing interval. This uncertainty can lead to testing either too frequently (Klungsøyr *et al.*, 2009) or too rarely (Anderssen, 2021), both of which undermine the program's effectiveness.

Although test results are expected to be communicated by the prescriber (typically the general practitioner), normal results are often not conveyed, leaving participants waiting anxiously for confirmation. When results are shared, there is no standardized communication method, as the choice is left to each prescriber despite CRN guidelines (*Informasjonsmateriell til leger*, 2024). Studies show that patients struggle to interpret their results, with unmet information needs (Monsonego *et al.*, 2011; Siegel, 2022; Klonteig, Li and Halvorsrud, 2023), and that better communication could reduce anxiety and improve compliance (Papa *et al.*, 2009; Markovic-Denic *et al.*, 2018).

Currently, screening participants in Norway have limited access to their test history, which requires a lengthy retrieval process. While Helsenorge provides public health services (*About Helsenorge*, 2023), it does not yet deliver cervical screening results or history.

To enhance participation and the effectiveness of the screening program, participants must have feasible access to their results and histories. We now present the development of a digital tool designed to (i) communicate test results, (ii) present test history clearly, and (iii) offer decision support based on the results.

Section 2 describes the development process from user requirements to a functional web application. Section 3 presents the digital tool, its features, and an early evaluation. Section 4 discusses design challenges, the use of AI in coding, limitations, and future work. Finally, Section 5 concludes with design recommendations and insights on using AI tools in development.

2. DEVELOPMENT OF THE DIGITAL TOOL

2.1 From User Requirements to Wireframe

A user-centered design approach (Salinas, Cueva and Paz, 2020) was applied in developing the digital tool. User requirements were collected from co-design workshops with screening participants (Klonteig, Li and Halvorsrud, 2023) and general practitioners, and unsystematic searches in the literature (Monsonego *et al.*, 2011; Marlow *et al.*, 2020; Siegel, 2022). In the workshops, several information needs and user requirements for a digital solution were identified. Information needs include the following:

- Screening methods: Explanation of the difference between HPV testing and Pap smear testing, and the reasons for selecting one or both.
- Screening algorithm: Clarification of why the interval between screenings may vary.
- HPV virus and genotypes: Description of common HPV genotypes and their association with cancer risk.
- Cervical cancer: Information on how long it typically takes for cervical cancer to develop and its potential symptoms.

The main user requirements are summarized as follows:

- Data Access: Users should easily access test results and history, preferably via an existing Health Portal.
- Clear Communication: Show the latest results first, use simple language, explain medical terms, and offer more details for abnormal results.
- Decision Support: Provide next steps based on results and contact options (chatbot or doctor).
- Notifications Preference: Allow users to track test status and choose reminder and communication preferences (SMS or email).
- Reliability: Link to trusted sources to prevent misinformation.

These user requirements were translated into four key features: (i) Latest screening result; (ii) Screening overview; (iii) My data; and (iv) Information page. The wireframe was designed using a wireframing tool called Figma. Furthermore, five use-cases with different test results and test history were identified and implemented in the web application.

The wireframe has since the co-design workshops gone through several iterations, based on feedback from experts from CRN, internal healthcare and design experts from SINTEF, as well as from our project group.

2.2 From Wireframe to Web Application

The final wireframe was translated into a functional web application with a frontend and backend. The frontend, built with TypeScript, ReactJS, Vite, TailwindCSS, and DaisyUI, supports flexible layouts from mobile to desktop. The backend, developed with FastAPI, manages the five use cases and aligns the data model with CRN's metadata[1] for future live data compatibility.

[1] https://metadata.kreftregisteret.no/help

To accelerate development, OpenAI's GPT-4o was used to generate the frontend code. Screenshots of the wireframe were provided with brief instructions, followed by iterative refinements. Once initial screens were completed, GPT-4o replicated the process for the remaining screens, with additional prompts for specific UI elements when needed. This approach quickly produced a working application closely aligned with the original design.

3. THE DIGITAL TOOL

3.1 The Digital Tool

The design of the digital tool employs intuitive navigation, and minimal medical jargon to ensure user-friendliness and accessibility. When opening the web application, user will reach the home page where 4 features were listed, as shown in Figure 1(a). It was intended to give user access to the information page only before login. The 4 key functionalities are listed as follows:

- Latest result: This page first and foremost communicates the most recent test result. It further aims to meet the patients' information needs and user requirements by communicating the specific test result in a clear but concise manner by highlighting the result (e.g.HPV not found), recommending a time for the upcoming test, organizing information into different categories, and putting detailed information behind a further click at the dropdown icon.
- Screening overview: Test result history and upcoming test are structured in reverse chronological order, with the latest test highlighted for quick access. The screening overview is visualized in a vertical timeline, commonly used to demonstrate event evolution over time (Scheer *et al.*, 2022; Yeshchenko and Mendling, 2024). Specifically, a scaled timeline was used to indicate the difference of time interval between two test results, with the possibility to discover pattern ('Chart Combinations: Overlaying Extra Data Onto Timelines', 2017).
- My data: The diagram visualize who can access user´s screening data to promote transparency. It al-so provides the user with data sharing choices.
- Information page: This feature tries to directly address the information needs identified in the co-design workshops by providing reliable and general information from the official website of CRN, reducing misinformation risks.

Front-end and back-end can be accessed with the following links: front-end: https://vismeg.sintef.cloud/, and back-end: https://vismeg-backend.sintef.cloud/docs.

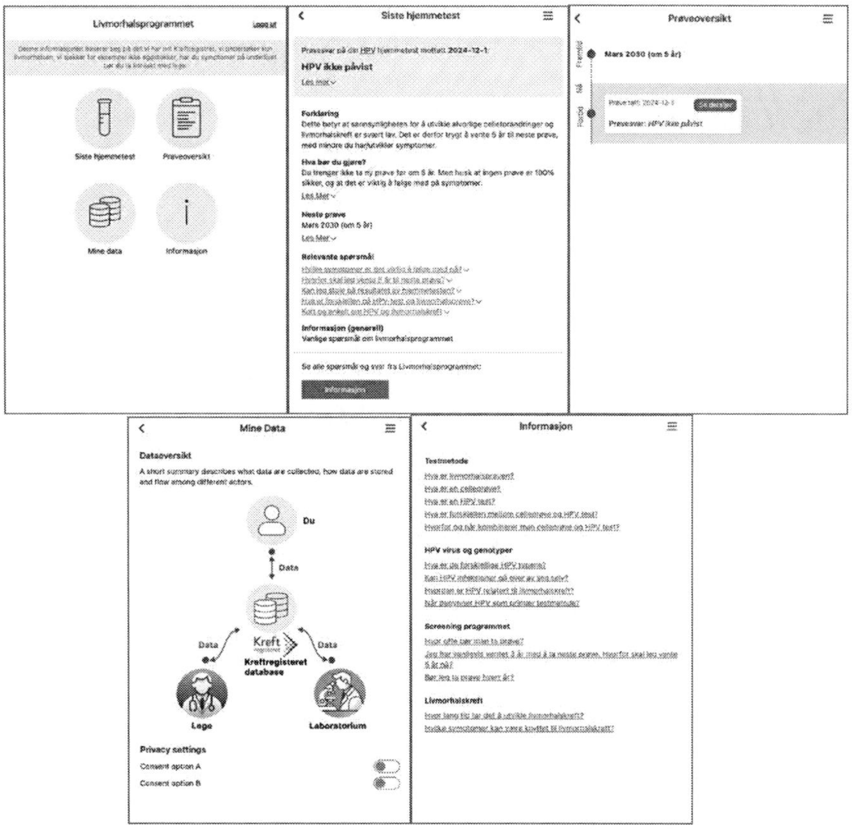

Figure 1. Screenshots of the digital tool, from left to right are (a) Home Page, (b) Latest Screening Result, (c) Screening Overview, (d) My Data, and (e) Information Page

3.2 Early Evaluations and Feedback

The screenshots shown in Figure 1 represent the latest version of the digital tool. Shortly after the design workshops, the first version was shown to general practitioners in two separate workshops. The doctors responded positively to the tool's potential to deliver test results and screening history, but raised concerns about wording (e.g., the term "cancer"), color choices (e.g., red), icons, and the risk of overwhelming patients with too much information (Hoffmann *et al.*, 2025).

Based on their feedback, the wireframe was revised and presented to domain experts from CRN. Discussions focused on wording and visual design, leading to changes such as removing test type and interval de-tails and renaming the “Screening History” page to “Screening Overview,” with the timeline split into past, present, and future.

Following further revisions, the updated wireframe was reviewed by design experts and health researchers from SINTEF. They recommended exploring non-text-based communication methods, using the waiting period to educate users, and improving risk communication.

4. DISCUSSION

4.1 Design Considerations and Challenges

There are several challenges during the design process of the digital tool.

4.1.1 Prioritizing Information Through Progressive Disclosure

The main challenge is delivering results in clear, easy-to-understand manner, while not overloading the user. We aimed for a good balance between providing the most important information while not overwhelming the user. Therefore, it´s important to decide on the priority of information, the more important it is, the more "visible" it should be. In the UX design field, gradual disclosure, or progressive disclosure (Spillers, 2004; Nielsen, 2006), has been a common practice to hide less important information or features in a secondary screen, to make the interface more user-friendly for novice users, while at the same time allows certain users to explore more advanced features. An example of how this was applied on our design is the page of the latest test result, where we only show a short summary of each category at first sight, and users need to click on the expand icon to view extra information (as shown in Figure 1 (b)).

4.1.2 Balancing Risk Communication and Emotional Impact

Risk communication is a critical aspect of delivering test results. Disclosing risk levels appropriately is challenging, as it must encourage users to act without causing unnecessary stress or anxiety. This is particularly important given that many screening participants either test too frequently or too infrequently compared to recommended intervals.

Various strategies for communicating risk were considered, including color coding test results (Shrivastava, Shrivastava and Ramasamy, 2014). Test results were categorized into four types, with severity ranked as (b) < (c) < (d), as shown in the table below. We initially used green for (b) (normal/safe), yellow for (c) (later changed to blue for a calmer feeling), orange for (d) (warning/danger), and grey for (a) (invalid or unavailable) (Braun, Mine and Clayton Silver, 1995; Oettinger *et al.*, 2009; *Material Design*, no date). These colors were primarily applied in the Screening overview page (as shown in Figure 1 (c)).

Table 1. Test Results and their color representatives

Test results	Color coding 1	Color coding 2
(a) Invalid	Grey	Grey
(b) HPV virus not found	Green	Green
(c) HPV virus found (not type 16)	Yellow	Blue
(d) HPV virus found	Orange	Orange

Although several participants expressed a desire to include risk-levels associated with a given test result, color coding or other risk communication formats were not used in the final design due to a concern that it will cause unnecessary stress for case (c) and (d).

4.1.3 Enhancing Patient Understanding Through Accessible Information

Research have found that increasing women´s knowledge of cervical cancer screening can help reduce stress and anxiety (Papa *et al.*, 2009; Markovic-Denic *et al.*, 2018). Our findings also revealed that many women had unmet information needs regarding primary screening methods, the screening algorithm, HPV viruses and genotypes, and cervical cancer itself (Klonteig, Li and Halvorsrud, 2023). To address this, we included an information page offering details about testing methods, HPV, the screening program, and cervical cancer, based on content from the CRN official website. Additionally, we integrated a section with frequently asked questions tailored to specific test results directly on the "Latest Screening Result" page, allowing users to easily find relevant information without having to search online.

4.2 Lessons Learned using GPT-4o for Design Replication

We identified several challenges and opportunities when using GPT-4o to generate front-end code out of the wireframe. Firstly, using GPT-4o can streamline the cumbersome and time-consuming process of translating complex UI designs into code. Manually implementing intricate wireframe design is often tedious and highly resource demanding. In practice, teams often face trade-off between time, effort, and design replicability and quality, leading to partial or simplified implementations of designs (Feng *et al.*, 2023). GPT-4o can help bridge this gap by generating code that rapidly and accurately reflect intricate designs, while maintaining a component-based structure with integrated UI components. Unlike Figma's own plugin that does not recognize UI elements and structure the code (Lin *et al.*, 2023). This is particularly useful for rapidly creating high-fidelity prototypes, where users still expect high-quality visuals, and a poor design can negatively impact the evaluation of the prototype (Anastassova, Mégard and Burkhardt, 2007; Sauer and Sonderegger, 2009). Another advantage is GPT-4o's ability to generate complete, testable React components along with coding explanations which makes it easy to quickly test the design and code within the system.

While using AI-models such as GPT-4o can accelerate development, sole reliance for code and interface generation has drawbacks. One key challenge is that the model does not consider the larger context of the code-base and problem-definition. As such, modifying the code becomes more time consuming. Additionally, the code generated for complex design elements was often highly specified, prioritizing exact visual replication over flexibility and readability. As a result, debugging errors, implementing interaction, or altering design elements may require significant effort. Another crucial point is interaction. While GPT-4o can infer basic interactivity cues - such as underlines indicating links - it often lacks sufficient context to fully implement functional UI behavior. Providing multiple frames improved its understanding of the high-level interaction flow, but additional manual adjustments were usually required to ensure proper functionality.

Additionally, we experienced that GPT-4o does not inherently use external libraries, which can be both beneficial and limiting (Constantinou and Stamelos, 2017). While avoiding dependencies eliminates the need to maintain libraries, it also results in more verbose and rigid code. For instance, instead of using icon libraries or component frameworks, GPT-4o would generate hardcoded SVGs and custom elements, resulting in messier and less maintainable code. However, this could be improved through more precise prompt engineering or by explicitly specifying preferred libraries.

4.3 Limitation and Future Work

Our work has a few limitations as followed: (i) Small sample size in the user requirement collecting process; (ii) Lack of user validation on the digital tool.

Firstly, the user requirements that were used in the development of the prototype were collected from a relatively small sample that is not representative. The sample mainly consists of women from the same re-search organization as the authors, whose needs might not represent the needs from women with other backgrounds. This can potentially have negative impact on the relevance and usefulness of the features in the prototype. Secondly, since the prototype hasn´t been through a formal validation with the end users, the usability and user satisfaction of the prototype is purely theoretical. The lack of direct user involvement in validating the prototype can result in usability issues uncovered and some user needs unaddressed. However, we have mitigated these challenges by involving general practitioners, health experts and design experts. The prototype has recently been tested with 11 screening participants and the results were still being analyzed.

Future work should further develop the prototype based on the feedback from user testing. We would also like to investigate more into the use of AI tools in assisting the coding process. For example, to explore the possibility to enhance the flexibility of the generated code through prompt engineering, and how AI tools can understand the interaction between different interfaces (e.g. by processing multiple screens of the wireframe simultaneously) in a single query to generate multiple React components.

5. CONCLUSION

We have developed a prototype to show the test results from cervical cancer screening to screening participants in Norway. The prototype has 4 main features: (i) Latest screening result; (ii) Screening overview; (iii) My data; and (iv) Information page. The wireframe was first developed with several design considerations in mind. And these suggestions are not just relevant for communicating cervical cancer screening results, but also beneficial for other digital communication platform for other medical test results.

- Prioritizing Information Through Progressive Disclosure: only show the most important messages at first glance and gradually disclose more information if needed.
- Balancing Risk Communication and Emotional Impact: there are different ways of communicating the test results and risk levels, which can have different emotional impact on different people. Make sure the communication methods have been tested with both users and health experts before release.
- Provide reliable and accessible information along the results, this can help them better understand the results and preparing for them.

The wireframe was then translated into a functional web application with the assistance of Open AI's GPT-4o model, which significantly reduced the time required for development. While using these tools offer considerable potential for accelerating prototyping, they also present certain advantages and limitations, summarized in the table below. Future work should involve a diverse sample of end users to test the effectiveness and usability of the prototype, as well as further explore strategies to mitigate the disadvantages of using AI tools in the process of converting designs into high-fidelity prototypes and, ultimately into commercial products.

Table 2. Advantages and disadvantages of using AI tools in coding

Advantages	Disadvantages
Quickly converts complex high-fidelity wireframe into usable code	Harder to understand or modify if you're not involved in writing the code
Capable of generating complete components (e.g., React)	AI generated code may be rigid and overly focused on visual accuracy
Eliminates need for some external dependencies	Poor handling of interactivity without context
	Avoiding libraries leads to bulkier, less maintainable code

ACKNOWLEDGEMENT

This work was (co-)funded by the Norwegian Research Council (project VisMeg), ES/321081), the Norwegian Cancer Registry, and the Norwegian Cancer Society.

DISCLOSURE OF INTERESTS

The authors have no competing interests to declare that are relevant to the content of this article.

REFERENCES

About Helsenorge (2023). Available at: https://www.helsenorge.no/en/about-helsenorge/ (Accessed: 6 March 2025).

Anastassova, M., Mégard, C. and Burkhardt, J.-M. (2007) 'Prototype Evaluation and User-Needs Analysis in the Early Design of Emerging Technologies', in J.A. Jacko (ed.) Human-Computer Interaction. Interaction Design and Usability. Berlin, Heidelberg: Springer, pp. 383–392. Available at: https://doi.org/10.1007/978-3-540-73105-4_42.

Anderssen, H. (2021) Kvinner som ikke har testet seg for livmorhalskreft på 8-10 år får tilbud om hjemmetest. Available at: https://www.healthtalk.no/alle-artikler/onkologi-tilbud-om-hjemmetest-for-livmorhalskreft-til-kvinner-som-ikke-har-testet-seg-på-8-10-år/ (Accessed: 1 July 2022).

Bjørge, T. et al. (2022) 'CervicalScreen Norway – A screening programme in transition', Norsk Epidemiologi, 30(1–2). Available at: https://doi.org/10.5324/nje.v30i1-2.4978.

Braun, C.C., Mine, P.B. and Clayton Silver, N. (1995) 'The influence of color on warning label perceptions', International Journal of Industrial Ergonomics, 15(3), pp. 179–187. Available at: https://doi.org/10.1016/0169-8141(94)00036-3.

'Chart Combinations: Overlaying Extra Data Onto Timelines' (2017) The Data Visualisation Catalogue Blog, 22 December. Available at: https://datavizcatalogue.com/blog/chart-combinations-timelines/ (Accessed: 18 April 2023).

Cohen, P.A. et al. (2019) 'Cervical cancer', Lancet (London, England), 393(10167), pp. 169–182. Available at: https://doi.org/10.1016/S0140-6736(18)32470-X.

Constantinou, E. and Stamelos, I. (2017) 'Identifying evolution patterns: a metrics-based approach for external library reuse', Software: Practice and Experience, 47(7), pp. 1027–1039. Available at: https://doi.org/10.1002/spe.2484.

Feng, S. et al. (2023) 'Designing with Language: Wireframing UI Design Intent with Generative Large Language Models'. arXiv. Available at: https://doi.org/10.48550/arXiv.2312.07755.

Hoffmann, V. et al. (2025) 'Perspectives, Dilemmas, and Guidelines in Patient-Centered Cervical Screening Communication', in. 2025 IEEE International Conference on Digital Health (ICDH), Helsinki, Finland. doi: In Press.

Informasjonsmateriell til leger (2024). Available at: https://www.kreftregisteret.no/screening/livmorhalsprogrammet/Helsepersonell/informasjonsmateriell-til-leger/ (Accessed: 28 April 2025).

Klonteig, S., Li, J. and Halvorsrud, R. (2023) '"My Heart Jumped. Do I Have Cancer?"—Results of a Co-design Study with Cervical Cancer Screening Participants', in M.A. Pfannstiel (ed.) Human-Centered Service Design for Healthcare Transformation: Development, Innovation, Change. Cham: Springer International Publishing, pp. 329–351. Available at: https://doi.org/10.1007/978-3-031-20168-4_19.

Klungsøyr, O. et al. (2009) 'Validity of Self-reported Pap Smear History in Norwegian Women', Journal of Medical Screening, 16(2), pp. 91–97. Available at: https://doi.org/10.1258/jms.2009.008087.

Lin, Z. et al. (2023) 'Figma2Code: Automatic Code Generation Method for Figma Design Drafts', Journal of Computer-Aided Design & Computer Graphics, 37(2), pp. 321–329. Available at: https://doi.org/10.3724/SP.J.1089.2023-00336.

Markovic-Denic, L. et al. (2018) 'Effects of Human Papillomavirus Awareness and Knowledge on Psychological State of Women Referred to Cervical Cancer Screening', Journal of Lower Genital Tract Disease, 22(3), pp. 178–183. Available at: https://doi.org/10.1097/LGT.0000000000000397.

Marlow, L. et al. (2020) 'Information needs among women taking part in primary HPV screening in England: a content analysis', BMJ Open, 10(12), p. e044630. Available at: https://doi.org/10.1136/bmjopen-2020-044630.

Material Design (no date) Material Design. Available at: https://m2.material.io/design/interaction/states.html#usage (Accessed: 28 April 2025).

Monsonego, J. et al. (2011) 'Psychological impact, support and information needs for women with an abnormal Pap smear: Comparative results of a questionnaire in three European countries', BMC women's health, 11, p. 18. Available at: https://doi.org/10.1186/1472-6874-11-18.

Nielsen, J. (2006) Progressive Disclosure, Nielsen Norman Group. Available at: https://www.nngroup.com/articles/progressive-disclosure/ (Accessed: 28 April 2025).

Oettinger, M.D. et al. (2009) 'Color-Coding Improves Parental Understanding of Body Mass Index Charting', Academic Pediatrics, 9(5), pp. 330–338. Available at: https://doi.org/10.1016/j.acap.2009.05.028.

Papa, D. et al. (2009) 'Assessing the role of education in women's knowledge and acceptance of adjunct high-risk human Papillomavirus testing for cervical cancer screening', Journal of Lower Genital Tract Disease, 13(2), pp. 66–71. Available at: https://doi.org/10.1097/LGT.0b013e31818a53f0.

Primær HPV-test - nå også for kvinner under 34 år (no date). Available at: https://www.kreftregisteret.no/screening/livmorhalsprogrammet/hpv/ (Accessed: 10 March 2025).

Salinas, E., Cueva, R. and Paz, F. (2020) 'A Systematic Review of User-Centered Design Techniques', in A. Marcus and E. Rosenzweig (eds) Design, User Experience, and Usability. Interaction Design. Cham: Springer International Publishing, pp. 253–267. Available at: https://doi.org/10.1007/978-3-030-49713-2_18.

Sauer, J. and Sonderegger, A. (2009) 'The influence of prototype fidelity and aesthetics of design in usability tests: Effects on user behaviour, subjective evaluation and emotion', Applied Ergonomics, 40(4), pp. 670–677. Available at: https://doi.org/10.1016/j.apergo.2008.06.006.

Scheer, Jan et al. (2022) 'Visualization Techniques of Time-Oriented Data for the Comparison of Single Patients With Multiple Patients or Cohorts: Scoping Review'. (J Med Internet Res 2022;24(10):e38041). Available at: https://doi.org/10.2196/38041.

Shrivastava, S.R., Shrivastava, P.S. and Ramasamy, J. (2014) 'Color coding: a tool to enhance the quality of health care in low resource settings', Healthcare in Low-resource Settings, 2(2). Available at: https://doi.org/10.4081/hls.2014.4772.

Siegel, L. (2022) Under Kontroll. The Oslo school of Architecture and Design.

Spillers, F. (2004) 'What is Progressive Disclosure? - ED', Frank Spillers, 25 March. Available at: https://frankspillers.com/progressive-disclosure-the-best-interaction-design-technique/ (Accessed: 28 April 2025).

Yeshchenko, A. and Mendling, J. (2024) 'A survey of approaches for event sequence analysis and visualization', Information Systems, 120, p. 102283. Available at: https://doi.org/10.1016/j.is.2023.102283.

QUALITY REQUIREMENTS FOR LARGE LANGUAGE MODELS IN CLINICAL SETTINGS: A SCOPING REVIEW

Theresa Thien Thanh Do Huynh, Uffe Kock Wiil and Ali Ebrahimi
SDU Health Informatics and Technology, The Maersk Mc-Kinney Moller Institute, University of Southern Denmark, Odense, Denmark

ABSTRACT

This scoping review investigates the current quality requirements for implementing large language models (LLMs) in clinical settings and evaluates the methods used to assess their performance. The review follows the Joanna Briggs Institute (JBI) methodology and the PRISMA-ScR guidelines. A systematic search is conducted across PubMed, Scopus, and Web of Science for studies published between January 2022 and October 2024. A total of twelve studies meets the inclusion criteria, and their findings are analyzed using structured data charting tables. Key findings highlight that diagnostic accuracy is the most used metric for evaluating LLM performance, alongside sensitivity, specificity, and adherence to clinical guidelines. Identified challenges include data biases, ethical concerns such as privacy risks, and issues related to model transparency, all of which complicate the integration of LLMs into clinical workflows. The review emphasizes the need for standardized evaluation frameworks to enable consistent, reliable comparisons of LLM performance across studies and healthcare contexts. Addressing ethical challenges, reducing bias, and improving real-world validation will ensure LLMs' safe and effective integration into healthcare practice.

KEYWORDS

Large Language Models, Quality Requirements, Healthcare Integration

1. INTRODUCTION

Recent innovations in healthcare have increasingly focused on generative Artificial Intelligence (AI) models, notably large language models (LLMs) trained on electronic health record (EHR) data. LLMs are built on multi-layer neural networks trained on a large amount of data and have quickly gained global popularity, especially since OpenAI launched ChatGPT in November 2022. These models can simulate human-like conversation, answer questions, summarize information, translate text, and even create stories (Raza et al., 2024; Clusmann et al., 2023). In healthcare, LLMs have already demonstrated their potential in writing research articles and interpreting complex EMR data, including patient notes, lab results, and billing codes. These models could improve predictive accuracy, simplify model development with fewer labeled data requirements, and reduce deployment costs (Raza et al., 2024). They offer promising capabilities in answering patient inquiries, summarizing medical research, and providing explanations in simpler language (Clusmann et al., 2023).

However, some challenges must be addressed before LLMs can be safely and effectively used in clinical practice. Concerns such as misinformation, privacy issues, biases in training data, and potential misuse require careful consideration (Clusmann et al., 2023). This scoping review systematically maps existing research on LLM implementation in clinical settings, with an emphasis on determining the quality requirements for its usage in healthcare. Existing research often focuses on specific applications of LLMs but lacks a comprehensive understanding of the quality requirements for effective implementation in healthcare. This gap in knowledge highlights the need for a scoping review to map existing evidence and identify key themes and challenges. The primary research question guiding this review is: *What quality requirements have been identified in recent studies for implementing LLMs in clinical settings?*

This review also explores the common evaluation methods and metrics used to assess LLMs' performance, reliability, and safety in healthcare contexts. Furthermore, it examines challenges such as data bias, ethical concerns, and the broader impact of LLMs on clinical workflow. By synthesizing evidence across these aspects, the review aims to provide a comprehensive overview of current research and highlights areas for further investigation.

2. METHODS

We used the scoping review methodology from the Joanna Briggs Institute (JBI), following five stages: (1) defining the research question, (2) conducting a comprehensive search for relevant studies, (3) selecting studies based on inclusion and exclusion criteria, (4) charting extracted data, and (5) summarizing and reporting the findings. This iterative, interpretative approach emphasizes relevance and credibility of evidence, suited for diverse sources, including qualitative studies.

2.1 Identifying Research Question and Relevant Studies

This review explores how recent research has addressed the quality requirements for implementing LLMs in clinical settings. It examines the scope, range, and methodologies of existing studies, with a particular focus on how LLMs are evaluated for accuracy, reliability, and safety, as well as how ethical considerations, such as bias, fairness, and patient safety, are addressed in healthcare applications. Therefore, the following research questions will be addressed:

1. *What requirements have been identified in recent research for implementing LLMs in clinical settings?*
2. *How are LLMs assessed for accuracy, reliability, and safety in healthcare?*
3. *What evaluation methods and metrics are most used to evaluate LLMs, and in what contexts?*
4. *What challenges and limitations are reported in the evaluations?*
5. *How are ethical issues such as bias, fairness, and patient safety addressed in these studies?*

We used systematic search strategies across multiple databases, applying MeSH and Boolean operators to refine terms: ("Large language models" OR "LLMs" OR "Generative AI" OR "Clinical AI" OR "chatbot" OR "ChatGPT" OR "Llama" OR "CoPilot") AND ("quality requirements" OR "safety" OR "effectiveness" OR "validation" OR "accuracy" OR "reliability" OR "risk management") AND ("hospital*").The databases PubMed, Scopus, and Web of Science were searched and included peer-reviewed articles in English published since 2022. Building on this foundation, the search strategies were drafted and refined in collaboration with an experienced librarian from the University of Southern Denmark.

Table 1. List of the Inclusion and Exclusion Criteria

Inclusion Criteria
1. The study's main goal must be evaluating LLMs in clinical settings.
2. The study should assess LLMs for diagnostic support, discharge summaries, clinical decision-making
3. The research must involve clinicians, clinical administrators, or patients.
4. LLMs must be validated using expert or gold-standard benchmarks.
5. The article must be a peer-reviewed publication in English between January 2022 and October 2024.
6. Preference should be given to studies on newer LLM versions (e.g., GPT-4) if many are available.
Exclusion Criteria
1. Studies in non-clinical settings or without clinical personnel or direct workflows (e.g., telehealth, predictive scoring).
2. Theoretical or background studies without practical and quantitative evaluation.
3. Studies focusing solely on a single clinical task without broader quality evaluation.
4. Studies on outdated LLM models with no relevance to current clinical practice.
5. Studies for which the full text is not available.

2.2 Data Screening, Selection, and Charting

This search resulted in 460 references, which were exported to the reference management tool EndNote for organization and deduplication. The remaining references were then systematically evaluated, beginning with title screening, followed by iterative abstract reviews using increasingly stringent exclusion criteria, and concluding with full-text assessments to identify studies of potential relevance. In cases of uncertainty, a second reviewer was consulted to assist in making final decisions. Table 1 provides a detailed overview of the criteria used for study selection. Figure 1, the PRISMA Flow Diagram was utilized to provides an overview of the selection process, including the number of studies screened, excluded, and included in the review.

Data charting tables were created to extract key information from articles aligned with the research objectives. The form was refined into three tables (2, 3, and 4) summarizing study references, design, objectives, examined technology or LLM, context, quality requirements, evaluation methods, findings, challenges, limitations, ethical considerations, data details, and availability.

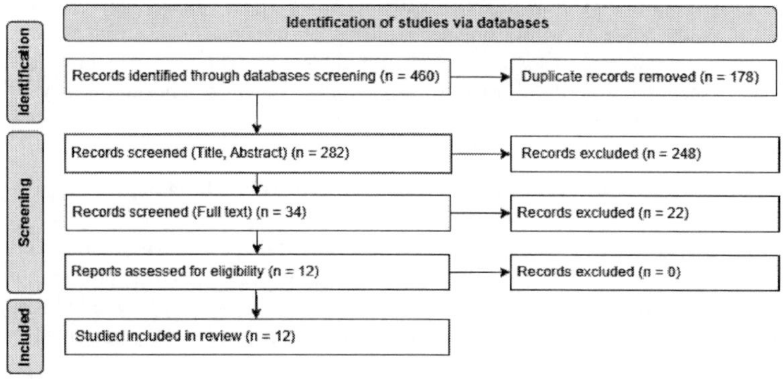

Figure 1. PRISMA Flow Diagram

3. RESULTS

The screening process identified twelve studies meeting the inclusion criteria, shown in the PRISMA flow diagram (Figure 1). Study characteristics are summarized in Tables 2 and 3, covering various designs, such as observational (Cho et al., 2024; Garcia et al., 2024), feasibility (Clough et al., 2023), comparative (Masanneck et al., 2024), and blinded evaluations (Guo et al., 2023; Yi et al., 2024). Key findings are organized into Diagnostic Support, Administrative Tasks, and Specialized Clinical Applications. Table 4 details the data types and availability, with datasets ranging from discharge summaries and EHRs to simulated data, some restricted due to privacy concerns (e.g., Cho et al., 2024) and others openly accessible (e.g., Sarvari et al., 2024). Several studies provided supplementary materials to enhance transparency (Guo et al., 2023; Matsler et al., 2024; Yi et al., 2024).

3.1 Diagnostic Support

Several studies explored LLMs for diagnostic support in emergency medicine (Hoppe et al., 2024; Masanneck et al., 2024), oncology (Cho et al., 2024), and other clinical settings (Levin et al., 2024; Sarvari et al., 2024), focusing on diagnostic accuracy, sensitivity, and specificity. In emergency medicine, Hoppe et al. (2024) found GPT-4 outperformed ED residents in accuracy, though bias and sample size concerns remained. In oncology, Cho et al. (2024) reported high accuracy in lung cancer staging but noted challenges like class imbalance. Sarvari et al. (2024) found GPT-4 had high sensitivity (93.9%) in comorbidity diagnosis from EHRs, despite integration and bias issues. Levin et al. (2024) showed LLMs like GPT-4 were faster than nurses in neonatal care, but less accurate, needing further improvement. Masanneck et al. (2024) reported GPT-4's substantial agreement in ED triage, though over-triage and limited generalizability were challenges. These findings highlight the potential of LLMs like GPT-4 in diagnostics, with challenges in integration, bias, and real-world validation.

3.2 Administrative Tasks

LLMs were explored for automating administrative tasks like discharge summary generation (Clough et al., 2023; Yi et al., 2024) and patient messaging (Garcia et al., 2024), focusing on usability, accuracy, and time efficiency. Clough et al. (2023) and Yi et al. (2024) found ChatGPT generated high-quality discharge summaries, outperforming junior clinicians without errors, but these findings were based on simulated records, requiring real-world validation. Garcia et al. (2024) reported that a HIPAA-compliant LLM improved patient messaging efficiency, reducing clinician burnout, though the study had a short duration and lacked reference standards. These results suggest LLMs could reduce clinician workload and improve efficiency, but real-world validation and integration remain challenges.

3.3 Specialized Clinical Applications

Several studies explored LLMs in specialized clinical contexts, showing their adaptability beyond general diagnostics. In chronic hepatitis C management, Kresevic et al. (2024) found GPT-4 Turbo and ChatGPT-3.5 improved adherence to clinical guidelines, though with a single-disease focus and limited comparison. In neurosurgical support, Guo et al. (2023) evaluated neuroGPT-X for point-of-care support, showing promise but struggling with rare conditions and hallucinations. In poison center calls, Matsler et al. (2024) found GPT-4 generated more readable and complete medical charts, though sample size and summarization reliability were concerns. These studies highlight LLMs' versatility, but limitations suggest they should complement, not replace, expert judgment.

Table 2. Summary of Study Design, Aim, Technology/LLM, and Context of Use

Article	Aim	Technology/ LLM	Context of Use	Challenges and Limitations
Cho et al. (2024)	Evaluate GLMs for pathology extraction.	6 GLMs, Llama-2, Mistral, Titan	Lung cancer staging support	Resource constraints, class imbalance, descriptor extraction
Clough et al. (2023)	Assess AI's potential for discharge summaries.	GPT	Discharge summary	Limited to mock vignettes, lack of real-world data, single GP rating per summary
Garcia et al. (2024)	Evaluate LLMs for patient message responses.	HIPAA-compliant LLM	Clinical messaging	Single-center, short study duration, lack of reference standards
Guo et al. (2023)	Compare context-enriched LLM performance with neurosurgeons	neuroGPT-X	Point-of-care neurosurgical support	Limited knowledge of rare conditions, potential hallucinations
Hoppe et al. (2024)	Compare ChatGPT and ED physicians' diagnostic accuracy.	GPT-3.5 & 4	Diagnostic support in emergency medicine	Small sample size, potential diagnosis bias
Kresevic et al. (2024)	Integrate LLMs for Hepatitis C decision support.	GPT-3.5 & 4 Turbo	Hepatitis C clinical guidelines	Single disease focus, limited LLM comparison
Levin et al. (2024)	Assess clinical reasoning of LLMs compared neonatal nurses	GPT-4, Claude-2.0	Clinical decision support in neonatal care	Small sample size, training bias, diagnostic limitations
Masan-neck et al. (2024)	Evaluate LLM performance in ED triage	GPT 3.5 & 4, Llama 3 70B, Gemini 1.5, Mixtral 8x7b	Triage support in emergency medicine	Overtriage by LLMs, undertriage by doctors, lack of gold-standard, limited generalizability, zero-shot approach
Matsler et al. (2024)	Evaluate ChatGPT's chart generation from poison center calls.	GPT-4	Poison center consultations	Poor intra-rater reliability, small sample size, no time savings evaluation
Sarvari et al. (2024)	Evaluate LLMs for diagnosing comorbidities	GPT-4, PaLM2	EHR-based diagnosis	Oversight, hallucination, integration, robustness
Sun et al. (2024)	Compare LLMs and ICD codes in detecting irAEs.	Mistral OpenOrca	Hospital admissions for ICI therapy	Single institution, need for external validation
Yi et al. (2024)	Compare GPT-4's discharge letters with junior clinicians.	GPT-4	Urology outpatient discharge letters	Fictional records, few assessors, no time measurement

Table 3. Summary of Quality Requirements, Evaluation Methods, and Key Findings

Article	Evaluated Quality Requirements	Evaluation Metrics	Key Findings	Ethical Consideration
Cho et al. (2024)	Accuracy and TN classification	F1-scores, Exact Match Ratio, Accuracy	High performance in information extraction and classification	Patient privacy in data use
Clough et al. (2023)	Healthcare providers' acceptance, Compliance, Completeness, clarity, Accuracy	Fisher's exact test, accuracy, sensitivity, specificity, PPV, Mann-Whitney U-test.	GPT produced summaries with high information completeness	Privacy in patient info generation, Risk of automation errors
Garcia et al. (2024)	Usability, Time usage, Impact on Clinician Well-Being	Quantitative and Qualitative evaluation.	High adoption, reduced burden, no time savings	Not explicitly mentioned
Guo et al. (2023)	Accuracy, coherence, relevance, thoroughness, and overall performance	Expert comparison, Consensus Evaluation, Time measurements	Context-enriched LLM showed promise as a point-of-care medical resource	Dataset bias, data security, and policies for LLMs
Hoppe et al. (2024)	Diagnostic Accuracy	Compared with ED physicians and gold standard; accuracy ANOVA.	GPT-4 surpassed ED residents in diagnostic accuracy, and excelled in specific disease	Privacy concerns; patient data was anonymized
Kresevic et al. (2024)	Accuracy, guideline adherence	expert review and text-similarity score	Improved accuracy with structured guideline reformatting	Concerns about incorrect responses from LLMs
Levin et al. (2024)	Accuracy, Completeness, and Response Time	Scenario questionnaire, expert scoring, descriptive stats, ANOVA.	Nurses outperformed LLMs; LLMs were faster. Claude2.0 surpassed GPT-4. Accuracy gaps persist.	Anonymized data, consent obtained
Masan-neck et al. (2024)	Agreement with trained professionals, over/under triage	Case vignettes, Manchester Triage comparison, Cohen's kappa, ANOVA.	Varying LLM results; GPT-4 showed strong agreement with professional raters	Bias in triage outcomes, fairness across staff levels.
Matsler et al. (2024)	Accuracy, Readability, Completeness, Data extractability	Summary grading and Kappa statistics	Preferred LLM summaries.	Privacy concerns with handling sensitive patient data
Sarvari et al. (2024)	Diagnostic Sensitivity and Accuracy	Ground truth comparison, Diagnostic & Aggregate Hit Rates.	GPT-4: 93.9% sensitivity; PaLM2 performed strongly.	Privacy concerns, Lack of transparency ("black box")
Sun et al. (2024)	Sensitivity, Specificity, Effectivity	Comparison with ICD, McNemar's test, Scatter text.	LLM surpassed ICD in sensitivity and efficiency.	Bias in Adverse Event Detection
Yi et al. (2024)	Clarity, Conciseness, Collegiality, Overall satisfaction	Standardized questionnaire, Mann-Whitney U-test	GPT-4 outperformed juniors, no hallucinations or errors.	Ethical concerns on automated patient communication, LLM hallucinations.

Table 4. Data charting table summarizing the data characteristics and availability from the included studies

Article	Data Type	Data Availability	Link to data
Cho et al. (2024)	Surgical pathology reports	Not publicly available	No public access
Clough et al. (2023)	Discharge summaries	Not publicly available	Contact author
Garcia et al. (2024)	Medical guidelines and LLM responses.	Not publicly available	No public access
Guo et al. (2023)	Web-scraped articles and abstracts, plus responses from neurosurgeons	Supplementary Table 4: Interview sample.	Supplementary Tables
Hoppe et al. (2024)	Electronic health records	Not explicitly stated	No public access
Kresevic et al. (2024)	Hepatology clinical guidelines	Not publicly available	Contact author
Levin et al. (2024)	Clinical descriptions of neonatal cases	Available upon request	Contact author
Masan-neck et al. (2024)	Simulated acute medical cases	Available upon request	Contact author
Matsler et al. (2024)	Real poison control call transcripts.	Supplementary material.	Supplemental material
Sarvari et al. (2024)	Electronic health records	Available at PhysioNet	MIMIC-IV
Sun et al. (2024)	Electronic health records	Not explicitly stated	Contact author
Yi et al. (2024)	Fictitious cases and summaries.	Available upon request	Contact author

4. DISCUSSION

Based on the results summarized in Section 3, this scoping review highlights the key quality requirements for using LLMs in clinical settings. The reviewed studies suggest that LLMs, show promise in improving clinical workflows, decision-making, and efficiency. However, challenges remain regarding consistent evaluation, integration into clinical practice, and overcoming limitations such as model bias and privacy concerns.

4.1 Accuracy and Evaluation Frameworks

The reviewed studies highlight diagnostic accuracy as the primary metric for evaluating LLMs in clinical applications, often complemented by sensitivity, specificity, and overall diagnostic performance. As shown in Section 3.A, GPT-4 demonstrated strong diagnostic support, surpassing clinicians in tasks like comorbidity diagnosis and cancer staging. However, variations in metric application across studies complicate comparisons. For example, while GPT-4 achieved high sensitivity for comorbidity detection from EHR data (Sarvari et al., 2024), challenges like data integration and bias remain. These findings underscore the need for standardized evaluation frameworks to ensure consistent, reliable assessments and uphold quality standards in healthcare.

4.2 Usability and Efficiency

Beyond accuracy, LLM usability and workflow efficiency emerged as key findings. Studies in Section 3.B showed that LLMs, such as ChatGPT, generated high-quality discharge summaries without errors, outperforming junior clinicians. This highlights their potential to reduce administrative burdens, streamline workflows, and alleviate clinician burnout. However, further research is needed to validate their effectiveness in real clinical settings, as many studies were conducted in simulated environments.

4.3 Challenges in Evaluation and Standardization

A significant challenge is the variation in evaluation methods across studies, including diverse metrics (e.g., F1 score, sensitivity, accuracy) and mixed assessment approaches. This inconsistency complicates cross-study comparisons, as seen in triage task variability in emergency medicine (Section 3.A). A standardized evaluation framework is essential for consistent, reliable LLM assessment and broader adaptability in healthcare settings.

4.4 Ethical Considerations

Ethical concerns, including data privacy, model bias, and decision-making transparency, were noted but lacked in-depth exploration. Using sensitive patient data poses privacy risks, while training biases can lead to unequal outcomes. These issues undermine clinician trust, particularly in high-stakes contexts like triage and diagnosis. Addressing these challenges through further research is crucial to ensure responsible LLM implementation that prioritizes fairness, privacy, and clinician confidence.

4.5 Data Access and Generalizability

Data sources varied widely, from EHRs to simulated datasets, complicating result comparison and generalizability. While some datasets were publicly available, most were restricted due to privacy concerns, limiting broader applicability. This highlights the need for diverse, representative data in future research and greater access to real-world datasets for effective LLM validation in clinical practice.

4.6 Recommendations for Implementation

To ensure safe and effective LLM integration in clinical settings, three key actions are recommended. First, standardized evaluation frameworks are needed to establish consistent quality metrics for cross-study comparisons and effective implementation. Second, ethical concerns must be addressed through policies prioritizing data privacy, transparency, and fairness to build clinician and patient trust. Finally, real-world validation through pilot studies with diverse datasets is essential to assess practical effectiveness and refine integration, ensuring LLM reliability across clinical environments.

4.7 Limitations of the Review and Future Research

The review is limited by using only three databases for the literature search. While this approach captured a wide range of studies, it may have excluded relevant research published in other sources. The exclusion of studies not published in English, and the absence of grey literature, may have omitted valuable evidence in different languages and restricted the inclusion of important non-peer-reviewed sources or from regions with varying healthcare contexts. Additionally, relevant articles published after October 2024 may have been missed, as the field is rapidly evolving.

5. CONCLUSION

This scoping review mapped existing research on LLM implementation in clinical settings, focusing on quality requirements for their evaluation. Findings highlight LLMs' potential, particularly models like GPT-4, in enhancing clinical workflows, decision-making, and efficiency. LLMs demonstrated strong diagnostic support, notably in cancer staging and comorbidity diagnosis, while also excelling in administrative tasks such as discharge summary generation and patient communication, reducing clinician workload and burnout.

However, significant challenges remain. The absence of standardized evaluation frameworks, along with inconsistent metrics like accuracy, sensitivity, and specificity, complicates cross-study comparisons. Ethical concerns, including data privacy, model bias, and decision-making transparency, further emphasize the need for responsible implementation.

This review identifies key quality requirements while addressing gaps in evaluation, ethics, and real-world validation. Future research should prioritize standardized frameworks, diverse datasets, and deeper exploration of ethical issues to ensure LLMs benefit both clinicians and patients. While LLMs hold promise for healthcare, responsible integration requires addressing these challenges to unlock their full potential and enable widespread adoption.

ACKNOWLEDGEMENT

The authors would like to express their gratitude to the Centre for Clinical Artificial Intelligence (CAI-X) at Odense University Hospital and the University of Southern Denmark for their support and collaboration in conducting this scoping review.

REFERENCES

Cho, H., Yoo, S., Kim, B., Jang, S., Sunwoo, L., Kim, S., Lee, D., Kim, S., Nam, S., & Chung, J.-H. (2024). Extracting lung cancer staging descriptors from pathology reports: A generative language model approach. *Journal of Biomedical Informatics*, *157*, 104720–104720. https://doi.org/10.1016/j.jbi.2024.104720

Clough, R., Sparkes, W. A., Clough, O. T., Sykes, J., Steventon, A., & King, K. (2023). Transforming healthcare documentation: Harnessing the potential of AI to generate discharge summaries. *BJGP Open*, *8*(1), BJGPO.2023.0116–BJGPO.2023.0116. https://doi.org/10.3399/bjgpo.2023.0116

Clusmann, J., Kolbinger, F. R., Muti, H. S., Carrero, Z. I., Eckardt, J.-N., Laleh, N. G., Löffler, C. M. L., Schwarzkopf, S.-C., Unger, M., Veldhuizen, G. P., Wagner, S. J., & Kather, J. N. (2023). The future landscape of large language models in medicine. *Communications Medicine*, *3*(1), 1–8. https://doi.org/10.1038/s43856-023-00370-1

Elsevier. (2023). *About Scopus | Abstract and citation database | Elsevier*. Beta.elsevier.com. https://www.elsevier.com/products/scopus

Garcia, P., Ma, S. P., Shah, S., Smith, M., Jeong, Y., Devon-Sand, A., Tai-Seale, M., Takazawa, K., Clutter, D., Vogt, K., Lugtu, C., Rojo, M., Lin, S., Shanafelt, T., Pfeffer, M. A., & Sharp, C. (2024). Artificial Intelligence–Generated Draft Replies to Patient Inbox Messages. *JAMA Network Open*, *7*(3), e243201–e243201. https://doi.org/10.1001/jamanetworkopen.2024.3201

Guo, E., Gupta, M., Sinha, S., Rössler, K., Tatagiba, M., Ryojo Akagami, Ossama Al-Mefty, Sugiyama, T., Stieg, P. E., Pickett, G. E., Madeleine de Lotbinière-Bassett, Singh, R., Lama, S., & Sutherland, G. R. (2023). neuroGPT-X: toward a clinic-ready large language model. *Journal of Neurosurgery*, 1–13. https://doi.org/10.3171/2023.7.jns23573

Hoppe, J. M., Auer, M. K., Strüven, A., Massberg, S., & Stremmel, C. (2024). ChatGPT With GPT-4 Outperforms Emergency Department Physicians in Diagnostic Accuracy: Retrospective Analysis. *Journal of Medical Internet Research*, *26*, e56110–e56110. https://doi.org/10.2196/56110

Kresevic, S., Giuffrè, M., Ajcevic, M., Accardo, A., Crocè, L. S., & Shung, D. L. (2024). Optimization of hepatological clinical guidelines interpretation by large language models: a retrieval augmented generation-based framework. *Npj Digital Medicine*, *7*(1), 1–9. https://doi.org/10.1038/s41746-024-01091-y

Levin, C., Kagan, T., Rosen, S., & Saban, M. (2024). An evaluation of the capabilities of language models and nurses in providing neonatal clinical decision support. *International Journal of Nursing Studies*, *155*, 104771–104771. https://doi.org/10.1016/j.ijnurstu.2024.104771

Masanneck, L., Schmidt, L., Seifert, A., Kölsche, T., Huntemann, N., Jansen, R., Mehsin, M., Bernhard, M., Meuth, S. G., Böhm, L., & Pawlitzki, M. (2024). Triage Performance Across Large Language Models, ChatGPT, and Untrained Doctors in Emergency Medicine: Comparative Study. *Journal of Medical Internet Research*, *26*, e53297. https://doi.org/10.2196/53297

Matsler, N., Pepin, L., Banerji, S., Hoyte, C., & Heard, K. (2024). Use of large language models to optimize poison center charting. *Clinical Toxicology*, *62*(6), 385–390. https://doi.org/10.1080/15563650.2024.2348107

PubMed. (2023, August 15). *PubMed Overview*. PubMed; National Library of Medicine. https://pubmed.ncbi.nlm.nih.gov/about/

Raza, M. M., Venkatesh, K. P., & Kvedar, J. C. (2024). Generative AI and large language models in health care: pathways to implementation. *Npj Digital Medicine*, *7*(1). https://doi.org/10.1038/s41746-023-00988-4

Sarvari, P., Zaid Al-fagih, Abdullatif Ghuwel, & Othman Al-fagih. (2024). A systematic evaluation of the performance of GPT-4 and PaLM2 to diagnose comorbidities in MIMIC-IV patients. *Health Care Science*, *3*(1), 3–18. https://doi.org/10.1002/hcs2.79

Sun, V. H., Heemelaar, J. C., Hadzic, I., Raghu, V. K., Wu, C.-Y., Zubiri, L., Azin Ghamari, LeBoeuf, N. R., Osama Abu-Shawer, Kehl, K. L., Grover, S., Singh, P., Suero-Abreu, G. A., Wu, J., Falade, A. S., Grealish, K., Thomas, M. F., Hathaway, N., Medoff, B. D., & Gilman, H. K. (2024). Enhancing Precision in Detecting Severe Immune-Related Adverse Events: Comparative Analysis of Large Language Models and International Classification of Disease Codes in Patient Records. *Journal of Clinical Oncology*. https://doi.org/10.1200/jco.24.00326

Web of Science. (2020, January 22). Sdunet. https://sdunet.dk/en/servicesider/analytics/analyseredskaber/web_of_science

Yi, J., Sunil Ravinder Gill, Gui, G., Yan, D., Ke, Y., Ting Fang Tan, Jin, L., Kabilan Elangovan, Chiat, J., Hairil Rizal Abdullah, Shu, D., & Tsung Wen Chong. (2024). Comparison of the Quality of Discharge Letters Written by Large Language Models and Junior Clinicians: Single-Blinded Study. *Journal of Medical Internet Research*, *26*, e57721–e57721. https://doi.org/10.2196/57721

Short Papers

COMPARING DEEP LEARNING MODELS USED FOR IMAGE CLASSIFICATION ON PATTERNS ON DRESSED CLOTHING

Marian Sousan, Sofia Azam and Nic Herndon
Department of Computer Science, East Carolina University East 5th Street, Greenville, NC, United States of America

ABSTRACT

Previous studies of image classification on fashion datasets focused on how accurate different items of clothing could be classified. However, many of these datasets involved isolated pieces of clothing with the same background rather than clothes worn by any individual in any setting. This study evaluated three convolutional neural networks, VGG, AlexNet, and LeNet, on image classification of fabric patterns of clothes worn by individuals with different background images, to enhance the understanding of fashion image classification and to test the limits of these models. This could be beneficial in a variety of ways, such as aiding in e-commerce with better recommendations or helping with mundane tasks such as categorizing clothes. Out of the three models evaluated, VGG was the best model with a classification accuracy of 68%.

KEYWORDS

Fashion Patterns, Deep Learning, Lenet, VGG, Alexnet, Image Classification

1. INTRODUCTION

In the digital age, e-Commerce success hinges on customer satisfaction, which directly impacts return rates. Online returns cost businesses $200 billion annually, with return rates outpacing sales growth (Guha, 2024). To retain diverse customers with varying preferences, businesses must improve recommendation systems (Helmy et al., 2023). One solution is to use deep learning to analyze online images and identify trending fabric patterns, enabling platforms to promote relevant products and reduce returns. Accurate image classification of clothes on people in diverse settings is key to enhancing automated recommendations, minimizing manual sorting, and improving customer satisfaction. This study investigates how image classification can address return-related revenue loss in e-Commerce. It emphasizes supporting, rather than replacing, fashion and retail professionals. To this end, the paper evaluates three CNNs: LeNet, AlexNet, and VGG. While VGG has been used in similar tasks (Xu et al., 2022), its performance on images with complex backgrounds has not been explored. LeNet and AlexNet had not been applied to clothed-person classification prior to this. All models were tested using the Fashion Dataset from Kaggle (Singh, 2019), which was selected due to the limited research on classifying clothed individuals in complex scenes.

The contributions of this work are as follows:

- We evaluated clothing classification using images with diverse outfits and backgrounds, unlike previous studies that used isolated clothing images.
- We compared the performance of LeNet, AlexNet, and VGG models, and found that VGG yielded the highest accuracy, 68.63%.
- We performed hyperparameter tuning, which increased the performance of these models, by up to 11.1%.

2. RELATED WORKS

Image classification is commonly performed with deep learning models, which are fine-tuned on specific datasets and optimized through hyperparameter tuning using methods like grid search (which will be explained in Section 3.1 of this paper). Fashion classification, a subfield of this area, often involves adapting models pre-trained on general datasets, like ImageNet, to evaluate their accuracy on fashion-specific data (Seo and Shin, 2018). One widely used dataset is Fashion- MNIST, which contains grayscale images of isolated clothing items. LeNet has shown strong performance on this dataset (Kayed et al., 2020), though Fashion-MNIST has been criticized for its lack of diversity and realism (Academy, 2023). To address these limitations, we incorporated color channels and images of individuals wearing clothing in more realistic settings.

Other studies have also focused on clothing worn by people. For instance, VGG models outperformed others in classifying worn apparel for e-commerce applications (Xu et al., 2022). Another study classified traditional clothing from Bangladeshi tribes using YOLO models, with YOLOv5 achieving the highest accuracy (Islam et al., 2022). Fashion classification has also been explored through specific attributes like color, fabric, and patterns. CNNs with algorithms like SIFT achieved 85% color detection accuracy (Ivanov et al., 2018), while SVMs and random forests showed limited success in pattern recognition, with accuracies around 35–38% (Bossard et al., 2013).

3. METHODOLOGY

3.1 Experimental Design

The dataset was first extracted from Kaggle, featuring images of clothed individuals outlined in red boxes against a complex background, including buildings, nature and other structures. Each image is labeled with a specific fabric pattern. Two categories, tribal and ikat, were excluded out of respect for Indigenous and Central Asian cultures as to not appropriate nor deal with images without permission from a representative of their culture. The data was then split into 60% for training, 20% for validation, and 20% testing. Image classification was performed using the original architectures of LeNet5, AlexNet, and VGG16. Each model was run twice on an NVIDIA GeForce RTX 4060 Ti: once with default parameters and once using grid search to optimize hyperparameters for better accuracy (Muhathir et al., 2023).

3.2 Data Collection and Preprocessing Data

For this dataset, we had to extract the images, since only their links were provided, using the Pandas, Pillow, and io libraries. We used a total of 13,512 images from this dataset where each image was 400 × 600 in size. There are 14 different pattern categories, where a sample of three images shown in Figure 1.

(a) Animal (b) Geometry (c) Stars

Figure 1. Examples of the images that were used and the categories they were respectively classified as.

3.3 CNN Architectures

We used LeNet5 (Figure 2), AlexNet (Figure 3), and VGG-16 (Figure 4) models for this study's experiments.

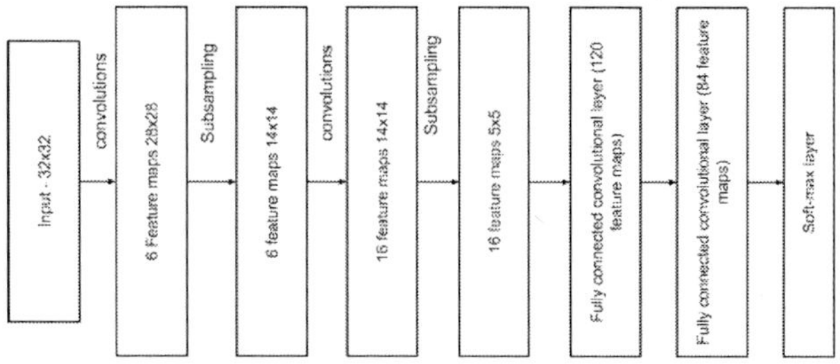

Figure 2. LeNet5 is a seven-layer CNN that has the architecture shown above. Image adapted from (Kayed et al., 2020)

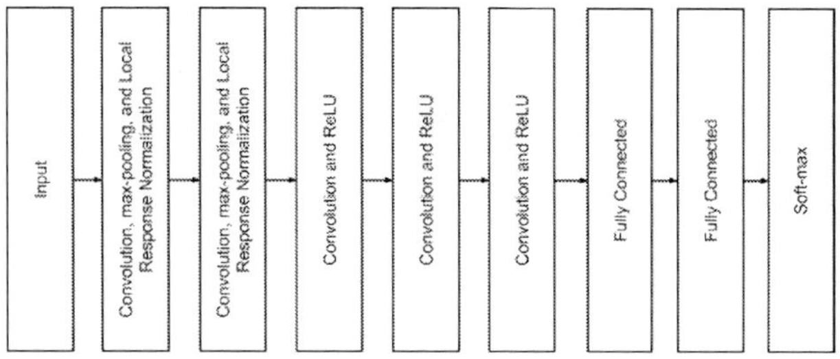

Figure 3. AlexNet is a CNN with a total of eight layers, with five convolutional and three fully connected layers. Image adapted from (Alom et al., 2018)

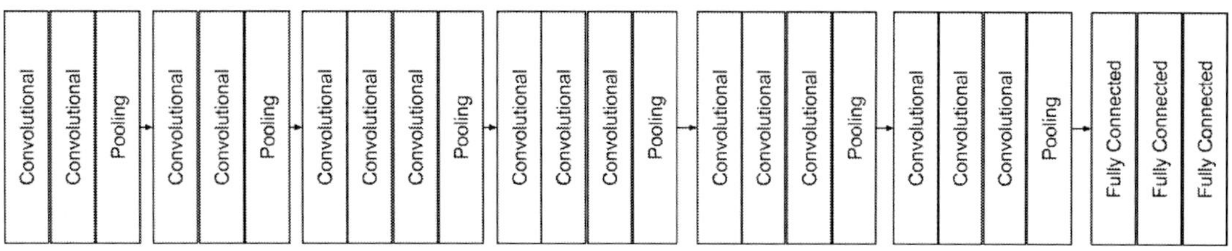

Figure 4. VGG-16 consists of 16 different layers – 13 convolutional layers and 3 fully connected layers (Xu et al., 2022)

Each layer must include a convolution filter, and ReLU (non-linear activation function) to speed up the process since all negatives are automatically counted as zero ($f(x) = \max(0, x)$), effectively leaving certain neurons inactive (Sun et al., 2016).

3.4 Hypertuning

We tested different hyperparameter values to see if they will improve the deep learning models by increasing their accuracy and performance. Considerations for epoch, batch size, and the optimizer hyperparameter values used for each model had priority, as related papers dealing with fashion classification commonly looked at these values (Bhoir and Patil, 2022; Kayed et al., 2020). The hyperparameter values that were tested for each model and those selected to obtain the most accurate classification are listed in Table 1.

4. RESULTS AND DISCUSSION

At the start of testing, the initial default hyperparameter values were a batch size of 64, ReLU for the activation function, 20 epochs, and an Adam optimizer. Afterwards, grid search was used to look for specific parameters that would work best, with the accuracy shown in Table 2. The row describing "after grid search" describes the results of the model after grid search using the selected values for each parameter described in Table 1. For each model, we also show the training and validation loss in Figure 5, with the hyperparameters that led to the best accuracy.

Table 1. Hyperparameters and their respective values tested for each model along with the selected values.

Parameter	Potential Values	Selected Values for LeNet	Selected Values for AlexNet	Selected Values for VGG16
Epoch	10, 20	20	20	10
Batch Size	32, 64, 128	128	128	32
Optimizer	Adam, SGD	Adam	Adam	SGD

Table 2. Accuracy before and after the use of grid search. The best accuracies are shown in **bold font**. Although AlexNet had the best accuracy with the default parameters, after grid search, the accuracy increased for LeNet and VGG but not for AlexNet. The best overall accuracy was with VGG with fine-tuned hyperparameters

	LeNet	AlexNet	VGG
Before Grid Search	56.53%	**67.22%**	57.53%
After Grid Search	57.82%	66.26%	**68.63%**

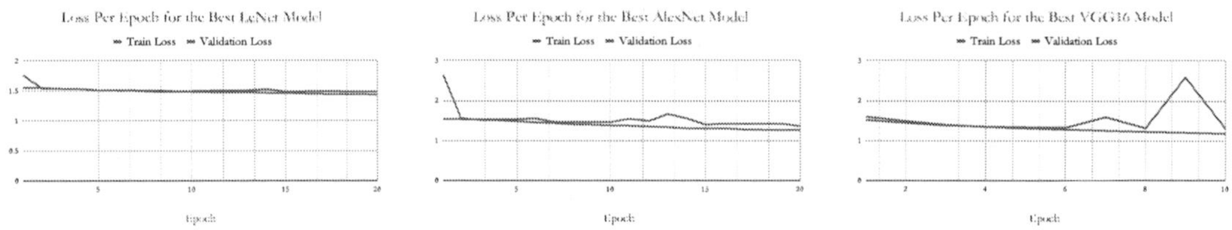

(a) LeNet (b) AlexNet (c) VGG

Figure 5. Training and validation loss for the best performing model per CNN. LeNet and AlexNet seem to have converged to the optimal weights early in the training, whereas VGG seems to have potential for higher accuracy as the validation loss varied in the last epochs

Initially, experiments were conducted on two different GPUs using the same models and data, yet accuracy results varied. Papers on deep learning reproducibility highlight this issue, noting that even with identical datasets, seeds, and GPUs, results can differ along some researchers proposing methods to reduce this variance (Pham et al., 2021). Another study described how GPU libraries such as cuDNN from NVIDIA can affect variance and decrease the chance of reproducibility for models due to stochastic factors (Piantadosi et al., 2019).

In this study, it's unclear whether accuracy differences stem from parameter choices or random variation across runs. While existing research addresses such factors, this study will defer their investigation to future work. With these findings, the CNNs were modified to include a seed and normalization of 0.5 to increase the chance of reproducibility. After modifications, the model that provided the highest accuracy was VGG with 68%. Grid search aided in increasing accuracy for most of the models except for AlexNet.

5. CONCLUSION AND FUTURE WORK

VGG with a batch size of 64, an epoch size of 10, and an SGD optimizer proved to be the most efficient model. However, while VGG has been shown to have achieved an accuracy of up to 93.2% in other fashion tasks (Meng et al., 2022), this study reached only 68%, underscoring the challenge of classifying fabric patterns in uncontrolled settings with varied lighting, poses, and backgrounds. This highlights the need for further research in classifying dressed garments in complex environments. Future work will explore more diverse datasets, including various clothing types, textures, and cultural styles to improve generalization. We also plan to assess the impact of visual artifacts and examine model reproducibility and variance to enhance reliability.

REFERENCES

Academy, E. (2023, August). What is the difference between the fashion-mnist dataset and the classic mnist dataset? (Cit. on p. 2).

Alom, M. Z., Taha, T. M., Yakopcic, C., Westberg, S., Sidike, P., Nasrin, M. S., Esesn, B. C. V., Awwal,

A. A. S., & Asari, V. K. (2018). The history began from alexnet: A comprehensive survey on deep learning approaches. (Cit. on p. 3).

Bhoir, S., & Patil, S. (2022). Transfer learning with deep neural networks for image classification in the e- commerce industry. *2022 IEEE 7th International conference for Convergence in Technology (I2CT)*, 1–8 (cit. on p. 3).

Bossard, L., Dantone, M., Leistner, C., Wengert, C., Quack, T., & Van Gool, L. (2013). Apparel classification with style. In K. M. Lee, Y. Matsushita, J. M. Rehg, & Z. Hu (Eds.), *Computer vision – accv 2012* (pp. 321– 335). Springer Berlin Heidelberg. (Cit. on p. 2).

Guha, I. (2024, August). Council post: Total cost of online returns breaks $200 billion-can retail afford it? (Cit. on p. 1).

Helmy, Y., Ashraf, M., & Abdelhamid, L. (2023). E-commerce challenges, definitions, solutions and evalua- tion. *Journal of Computer Science*, *19*(9), 1087–1097 (cit. on p. 1).

Islam, A., Chowdhury, T., Hossain, M., Nahid, N., & Rifat, A. I. (2022). An automatic system for identifying and categorizing tribal clothing based on convolutional neural networks. *2022 Fourth International Con- ference on Emerging Research in Electronics, Computer Science and Technology (ICERECT)*, 1–6 (cit. on p. 2).

Ivanov, A. Y., Borzunov, G. I., & Kogos, K. (2018). Recognition and identification of the clothes in the photo or video using neural networks. *2018 IEEE Conference of Russian Young Researchers in Electrical and Electronic Engineering (EIConRus)*, 1513–1516 (cit. on p. 2).

Kayed, M., Anter, A., & Mohamed, H. (2020). Classification of garments from fashion mnist dataset using cnn lenet-5 architecture. *2020 International Conference on Innovative Trends in Communication and Computer Engineering (ITCE)*, 238–243 (cit. on pp. 2, 3).

Meng, N., Tianyu, T., & Mingming, Z. (2022). Research on computer image recognition and intelligent analysis technology of water and land painting clothing. *2022 IEEE Conference on Telecommunications, Optics and Computer Science (TOCS)*, 946–952 (cit. on p. 4).

Muhathir, Khairina, N., Rehia Karenina, I. B., Ula, M., & Sahputra, I. (2023). Preserving cultural heritage through ai: Developing lenet architecture for wayang image classification. *International Journal of Ad- vanced Computer Science and Applications*, *14*(9) (cit. on p. 2).

Pham, H. V., Qian, S., Wang, J., Lutellier, T., Rosenthal, J., Tan, L., Yu, Y., & Nagappan, N. (2021). Problems and opportunities in training deep learning software systems: An analysis of variance. *Proceedings of the 35th IEEE/ACM International Conference on Automated Software Engineering*, 771–783 (cit. on p. 4).

Piantadosi, G., Marrone, S., & Sansone, C. (2019). On reproducibility of deep convolutional neural networks approaches. In B. Kerautret, M. Colom, D. Lopresti, P. Monasse, & H. Talbot (Eds.), *Reproducible research in pattern recognition* (pp. 104–109). Springer International Publishing. (Cit. on p. 4).

Seo, Y., & Shin, K.-s. (2018). Image classification of fine-grained fashion image based on style using pre-trained convolutional neural network. *2018 IEEE 3rd International Conference on Big Data Analysis (ICBDA)*, 387–390 (cit. on p. 2).

Singh, N. (2019). Fashion dataset: Over 15000 images with various dress categories. (Cit. on p. 1).

Sun, J., Cai, X., Sun, F., & Zhang, J. (2016). Scene image classification method based on alex-net model. *2016 3rd International Conference on Informative and Cybernetics for Computational Social Systems (ICCSS)*, 363–367 (cit. on p. 3).

Xu, J., Wei, Y., Wang, A., Zhao, H., & Lefloch, D. (2022). Analysis of clothing image classification models: A comparison study between traditional machine learning and deep learning models. *Fibres & Textiles in Eastern Europe*, *30*(5), 66–78 (cit. on pp. 1–3).

A DESIGN OF ELDERLY CARE MONITORING SYSTEM USING MULTI-MODAL SENSORS

Sumin Kim, Sangwon Nam, Jimin Seo, Seohee Cho and Junho Jeong
Dep. Of Computer Science and Engineering, Dongguk University
30, Pildong-ro 1-gil, Jung-gu, Seoul, 04620, Republic of Korea

ABSTRACT

This study proposes a system and mobile application UI/UX that enables caregivers and key stakeholders to monitor the daily living data of elderly individuals living alone, collected through eight types of non-contact sensors, in an effort to address social issues such as the prevention of solitary deaths. Daily living patterns are analyzed based on data related to outings, sleep, and physical activity, and a risk classification method is proposed accordingly. The study was conducted using anonymized data collected over a two-week period from ten participants. Based on these findings, a mobile application was designed to allow caregivers to access monitoring reports, thereby aiming to enhance the efficiency of health management for elderly individuals living alone.

KEYWORDS

Elderly Living Alone, Contactless Sensor, Lifestyle Pattern Analysis, Anomaly Detection, Machine Learning, Health Monitoring System

1. INTRODUCTION

As society becomes super-aged, managing the health of elderly individuals living alone and preventing solitary deaths have emerged as critical social issues. Traditional monitoring systems using wearable, contact-based sensors (Lee, 2020; Lee and Cha, 2021) face challenges like user discomfort and device loss. Research with non-contact sensors aimed to enhance accuracy by increasing sensor numbers (Yoon et al., 2022; Chang et al., 2023), but high installation and maintenance costs remain major hurdles. Recent studies have explored hidden patterns through unsupervised learning. Non-contact sensors have been used to monitor activities and physiological states via digital exhaust and digital behaviorome analysis (Schütz et al., 2022a), and to predict fall risks, mild cognitive impairment, and social isolation based on visit frequency (Schütz et al., 2022b).

This study designs an application using eight types of non-contact sensors, providing algorithms and visualizations suitable even for limited resources, enabling caregivers to easily monitor daily living patterns and respond appropriately. We introduce the proposed method in Sections 2 and 3. Finally in Section 4, we discuss the conclusion and future work.

2. DESIGN OF ELDERLY CARE MONITORING METHOD

The proposed system consists of five layers: the edge layer, data transmission layer, data processing and storage layer, data analysis layer, and user interface layer, as shown in Figure 1. The edge layer collects raw data from eight types of non-contact sensors such as PIR (Passive infrared sensor), radar, heartbeat, humidity, respiration, temperature, illuminance, and door opening detection. Then sensors transmit it via a gateway, the data transmission layer, to a server for analyzing elderly living patterns. In the data processing and storage layer, the data is stored in a time-series format within a database. The data analysis layer identifies daily living patterns such as sleep, outings, and physical activity, and classifies risk groups. The final analysis results are visualized and delivered to caregivers and social workers through a mobile application.

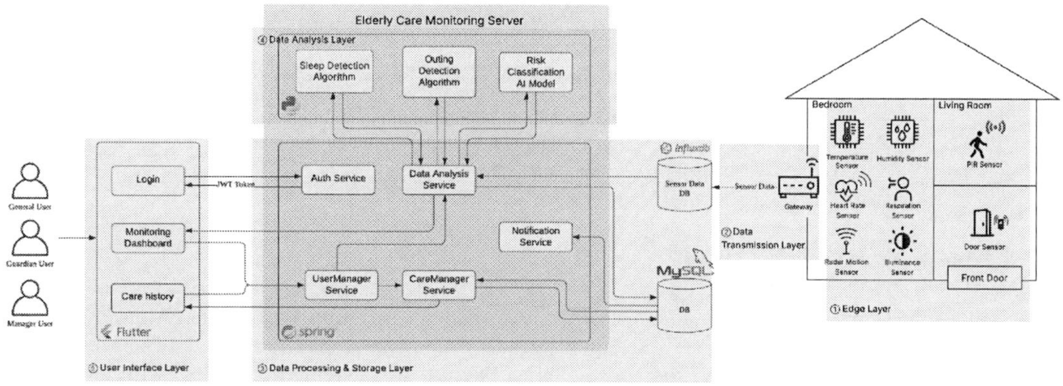

Figure 1. System Architecture for Data Collection, Analysis, and Service Provision

2.1 Outing Detection Method

To compensate for potential errors from the gateway and door sensors, outing status was determined by combining door open/close records with indoor activity data. If activity levels remained below a threshold for a set time after the door closed, the system classified the event as an outing; if door openings occurred during an outing or indoor activity exceeded a threshold, it was considered a return. Algorithm 1 presents pseudocode for outing pattern detection based on PIR, radar, heart rate, respiration, and door sensors. Three thresholds are defined to determine outing start and end points, and thresholds are dynamically updated based on recent activity data. Initial thresholds are set using the average activity level over a predefined period.

Algorithm 1.

Algorithm 1 Outing Analysis Algorithm

Require: Door event data and Sensor time-series data
Ensure: Outing status with timestamps (1: Outing, 0: Not Outing)
Initialize $is\_outside = False$
Set thresholds: T_{hb} (heart rate/breath), T_{rp} (radar/PIR), T_{ha} (home activity)
for each door event **do**
  if door closed (E0101) **then**
    Compute heart rate and breathing activity
    if activity $\leq T_{hb}$ **then**
      Check radar/PIR activity
      if activity $\leq T_{rp}$ **then**
        $is\_outside = True$
        Update T_{hb}, T_{rp}
        Log outing start time
      end if
    end if
  else if door opened (E0100) and $is\_outside == True$ **then**
    Compute home activity (30-min intervals)
    if activity $\geq T_{ha}$ **then**
      Log return time
      Update T_{ha}
    end if
    Ignore door events within 3 min of return
    $is\_outside = False$
  end if
end for

2.2 Activity Analysis Method

Activity levels were analyzed using data from PIR and radar sensors. Algorithm 2 outlines pseudocode for aggregating sensor data over a given period, calculating, and storing daily activity totals for each sensor.

Algorithm 2.

```
Algorithm 2: Sensor Activity Analysis Algorithm
Require: Sensor time-series data
Ensure: Total PIR and radar activity over the specified period
Retrieve PIR and radar activity data;
Convert timestamps to date format;
Sum all activity values within the specified period;
Store the total activity for the period;
```

2.3 Sleep Analysis Method

The subject's actual sleeping location was first identified using multi-sensor data. Algorithm 3 presents pseudocode for sleep pattern analysis based on PIR, radar, heart rate, respiration, and light sensors. Sleep onset and wake-up times are estimated from physiological signals or movement changes, with adjustments for mid-sleep awakenings and outings using sensor data.

Algorithm 3.

```
Algorithm 3 Sleep Analysis Algorithm
Require: Sensor time-series data (HR, Resp, Radar, PIR, Light)
Ensure: Sleep state per minute (1: sleep, 0: awake)
Initialize variables;
Set thresholds: wake_window ← 10, T_interrupt ← 60;
                sensor_zero_threshold ← 0.9;
Determine room type based on sensor activity;
for each date do
    Filter night data based on low Light values;
    if room type is "bedroom" then
        Detect sharp drop in HR + Resp + Radar;
    else
        Detect sharp drop in PIR;
    end if
    if drop detected then
        Save as sleep_start[date];
    end if
end for
for each next date do
    Filter morning data based on high Light values;
    if room type is "bedroom" then
        Check HR, Resp, Radar = 0 for wake_window minutes;
    else
        Check PIR = 0 for wake_window minutes;
    end if
    Save as wake_start[date];
end for
for each sleep interval do
    Set sleep_state = 1 between sleep and wake;
    if any awake segment ≥ T_interrupt then
        Mark as interrupted: sleep_state = 0;
    end if
end for
```

2.4 Risk Group Classification Method

Isolation Forest was applied to detect risk groups based on behavioral changes. Algorithm 4 presents pseudocode for anomaly detection using six features: outing frequency, outing duration, sleep duration, mid-sleep awakenings, and PIR/radar activity levels. Weekly averages and standard deviations were calculated, data were normalized using Z-scores, and change rates were computed. Combined vectors from Week 2 Z-scores and change rates were input into the Isolation Forest to calculate anomaly scores. Users identified as anomalies were classified into risk groups and stored in the database.

Algorithm 4.

Algorithm 4 Risk Group Detection via Isolation Forest

Require: Behavioral pattern data from week 1 and week 2
Ensure: `anomaly_score` and `risk_label` for each user
Load behavioral data into W_1 (week 1) and W_2 (week 2)
Select 6 features: outing count, outing duration, sleep duration, mid-sleep interruptions, PIR activity, radar activity
Compute z-scores:
  $Z_1 \leftarrow \frac{W_1 - \mu_1}{\sigma_1}$
  $Z_2 \leftarrow \frac{W_2 - \mu_2}{\sigma_2}$
Compute z-score change: $D_z \leftarrow Z_2 - Z_1$
Construct input matrix: $X \leftarrow [Z_2 \mid D_z]$
Train Isolation Forest model on X
Compute `anomaly_score`: $-$`decision_function`(X)
Predict anomaly labels: `label` $\leftarrow$ `predict`(X)
Assign `risk_label`: if label $=$ `-1`, then `HighRisk`; else `Normal`

3. PROPOSED APPLICATION UI/UX DESIGN

The proposed system visualizes the analysis results generated by the server and delivers them to caregivers and welfare managers through a mobile application. Figure 2 present the key UI/UX designs of the system. Figure 2 (a) is a dashboard interface that shows the results of analyzing the user's weekly activity and environmental data. Users can check key indicators such as average number of outings, sleeping time, indoor activity, temperature and humidity by setting the analysis period. Figure 2 (b) is an interface that allows users to view the list of management targets on the administrator page. Administrators can quickly view only the specific people they want through the risk group classification filter and long-term absentee filter. Figure 2 (c) allows users to select a person whose detailed information they need from the list of management targets and check detailed information such as the contact information and guardian information of the person in question.

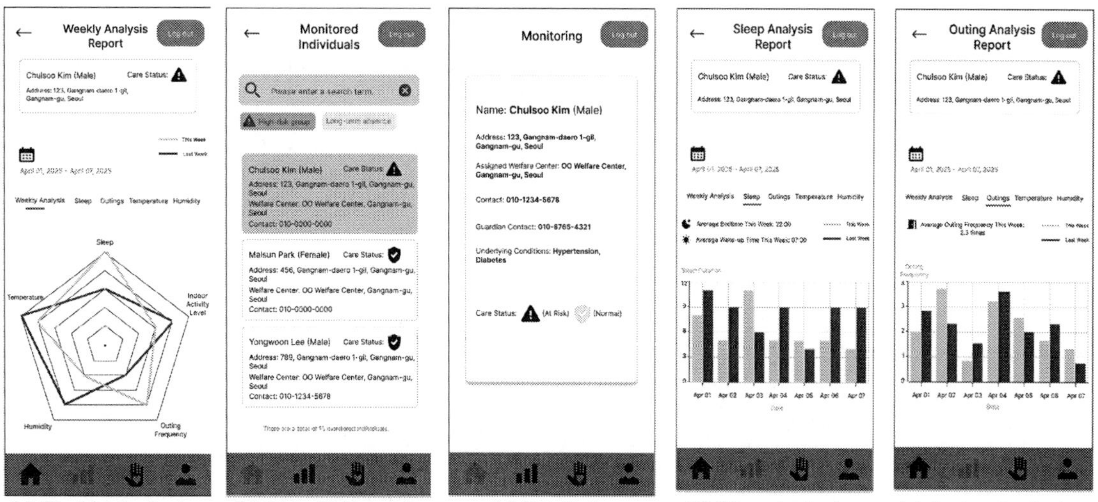

(a) Weekly Analysis Report (b) List of Managed Users (c) Detailed Information (d) Sleep Analysis Report (e) Outing Analysis Report

Figure 2. Overview of the Main UI/UX Components

Figure 2. (d) (e) are the UI/UX for a report that analyzes users' lifestyle patterns. Figure 2 (d) is an interface that shows a sleep analysis report of the subject of management, where you can check the average bedtime and average wake-up time for the corresponding week and easily understand the change in sleep time from last week to this week through a graph. Figure 2 (e) is an interface that shows an outing analysis report of the subject of management, where you can check the average number of outings for the corresponding week, and easily understand the change in outing time from last week to this week through a graph.

4. CONCLUSION

This study designs an elderly monitoring system based on 8 IoT sensors and data analysis. By applying anomaly detection algorithms and classifying risk groups, the system aims to overcome the limitations of manual monitoring and enhance safety for elderly individuals while improving convenience for caregivers and managers. Although the system faces challenges such as limited data collection, reduced accuracy from minimal non-contact sensors, and the absence of labeled data, these limitations are mitigated by introducing the Isolation Forest technique to automatically detect new anomaly patterns.

However, privacy concerns limit data collection, making it difficult to quantitatively assess accuracy. Future research will focus on final implementation and performance validation through real-world testing.

ACKNOWLEDGEMENT

The following are results of a study on the "Convergence and Open Sharing System "Project, supported by the Ministry of Education and National Research Foundation of Korea. And this research was supported by the MSIT (Ministry of Science, ICT), Korea, under the National Program for Excellence in SW), supervised by the IITP (Institute of Information & communications Technology Planning &Evaluation) in 2024(2023-0-00049).

REFERENCES

Chang, K. et al., (2023). A Study on an IoT Sensor-based Monitoring Algorithm for Caring for the Elderly Living Alone. *Journal of Digital Contents Society*, Vol. 24, No. 9, pp. 2147–2156.

Lee, J., (2020). A Study on Emergency Prediction Model for The Elderly Living Alone Using Big Data and Machine Learning. PhD thesis, Kongju National University.

Lee, J. and Cha, S., (2021). A Long-term Monitoring Demonstration of Smart Home System for the Elderly. *KIBIM Magazine*, Vol. 11, No. 3, pp. 75–90.

Schütz, N. et al., (2022[a]). A systems approach towards remote health-monitoring in older adults: Introducing a zero-interaction digital exhaust. *npj Digital Medicine*, Vol. 5, p. 116.

Schütz, N. et al., (2022b). A sensor-driven visit detection system in older adults' homes: Towards digital late-life depression marker extraction. *IEEE Journal of Biomedical and Health Informatics*, Vol. 26, No. 4, pp. 1560–1569.

Yoon, Y. et al., (2022). Development and Operation of Remote Lone-Senior Monitoring System Based on Heterogeneous IoT Sensors and Deep Learning. *Journal of the Korea Convergence Society*, Vol. 13, No. 1, pp. 387–398.

ARTIFICIAL REASONING FOR DIAGNOSIS SYSTEMS

Yamine Klioui[1], Mohamed Frendi[2], Pierre-Jean Charrel[3] and Malik Si-Mohammed[1]
[1]*University Mouloud Mammeri of Tizi-Ouzou, Tizi-Ouzou, Algeria*
[2]*Shared Services, Federal Government of Canada, Ottawa, Canada*
[3]*University of Toulouse II Jean Jaurès, Toulouse, France*

ABSTRACT

We present in this paper a domain-agnostic reasoning framework that allows to represent and structure expert knowledge and reasoning through ontological representations and graph-based inference. This approach mainly introduces two models: A knowledge formalisation model, that maintains domain semantics, and a human-like graph reasoning model that maintains logical consistency during problem-solving. Our system has been tested on diagnosis problem in several expert domains, mainly on clinical use cases, what allowed us to confirm its interest. In addition to its performance on health diagnosis, the system design itself confirms its generalisability to other expert domains.

KEYWORDS

Ontologies, Diagnosis, Artificial Reasoning, Conceptual Graphs

1. INTRODUCTION

Artificial Intelligence (AI) seeks to simulate human cognitive abilities, such as learning, reasoning, and problem-solving, within computational systems (Sarker, 2022). While Machine Learning (ML) and Natural Language Processing (NLP) enhance pattern recognition, they often lack interpretability in decision-making. Knowledge-Based Systems (KBS) bridge this gap by integrating structured knowledge representation: like rules and ontologies, with logical reasoning, enabling expert-like solutions that emphasise transparency. However, existing models struggle to balance adaptability with cognitive flexibility. Rule-based systems lack versatility, while statistical approaches fail in explainability. This challenge underscores the need for a reasoning framework that preserves both structural rigor and semantic richness. Domain ontologies offer a transformative advantage by organising concepts, relationships, and constraints into reusable frameworks, providing a basis for computational reasoning that mimics human thought processes. (Igoche B.and Matthew, 2024). This study aims to develop an ontology-based artificial reasoning model that enhances diagnostic decision-making by structuring expert knowledge in a machine-interpretable framework. Unlike classification tasks, which cluster input data, reasoning dynamically derives conclusions based on domain constraints and evolving knowledge patterns. The proposed model leverages declarative semantic graphs, ensuring adaptable inference mechanisms independent of predefined reasoning engines. The structure of the paper is as follows: Section II provides an overview of some related research, detailing existing gaps in AI reasoning. Section III and IV respectively introduce our ontology-based reasoning model and the reasoning process, before providing some details on the implementation framework our model.

2. RELATED WORK AND BACKGROUND

Ontologies, as formal specifications of domain knowledge (Gruber, 2018), facilitate structured representation and enhance communication among stakeholders. Their hierarchical nature enables precise conceptual definitions and semantic relationships (Igoche B.and Matthew, 2024). Recent research highlights ontology applications in expert systems, decision support, and semantic search. Examples include ontology-based smart search algorithms for E-commerce (Guskov Gleband Zarayskiy, 2022), ontology-driven wildlife data integration (Hamed *et al.*, 2025), cybersecurity frameworks (Pastuszuk, Burek and Ksiezopolski, 2021), business model knowledge representation (Hamrouni *et al.*, 2021) and IoT applications (Li *et al.*, 2019).

Despite these advancements, existing works emphasise ontology construction rather than reasoning within problem-solving. Our approach integrates ontologies with artificial reasoning to model human diagnostic processes, leveraging structured knowledge representation and inference mechanisms. Unlike rigid rule-based systems, our declarative semantic graph enables dynamic reasoning strategies, ensuring transparency and adaptability while capturing the evolving nature of knowledge acquisition.

3. KNOWLEDGE REPRESENTATION AND REASONING PROCESS

Reasoning process commonly consists in operating inferences and drawing conclusions from available (acquired) knowledge. Consequently, emulating this human ability is crucial for Knowledge-Based Systems, in their process to derive new "elements" of knowledge (facts). By combining existing elements of knowledge through inference mechanisms, new facts will enrich a pre-existing knowledge base. In the case of a diagnosis process, facts are generally obtained by requesting additional information on the considered subject, through answers to selected and relevant questions. We call this functionality "*fill in the gaps*".

Our main argument is that any diagnosis process can be modelled as a comparison process between two sets: the one describing the status of the object of diagnosis, and the one (if available) containing all known and potential situations (in the related domain), which in fact constitutes the knowledge of the expert in the considered domain. The problem we aim to explore in this paper is how domain ontologies and their characteristics can be used to represent this comparison process, and how this provides a way to simulate human reasoning, mainly in diagnosis processes.

Diagnosis follows a structured process where an expert gathers and analyses data before formalising reasoning. Initially, deductions emerge passively until enough relevant information is assembled. At this point, the expert shifts to targeted inquiry, using structured questions and diagnostic tools to refine hypotheses and prioritise plausible diagnostic paths. Throughout this process, knowledge is incrementally formalised into an ontological model, structuring observed data while progressively linking it to expert knowledge. This transition marks the shift from unstructured reasoning to structured problem-solving, where reasoning dynamically adapts to acquired knowledge. This mechanism aligns with our ontology-based approach, mirroring the expert's cognitive framework or "mental model". Since an ontology captures only a subset of domain expertise, diagnosis consists of retrieving and identifying the relevant portion within this structured representation.

Our approach is built upon an adapted version of the described meta-model in (Otmani *et al.*, 2018), which serves as a foundational meta-ontology for validating problem instances (e.g., diseases or faults) through structured diagnostic criteria. While our framework is inherently domain-agnostic, we demonstrate its validity through a use case in the medical domain, with a specific focus on cardiology diagnosis in (Klioui *et al.*, 2025). This dynamic approach continuously refines its conclusions through trial and error, readily adjusting initial assumptions when confronted with new evidence. Through this adaptive cycle, the range of potential diagnoses is systematically narrowed down to the most plausible options, as represented in Figure 1. To enable consistent knowledge representation across various domains (e.g., healthcare, engineering, cybersecurity), we add a generalised ontological meta-model comprising the following core classes described in Table 1. The unchanged class are not mentioned.

Table 1. Generalised classes for artificial reasoning

Generalised (Cross-Domain)	Original (Medical)	Definition
Observation	Sensation	Perceivable data
Component	Organ	Functional unit
Action	Treatment	Intervention to resolve the issue
Area	Part of Body	Structural container
Agent	Person	Entity performing/describing

4. IMPLEMENTATION

Our diagnostic model is implemented around two core components that interact dynamically: Knowledge Module, that serves as the structured repository of domain expertise, and the Inference Module, that performs reasoning over the first one using forward chaining, backward chaining and adapts its strategy iteratively based on data source (human input, sensor streams…etc.).

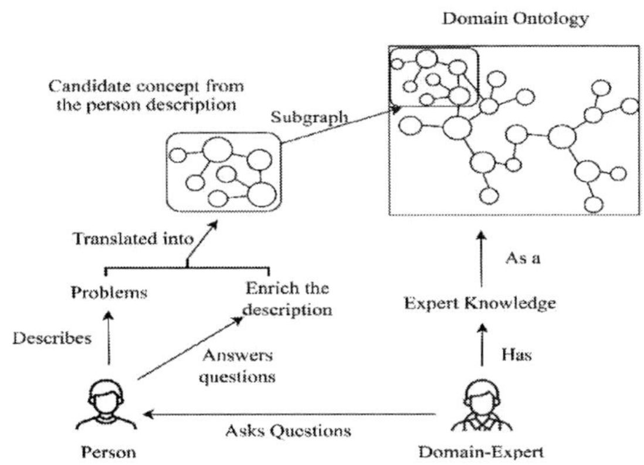

Figure 1. Diagnosis representation using ontologies

Our approach consists of instantiating our ontological framework for cardiology diagnosis, a domain chosen for its well-defined clinical guidelines (Visseren *et al.*, 2021), and a rich ontological foundation in alignment with SNOMED CT (SNOMED International, 2023). The Knowledge Module is encoded in Web Ontology Language (OWL2) (Grau *et al.*, 2008) and developed using Protégé editor version 5.5.0 (Musen and Team, 2015) . This module provides access for data retrieval and updates, including concepts like Observations (ChestPain, Dyspnea…etc), Diagnosis (Myocardial Infarction, Angina Pectoris…etc); Relations and Semantic Web Rule Language (SWRL) rules (Horrocks *et al.*, 2004). The OntoGraf Protégé's plugin facilitates interactive exploration of relationships within our ontology.We developed a client-server application, where the server-side is an API implemented with .Net 8 (Cvijić and Ranilović, 2024), since this version integrates AI/ML capabilities. We used the dotNetRDF (Leki and Mahoro, 2020) library to handle SPARQL Protocol and RDF Query Language (Arenas, Gutierrez and Pérez, 2010) queries like:

$$SELECT\ ?diagnosis\ WHERE\ \{:\ ChestPain\ :\ leadsTo\ ?diagnosis\ .?diagnosis\ a\ :\ CoronarySyndrome\}$$

The SPARQL query is performed on the ontology to search for the node patterns (a combination of inputs) that lead to a potential diagnosis. The results are then displayed to the user in a graph form using GoJS Library (Software, 2023). Additionally, the Inference Module, via Protégé's embedded reasoners, ensures consistency of the results by applying reasoning techniques to the knowledge base to derive conclusions.

We tested our approach on real cardiology cases from specialised symptomatology sources, assessing diagnostic capability across diverse symptom patterns. The first phase includes fifty anonymised cases, validated by a cardiologist. Each case was analysed using our ontology-based framework, extracting distinctive symptoms from full clinical data. The evaluation is being expanded to a larger anonymised dataset to ensure robust assessment, Table 2., our iterative diagnostic approach then progressively refined the diagnosis at each iteration until arriving at a final result Table 3. Finally, a study case application using our approach is presented in Table 4.

Table 2. Conditions with distinctive symptoms

Condition	Symptomatology	Distinctive Symptoms
Angina Pectoris (01)	Paleness, Weakness, Dizziness, Shortness of breath, Sweeting, Chest pain, Crushing feeling, Vomiting, Arm or shoulder pain.	*Sweating, chest pain, crushing feeling, vomiting, arm or shoulder pain.*

Myocardial Infarction (02)	Paleness, Weakness, Dizziness, Shortness of breath, Chest pain, Arm or shoulder pain, Nausea, Cold sweats, Anxiety, Syncope.	*Chest pain, arm or shoulder pain, cold sweats, Syncope.*
Dilated cardiomyopathy (03)	Paleness, Weakness, Dizziness, Shortness of breath, Abdomen Swelling, Lack of appetite, Sudden weight gain, Persistent cough, Syncope, Palpitations.	*Abdomen Swelling, Lack of appetite, Sudden weight gain, persistent cough.*
Hypertrophic Cardiomyopathy (04)	Cyanosis, Chest pain, Palpitations, Dizziness, Syncope, Heart murmurs, Shortness of breath, Headaches, Swelling, Cold extremities.	*Heart murmur, Rapid heartbeat post-exercise, swelling, syncope, Palpitations.*
Coronary Artery Disease (05)	Chest pain, Radiating pain, Shortness of breath, Weakness, Dizziness, Palpitations, Erectile dysfunction.	*Chest pain, Radiating pain, Erectile Dysfunction.*
Heart Failure (06)	Shortness of breath, Fatigue, Swelling, Coughing, Rapid heartbeat, Weight gain, confusion, exercise intolerance, loss of appetite.	*Swelling, coughing, rapid heartbeat, exercise intolerance.*
Infective Endocarditis (07)	Fatigue, Fever, Chills and sweats, Heart murmur, Joint pain Muscle aches, Shortness of breath, Cough, Petechiae, Weight loss.	*Fever, Chills and sweats, Heart Murmur, Joint pain, Muscle aches Cough, Petechiae, Weight loss.*

Table 3. Model accuracy on our dataset

Condition Number	Number of iterations	Our Model Accuracy (%)
01	05	*100*
02	06	*98*
03	04	*95*
04	08	*94*
05	08	*93*
06	06	*88*
07	05	*80*

Table 4. An example of application of our model on study case

Study case	Symptomatology Set	Our Tool Questions	Possible Diagnosis
Angina Pectoris	{Weakness, Paleness, Dizziness, Nausea, Shortness of breath, **Chest pain, Sweating, Arm or Shoulder pain, crushing feeling, vomiting**}	*Do you have **Chest pain**? Do you have **Arm or Shoulder pain**? Do you have **Sweating**? Do you **feel a heavy weight on your chest?** Have you had any **nause** or **vomiting***	*Angina Pectoris, Myocardial infarctio Ventricular Tachycardia, ... **Angina Pectoris***

5. CONCLUSION AND FUTURE WORK

This paper presents an ontology-based approach to simulating human reasoning in diagnosis, applied to cardiology (Klioui *et al.*, 2025), it facilitates expert interactions and formalises the diagnostic process using structured knowledge. We demonstrate that ontology-based reasoning enables adaptive knowledge acquisition while avoiding rigid inference mechanisms. Rather than automating diagnosis, our goal is to model expert knowledge and reasoning through ontology-driven frameworks. One of our future goals is to combine our ontology-based approach and large language models (LLMs) into a solution, where the domain ontology supports artificial reasoning, while the LLM facilitates continuous enrichment of the ontology and enables natural language interaction.

ACKNOWLEDGMENTS

The authors express their sincere gratitude to Dr. Rachid ACHAIBOU, cardiologist, for his valuable assistance and for the interest he has shown in this work.

REFERENCES

Arenas, M., Gutierrez, C. and Pérez, J. (2010). 'On the semantics of SPARQL' In *Semantic Web Information Management: A Model-Based Perspective*. Springer Berlin Heidelberg, pp. 281–307.

Cvijić, B. and Ranilović, P. (2024). 'From .NET Core to .NET 8: A Comprehensive Analysis of Performance, Features, and Migration Pathways', JITA - Journal of Information Technology and Applications (Banja Luka) APEIRON, 24(1).

Grau, B. C. *et al.* (2008). 'OWL 2: The next step for OWL' In *Journal of Web Semantics*, 6(4), pp. 309–322.

Gruber, T. (2018) *Ontology Adapted from Barry Smith's draft @ What is an Ontology?*

Guskov, G., Zarayskiy, V., Filippov, A. and Romanov, A. (2022). 'An Approach to Ontology Based Smart Search in Ecommerce', in V. and G.V. and S.D. Golenkov Vladimir and Krasnoproshin (ed.) *Open Semantic Technologies for Intelligent Systems*. Cham: Springer International Publishing, pp. 361–372.

Hamed, N. *et al.* (2025). 'FooDS: Ontology-based Knowledge Graphs for Forest Observatories', *ACM J. Comput. Sustain. Soc.*, 3(1).

Hamrouni, B. *et al.* (2021). 'Explainable ontology-based intelligent decision support system for business model design and sustainability', *Sustainability (Switzerland)*, 13(17).

Horrocks, I. *et al.* (2004). 'SWRL: A semantic web rule language combining OWL and RuleML.

Igoche B. and Matthew, O. and B.P. and G. A. (2024). 'Ontology-Driven Approach for Knowledge Discovery in Academic Databases', in R. and C.Y.-L. Pichappan Pit and Rodriguez Jorge (ed.) *Advances in Real-Time Intelligent Systems*. Cham: Springer Nature Switzerland, pp. 316–327.

Klioui, Y. et al. (2025). 'An Ontological Model for Artificial Reasoning', International Journal of Online and Biomedical Engineering (iJOE), 21(07), pp. 46–60. Available at: https://doi.org/10.3991/ijoe.v21i07.54361.

Leki,' and Mahoro, J. (2020). *A Comparative Study of dotNetRDF And Semweb.NET Semantic Web Libraries.*

Li, W. *et al.* (2019). 'Review of Standard Ontologies for the Web of Things'

Musen, M.A. and Team, the P. (2015). 'The Protégé Project: A Look Back and a Look Forward', *AI matters*, 1(4), p. 4.

Otmani, N.A. *et al.* (2018). 'Ontology-based approach to enhance medical web information extraction', *International Journal of Web Information Systems*, 15(3), pp. 359–382.

Pastuszuk, J., Burek, P. and Ksiezopolski, B. (2021). 'Cybersecurity Ontology for Dynamic Analysis of IT Systems', *Procedia Computer Science*, 192, pp. 1011–1020.

Sarker, I.H. (2022). 'AI-Based Modeling: Techniques, Applications and Research Issues Towards Automation, Intelligent and Smart Systems', *SN Computer Science*, 3(2).

SNOMED International (2023). 'SNOMED CT: The global language of healthcare'.

Software, N. (2023). 'GoJS: Interactive JavaScript Diagrams for the Web'.

Visseren, F. *et al.* (2021). '2021 ESC Guidelines on cardiovascular disease prevention in clinical practice', *European Heart Journal*. Oxford University Press, pp. 3227–3337.

INTEGRATING LOCAL CLIMATE ZONES AND 3D URBAN DATA TO ANALYZE THE SPATIAL DISTRIBUTION OF POTTED GREENERY IN CHIAYI CITY, TAIWAN

Yu Cheng Chen[1] and Jia Chen Zou[2]
[1]*Assistant Professor*
[2]*Master student*
Department of Architecture, Nanhua University, No. 55, Sec. 1, Nanhua Rd., Dalin Township, Chiayi County 62249, Taiwan

ABSTRACT

This study investigates the spatial distribution of potted greenery within the urban streetscapes of Chiayi City and its relationship with the urban environment. Utilizing the Local Climate Zone (LCZ) classification framework, the research integrates satellite imagery and three-dimensional urban data to categorize the city's urban morphology at a 200-meter grid resolution. These geospatial datasets enable the identification of LCZ types and the extraction of key urban features such as building density, surface coverage, and street geometry. Field surveys were conducted to map the location and density of potted plants across various urban zones, while environmental variables—including distance to roads and sidewalks, sky view factor (SVF), and land use—were collected using GIS-based analysis. Regression and multivariate analyses were applied to examine the underlying factors influencing the distribution patterns of potted greenery. Findings reveal that the presence of potted plants in Chiayi's street environments is more significantly correlated with urban and architectural physical factors than with socioeconomic indicators. Notably, differences in plant distribution patterns were observed between LCZ 2, 3&5 under distinct urban textures. By combining remote sensing, 3D city data, and micro-scale landscape observation, this study demonstrates a scalable methodology for analyzing vegetation presence in dense urban environments. The results provide valuable insights for local governments to refine urban design regulations, promote sustainable streetscape strategies, and support the development of climate-adaptive urban policies.

KEYWORDS

Local Climate Zone, 3D Urban Data, Potted Greenery Distribution, Urban Morphology, Sustainable Urban Design

1. INTRODUCTION

In urban settings, potted greenery reflects the interaction between residents and their environments. On a micro-social level, it may stem from gardening interests, leisure, plant exchanges, or cultural traditions such as warding off misfortune. Spatially, it signifies territorial awareness, personalization of semi-public areas, or practical uses like deterring illegal parking. Historically, urban greenery has long symbolized human efforts to connect with nature (Jin et al., 2021; Suárez et al., 2022; Bhatti et al., 2023).

Nowadays, potted plants serve as flexible, low-cost, and easily deployable greening tools used by households, shops, and communities to enhance their immediate surroundings. Government-led initiatives—like rooftop gardens and street-corner planting—highlight their evolving role as dynamic environmental infrastructure (Tan et al., 2021). As urbanization accelerates, green infrastructure becomes vital for public health, biodiversity, and climate resilience (Haaland & van den Bosch, 2020; Kabisch et al., 2020). Yet, most studies focus on fixed green spaces (e.g., parks, green roofs), overlooking mobile and informal elements like potted plants (Zhou et al., 2023; Pearlmutter et al., 2021).

This study explores the spatial distribution of potted greenery in Chiayi City, using LCZ framework. By integrating satellite imagery, 3D urban data, and GIS analysis using urban environmental factors and social economic factors, we examine how urban morphology relates to plant density. LCZ classification, based on physical factors like building height, street layout, and surface cover, offers greater precision than traditional land-use zoning (Stewart & Oke, 2012; Alexander & Mills, 2023).

The research has three goals: (1) to compare potted greenery distribution in residential and commercial areas under varied urban textures; (2) to analyze how physical factors such as SVF, road width, and surface types influence plant presence; and (3) to build a spatial-statistical model linking urban form to greenery distribution, supporting sustainable design policies. This study fills a research gap by highlighting informal, micro-scale greenery as part of resilient urban planning.

2. METHOD

2.1 Study Area

This study focuses on Chiayi City's to examine the spatial characteristics of non-permanent greenery within a complex urban fabric (Figure 1). The city exhibits a diverse range of architectural forms shaped by different historical periods—from Japanese colonial street grids and post-war modernism to contemporary commercial developments—all coexisting within the same urban context. Such layered spatial characteristics result from a gradual process of urban evolution, influenced by heritage preservation and uneven redevelopment strategies. Given that conventional land use zoning fails to fully capture this complexity, the study adopts the Local Climate Zone (LCZ) classification system as an analytical framework to more accurately interpret and quantify urban morphology.

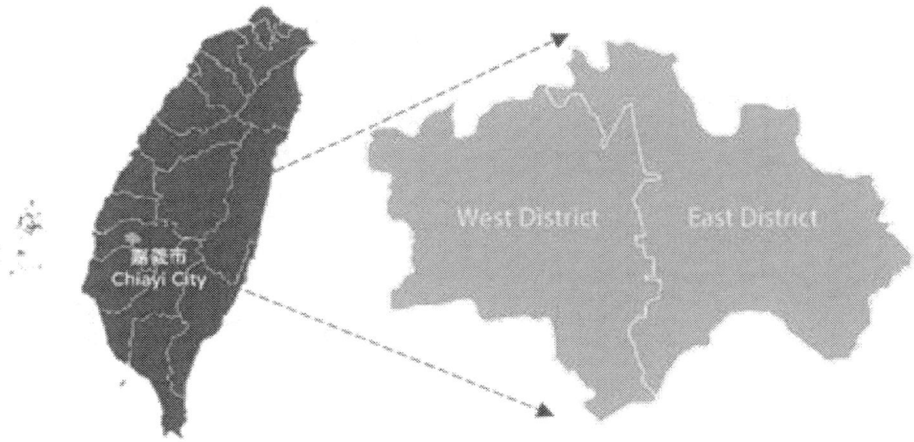

Figure 1. Study area- Chiayi City

2.2 Selecting Criteria for Sampling Sites

To investigate the spatial distribution of potted greenery across different urban settings, this study established a systematic sampling strategy based on four key principles to ensure data reliability (Figure 2):

1. Stratified Sampling Principle: Representative residential and commercial zones were selected based on land-use zoning maps. Population density within each zone was further considered to ensure statistical relevance. 2. Even Coverage Principle: A uniform 200m × 200m grid was overlaid on the selected zones. One street per grid cell was chosen as an observation point, with preliminary observations identifying areas with notable potted plant presence or influence. 3. Feasibility Optimization Principle: Final sampling locations were determined based on street accessibility, pedestrian flow, and proximity to urban centers. Each site was defined with a 200-meter radius of influence, corresponding to the typical double-block scale in Chiayi City. 4. Dynamic Adjustment Principle: Sites such as schools and commercial districts were preliminarily observed to assess the actual influence of non-permanent greenery on pedestrian activity. Sampling locations were flexibly adjusted based on these real-time conditions.

After site selection, the LCZ classification was applied to analyze variations in urban form, allowing a structured assessment of potted greenery distribution and its environmental implications. Based on the above criteria and sampling principles, a total of 60 sampling points were selected within the overlapping areas of stratified sampling zones, covering both residential and commercial areas. These points were used for classifying urban spatial types through LCZ analysis, while potted plants with container diameters of 20 cm or larger were included as valid samples in the survey.

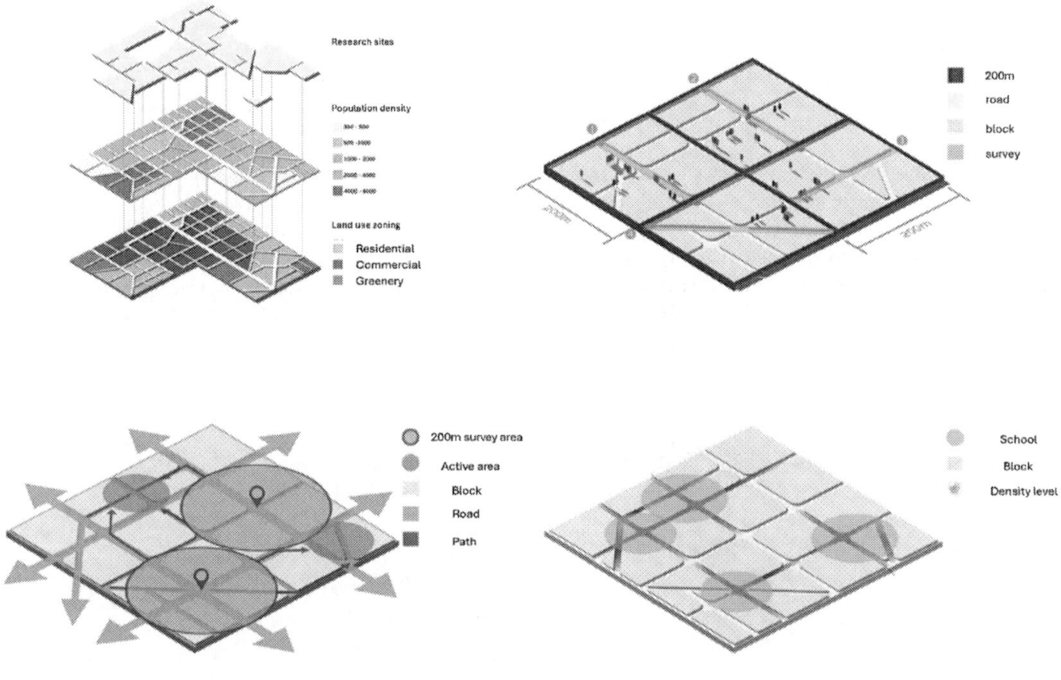

Figure 2. Research Sample Selection Process

2.3 Urban Morphology Classification Method

The LCZ classification system, proposed by climatologists Stewart and Oke in 2012, represents an innovative framework for urban climate research that surpasses the limitations of traditional urban–rural spatial analysis (Figure 3). Designed for use at a 2–5 km spatial scale, LCZs are defined through multidimensional spatial characteristics, offering a more structured understanding of urban morphology. The classification integrates aerial imagery, field surveys, and geospatial data to systematically collect indicators such as land use, building form, and surface cover. These are categorized into 10 built types.

Unlike conventional binary urban–rural models, LCZs adopt a panoramic view of spatial environments, including not only built-up areas but also agricultural land, forests, and water bodies. This allows for more accurate microclimate analysis, especially in geographically complex urban contexts like Taiwan. As a standardized system, LCZ enables cross-city comparisons and provides a science-based foundation for urban planning, offering valuable insights for addressing climate change and urban heat island challenges.

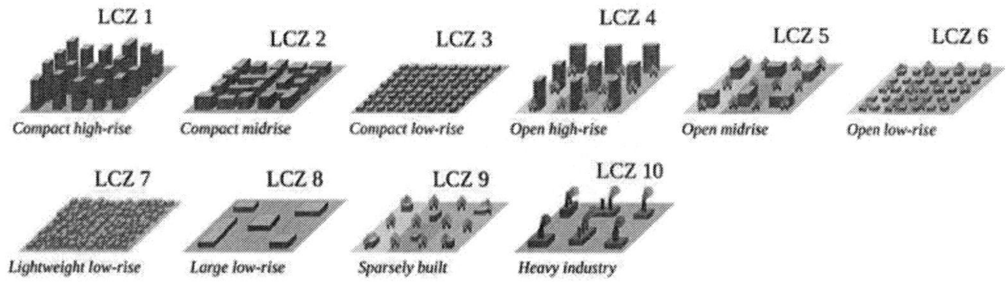

Figure 3. Research Sample Selection Process (Stewart & Oke, 2012)

2.4 Research Tools and Field Measurement Methods

This study focuses on Chiayi City, using a 200m × 200m grid system to define the study area. Within each of the 60 selected grid cells, one primary street was identified for field observation. The research emphasizes two key urban morphological parameters: SVF and Building Aspect Ratio (H/W). Sky View Factor is a dimensionless index ranging from 0 to 1 that quantifies the proportion of visible sky from a given point on the ground. An SVF value of 1 indicates a completely open environment with full sky exposure, while a value of 0 represents a fully obstructed space.

SVF is influenced by surrounding building heights, street widths, and other obstructions, making it a key parameter in assessing urban microclimates. It is commonly used to evaluate solar access, longwave radiation exchange, and ventilation potential in street canyons. In this study, SVF serves as a morphological indicator to understand how spatial openness affects the presence of potted greenery.

In this study, fisheye images were captured on-site using an Insta360 camera, and SVF values were calculated using the RayMan microclimate simulation tool.

Simultaneously, street width and building height were measured in the field using a distance wheel and measuring tape, allowing for precise calculation of the street aspect ratio. Additional environmental features, such as building heights and surface materials, were also recorded to better describe the spatial structure of each sample area. These data served as the basis for LCZ classification and further analysis of how urban form influences the distribution of potted greenery (Figure 4).

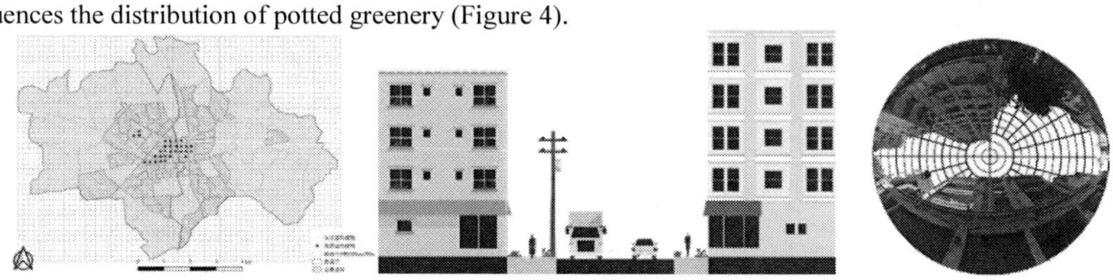

Figure 4. Schematic Map of Sampling Points and Measured Variables

3. RESULTS

This study analyzed 60 samples from LCZ 2, 3, and 5 using Excel-based regression to explore how urban physical factors affect potted greenery distribution (Table.1). Key variables included total road width and SVF. Regression results showed a strong correlation and good explanatory power ($R^2 = 0.641$), with both factors significantly influencing greenery quantity. Residential areas consistently had more potted plants than commercial ones, even within the same LCZ type. For example, in LCZ 5, residential zones had 3.46 times more greenery than commercial zones. This suggests land use strongly impacts greening patterns. Residential areas benefit from flexible space and residents' concern for quality of life, while commercial areas often limit greenery due to spatial constraints and business considerations.

Table 1. Regression analysis results of each urban spatial pattern

	LCZ 2	LCZ 3	LCZ 5
Lane width (m)			
Distant to sidewalk(m)	■		
Total road width(m)	■		■
Building height(m)			■
SVF	■	■	■
H/W			■
Landuse	■		

■Positive ■Negative

4. CONCLUSION

This study examined the spatial distribution of potted greenery in Chiayi City using the LCZ framework, integrating satellite imagery, 3D urban data, and field-based measurements. Among 60 sampling points across LCZ 2, 3, and 5, key physical variables such as SVF, road width, and aspect ratio were analyzed. Regression results confirmed that SVF and total road width significantly influence the number of potted plants in urban areas. Notably, residential areas consistently showed higher potted plant density than commercial zones, even within the same LCZ type. This suggests that land use exerts a considerable impact on informal greenery distribution, with residential zones offering greater flexibility for greening efforts.

In contrast, commercial areas are more limited due to regulatory constraints and functional priorities. The findings offer practical implications for urban planners and policymakers, emphasizing the value of integrating small-scale greening strategies into broader sustainability and resilience planning—especially in high-density, mixed-use urban environments.

To strengthen the robustness and generalizability of the results, future work will incorporate a broader range of LCZ types and expand the sample size. This extension will enhance the analytical precision and facilitate the transferability of the proposed methodology to cities with different morphological and cultural contexts.

REFERENCES

Alexander, L., & Mills, G. (2023). Advances in LCZ classification methods using remote sensing and urban form metrics. *Urban Climate*, 46, 101332.

Bhatti, M. A., Qureshi, S., & Breuste, J. H. (2023). Informal green practices and everyday environmentalism in South Asian cities. *Urban Forestry & Urban Greening*, 85, 127839.

Haaland, C., & van den Bosch, C. K. (2020). Challenges and strategies for urban green-space planning in cities undergoing densification: A review. *Urban Forestry & Urban Greening*, 48, 126552.

Jin, Y., Kong, F., Li, X., & Yin, H. (2021). Urban residents' engagement with small-scale greenery: A study on potted plants in dense neighborhoods. *Cities*, 111, 103097.

Kabisch, N., Korn, H., Stadler, J., & Bonn, A. (2020). Nature-based solutions to climate change adaptation in urban areas. *Springer Nature*, 1st ed.

Pearlmutter, D., Correa, E., & Hughes, M. (2021). The role of green infrastructure in urban microclimate: A review of climate adaptation potential. *Climate*, 9(9), 132.

Stewart, I. D., & Oke, T. R. (2012). Local Climate Zones for urban temperature studies. *Bulletin of the American Meteorological Society*, 93(12), 1879–1900.

Suárez, M., Páez, A., & Rodríguez, D. A. (2022). Micro-greening and well-being: Linking green interventions to urban form and health benefits. *Landscape and Urban Planning*, 218, 104280.

Tan, P. Y., Wang, J., & Sia, A. (2021). Perspectives on five decades of urban greenery in Singapore. *Cities*, 113, 103167.

Zhou, M., Lin, B. B., & Kendall, D. (2023). Informal urban green infrastructure: Understanding spontaneous and small-scale greening practices. *Urban Studies*, 60(2), 375–393.

IMPROVING CITIZEN PARTICIPATION IN SMART CITIES: A LEGAL PERSPECTIVE ON THE NEED FOR ADAPTING PROCESSES AND THE ROLE OF EMERGING TECHNOLOGIES

José Miguel Lucas
PhD student at the Faculty of Law – University of Coimbra
Pátio da Universidade, 3004-528 Coimbra, Portugal

ABSTRACT

According to several studies, despite being legally established, many mechanisms for citizen participation were rarely used, had poor adherence, or were unsatisfactory for participants. In contrast, the Organisation for Economic Co-operation and Development (OECD) has already catalogued effective deliberative practices across jurisdictions, revealing that well-designed mechanisms can increase legitimacy.

This study blends a targeted literature review with an argumentative legal analysis sustaining that emerging technologies (e.g., Artificial Intelligence (AI), Internet of Things (IoT), and blockchain) may enhance citizen participation in smart cities, addressing inclusivity and transparency. It is also analyzed the possibility of introducing a framework that improves regulatory instruments, evolving legal frameworks to ensure moving from symbolic to substantive participation, thereby aligning with Sustainable Development Goals 11 and 16. For instance, improving qualitative analysis via AI semantic analysis, identifying risks and minority concerns, boosting engagement with feedback loops, and ensuring transparency through explainable AI (namely by the own AI), blockchain logging, and audits. Thus, legal frameworks, aligned with OECD guidelines, should defend namely citizens' input integration and universal design, tailored to AI/IoT-rich or resource-limited settings, prioritising, notably, marginalised groups and the elderly. Lastly, effectiveness can be evaluated via participation rates, response times, and inclusivity, using dashboards and pilot projects.

KEYWORDS

Citizen Participation, Smart Cities, Artificial Intelligence

1. INTRODUCTION

We start by asking the same question as (Boussaguet, 2015) (although the Author does so concerning the European Union): has there really been a participatory shift so far, or have we witnessed a rhetorical shift without concrete implications? The problem we identify is that several mechanisms for citizen participation (such as citizens' conferences, deliberative surveys or citizen consultations, despite being normatively established in their respective legal systems, have rarely been used, have not had excellent adherence or have been unsatisfactory for participants, appearing to be more important in theory than in practice. This discrepancy, referred to by some authors as the "participation fallacy" (Delicado, 2018), reveals a critical gap between the normative value of participation and its practical effectiveness (Cardullo & Kitchin, 2017), (Martinez et al., 2019), (Levenda et al., 2020, p. 4), and (Chen et al., 2022).

In this study, we intend to answer the following research questions:

a) How can technology enhance citizen participation while addressing inclusivity and transparency?
b) How can legal frameworks evolve to ensure substantive citizen participation in smart cities?
c) How can participation mechanisms be evaluated for effectiveness?
d) What are the practical and regulatory challenges to implementing this framework, and how can they be addressed?

Previously, in (Lucas, 2025), drawing the idea from ANDREAS KLINKE and ORTWIN RENN, we have already proposed that participatory procedures should vary in scope, depth, and method depending on whether a decision is strategic or operational, addressing how administrative procedures can be adapted to enhance citizen participation in smart cities and how can smart city technologies support tailored participation by enhancing citizens' engagement, inclusivity, and decision-making. In both works, the concepts of "improving" and "greater effectiveness" of citizen participation are closely interconnected, aiming to address the "participation fallacy" by enhancing the practical impact, inclusivity, and transparency of participation processes in smart cities. The difference in terminology reflects nuanced emphases: "improving" highlights the transition from symbolic to substantive participation, while "greater effectiveness" emphasises optimising procedural outcomes.

More precisely, in (Lucas, 2025), we have defined the framework for enhancing citizen participation in smart cities, using six criteria to categorise municipal decisions and determine when and how to engage citizens: economic value, number of people affected, territorial area, impact duration, irreversibility, and qualitative factors (e.g., political sensitivity). Decisions are strategic, impactful, or sensitive if any criterion reaches Tier 3 or above, triggering tailored participation mechanisms. Routine decisions (Tier 1, e.g., pothole repairs) utilise simple digital forms, while higher tiers (3–5, e.g., transit or climate plans) employ workshops, citizen assemblies, or AI-driven platforms, engaging experts, Non-Governmental Organizations (NGO), or the broader public based on the complexity, uncertainty, or ambiguity of the issue. Qualitative decisions require public forums to resolve value conflicts.

In this work, we intend to focus on the legal aspects. Citizen participation should be recognised as a substantive legal right, not merely a symbolic or formal gesture, to ensure meaningful engagement in municipal governance. Grounded in the constitutional principles of democratic participation and the good administration of administrative law, this right entails citizens' entitlement to influence outcomes, not just be heard. Unlike symbolic mechanisms that prioritise visibility over impact, substantive participation requires legal obligations for public authorities to treat and incorporate citizen input, provide transparent feedback, and justify decisions. By regulating participation processes to ensure inclusivity, responsiveness, and accountability (supported by AI-driven tools for processing input) the proposed approach aims to transform participation into a cornerstone of legitimate governance, aligning with the vision of Sustainable Development Goals (SDG) 11 and 16.

As referred to in (United Nations (UN), 2015), Goal 11 of the SDG intends to "make cities and human settlements inclusive, safe, resilient, and sustainable". More precisely, the focus is to ensure that by 2030, all have access to several essential goods and services, like affordable housing, sustainable transport systems and what we would like highlight what we consider to be a holistic concept of sustainability (joining several areas, like environment, social and governance): “participatory, integrated, and sustainable human settlement planning and management in all countries” and “reduce the adverse per capita environmental impact of cities”, “support positive economic, social, and environmental links between urban, peri-urban, and rural areas”, namely by adopting and implementing integrated policies and plans toward inclusion and resource efficiency. Furthermore, regarding Goal 16, the purpose is to promote inclusive societies for sustainable development and build effective, accountable, and inclusive institutions at all levels, which includes ensuring responsive, inclusive, participatory, and representative decision-making at all levels.

Given the scope of this work, it will not be possible to address several topics, such as political participation, referendums, and specific risks like environmental or cyber issues. There will not be a detailed assessment of European regulations, comparative law options, or the strategies and regulations of international organisations, especially the potential implications for data protection. Likewise, given that the topic of participation is multi and interdisciplinary, it will not be possible to cover all realities within the scope of this work, in particular *(i)* the discussion on whether the excessive use of social networks could lead to an approximation of a regime of direct democracy; *(ii)* prior hearing when forming administrative acts; and *(iii)* popular participation in specific matters, such as territorial and environmental planning instruments.

Regarding the methodology, this work employs an exploratory, qualitative approach that combines a targeted literature review with an argumentative legal analysis to propose a specific approach to citizen participation in smart cities. Scientific articles were collected from reputable library databases, including EBSCO, IEEE, ResearchGate, and repositorium.sdum.uminho.pt, as well as Google Scholar. Rather than aiming for an exhaustive summary of all literature – a hallmark of scientific systematic reviews – this work prioritises pertinent sources that either bolster the proposal or illuminate counterarguments, aligning with qualitative research principles. This selective approach reflects a core tenet of legal scholarship, where persuasion takes precedence over comprehensiveness, allowing the analysis to construct a compelling case rather than merely describe the field. For references, the "American Psychological Association (APA), 17th edition" style and Zotero were used.

2. PROPOSED SOLUTION

2.1 Enhancement of Citizen Participation by Emerging Technologies

We are aware that, although smart cities are increasingly becoming a reality, their widespread implementation is uneven globally. We follow the concept established jointly by the United Nations Economic Commission for Europe (UNECE) and International Telecommunication Union (ITU), according to which a smart sustainable city "*is an innovative city that uses ICTs and other means to improve quality of life, efficiency of urban operation and services, and competitiveness, while ensuring that it meets the needs of present and future generations concerning economic, social, environmental as well as cultural aspects*" (UNECE (United Nations Economic Commission for Europe), 2018), to which we add a legal perspective (Lucas, 2023).

The document focuses on smart cities because they leverage emerging technologies (e.g., AI, Internet of Things (IoT), blockchain) to enhance governance, making them ideal for testing advanced participation mechanisms. In smart cities, traditional mechanisms like citizens' conferences, deliberative surveys, and citizen consultations are not entirely replaced. However, they are often transformed or integrated into more advanced, technology-enhanced mechanisms to align with the digital governance framework. These traditional methods, which rely on in-person or paper-based participation, are still used where digital access is limited or for specific, high-impact decisions that require in-depth dialogue. They are usually converted into hybrid or fully digital formats, such as the platform Go Vocal (www.govocal.com).

We defend that the emerging technologies (e.g. AI, IoT, and blockchain) may enhance citizen participation in smart cities by:

a) Improving qualitative participation and risk identification, processing large-scale qualitative data and using semantic analysis with AI, allowing a timely response once submissions have been received (de Vries, 2000) and (Ranchordas, 2021), and to more easily identify risks and minority relevant situations. In the example of changing the pavement to stone pieces on a street in Coimbra where buses pass, AI could analyse qualitative feedback, such as a citizen alerting that buses might break the stones. By cross-referencing this input with historical data, urban engineering insights, or similar cases, AI could flag the risk of structural damage and associated expenses (Câmara Municipal de Coimbra, 2022), preventing wasteful spending. Additionally, AI could identify minority-relevant risks – such as accessibility concerns for elderly residents or those with mobility issues due to uneven stone surfaces – by detecting less frequent but critical feedback from underrepresented groups, ensuring inclusive risk assessment and decision-making.

b) Increasing engagement and feedback validation with platforms that allow feedback loops to validate citizen impact (as exemplified by platforms like Decidim in Barcelona), establishing clear channels for participants to see how their input shapes outcomes (e.g., post-decision reports), and reducing dissatisfaction by demonstrating influence.

c) Adding transparency and accountability through explainable AI processes, supported by explainability clauses for AI-driven recommendations, system auditability (with public access to algorithms and training data), and public registers of participatory tools, alongside dashboards, cost-benefit projection, data visualisation, annual reports, independent oversight (addressing systemic barriers) and blockchain logging of all inputs and decisions with law or municipal regulations mandating public access, with regular audits to mitigate bias and adherence to data protection principles (e.g., anonymisation).

d) Promoting inclusivity and accessibility by providing hybrid mechanisms, universal design standards (accessible to all, intuitive, and easy to use), as well as considering offline access and multilingual interfaces/platforms with real-time translation and mediators to engage marginalised groups. IoT sensors can also provide complementary real-time urban data (e.g., traffic or air quality updates) via accessible mobile apps, enabling citizens, including those with limited mobility or resources, to participate informedly in decision-making processes by accessing more data and/or submitting location-specific feedback. People, if they are interested, can participate regardless of their availability or location, needing only to worry about the closing time for accepting proposals, and no longer about the closure of services.

e) Increasing decision-making efficiency and adaptability by tailoring mechanisms to the size of the municipality, the type of decision at stake (streamlining decision-making in certain types of decisions), and the platforms' ability to allow citizens to choose their preferred participation methods. The rule-setting process begins with public consultations (and, if necessary, expert panels) based on the established criteria (e.g., economic value, irreversibility) for tiered participation. Digital Twin technology can be used to create virtual replicas of urban environments, enabling citizens to simulate and evaluate the impact of proposed policies (e.g., new infrastructure projects) on a collaborative online platform, thereby accelerating consensus-building and ensuring that long-term urban planning aligns with citizen input. Internet and ICT have enabled new forms of citizen participation, as well as communication between citizens and institutions, such as notifications that can now be sent automatically by email or generated by a system incorporated into a website belonging to the competent body's service or the electronic one-stop shop. It is now crucial to consider the impact that participation mechanisms have or may have on administrative structures, particularly those of smaller organisations, which, by nature, have fewer human and financial resources.

f) Strengthening community empowerment and education by employing 5G networks to facilitate high-speed, low-latency communication for virtual town halls and live-streamed consultations, empowering geographically dispersed or marginalised communities to engage actively in real-time governance discussions and vote on key issues. Launching targeted public awareness campaigns and education initiatives, involving people (which seems like the most relevant) to highlight participation benefits and technology use, building trust and awareness (e.g., workshops on AI tools), and addressing low utilisation and simultaneously, simplifying interfaces and training public servants/citizens on technology use, mitigating complexity and enhancing satisfaction.

Legal frameworks (and municipal regulations) should evolve to ensure substantive citizen participation in smart cities by establishing universal principles, aligned with international standards like (OECD, 2022), especially in strategic/impactful/sensitive decisions (Lucas, 2025). This must consider existing participation laws, may involve crafting municipal regulations that mandate the abovementioned references, namely for integration of citizen input into decisions, transparent feedback, and justification of outcomes, especially for marginalised communities and the elderly (even those whose unique perspectives are critical), namely through weighted input mechanisms or mediators.

In 2019, SIMONOFSKI presented the CitiVoice Frameworks, a governance tool designed to help define a citizen participation strategy and serve as a comparison and creativity tool for several cities, facilitating the design of new means of participation (Simonofski et al., 2019). Municipalities could deploy platforms like Decidim (Bria, 2019), (Guardiola, 2019), (Leal Garcia et al., 2023) and (Barandiaran et al., 2024), customised with AI algorithms to analyse feedback in real-time, categorise inputs by theme, and prioritise issues based on sentiment analysis, as piloted in Barcelona's participatory budgeting.

It is necessary to reach those who need it most, primarily through inclusive technology and improved digital literacy, as well as to strengthen civil society channels through focused, accessible, and well-publicised platforms that increase awareness and trust. (Savaget et al., 2019) and Sieber & Brandusescu (2021) have analysed the role of AI in citizens' participation. AI may amplify the voices of citizens like never before.

Our proposed approach has several limitations, namely that it relies on municipal capacity to implement sophisticated technologies and regulatory frameworks, which may be challenging for smaller or resource-constrained cities (Martinez et al., 2019). Furthermore, the proposed tiered criteria (e.g., economic value, population affected) require local adaptation, risking inconsistency across jurisdictions due to subjective interpretations (OECD (Organisation for Economic Co-operation and Development), 2024). Lastly, the exploratory nature of the research lacks empirical validation, necessitating future testing to confirm the framework's effectiveness in diverse smart city contexts.

2.2 Metrics and Indicators for Evaluating Participation

To assess whether citizen participation in smart cities is genuine and impactful, clear metrics aligned with tiered decision-making frameworks are essential. Instead of assessing the general feeling of satisfaction namely through questionnaires, considering our previous work (Lucas, 2025), we sustain that citizen participation mechanisms can be evaluated for effectiveness focusing on *(i)* participation rates, *(ii)* response times, and *(iii)* inclusivity, bearing in mind that tiers should still be defined in concrete by municipalities (nevertheless, we suggest defining values, for instance, considering that a decision is strategic/impactful/sensitive if any quantitative criterion reaches Tier 3 or above or the qualitative criterion is triggered).

For instance, for Tier 1 decisions (e.g., pothole repairs), participation may be measured by response rates to digital forms, with a threshold of greater than 70% engagement indicating accessibility, and qualitative feedback ensuring proportionality. These values are a mere suggestion and should be analysed and proposed by the municipalities. For tier 2 (e.g., park upgrades), survey completion rates (>50%) and the diversity of respondents (e.g., >30% from marginalised groups) can gauge inclusivity. Tiers 3–4 (e.g., transit plans) may require workshop attendance (with>60% stakeholder representation) and actionable outcomes (e.g., >50% of proposals adopted), reflecting their influence. For Tier 5 (e.g., climate plans), hybrid assembly participation (>80% city-wide engagement) and blockchain-verified voting integrity (e.g., <1% discrepancies) could ensure transparency. Additional indicators include decision implementation rates (greater than 75% for influential participation) and feedback loop completion (e.g., public reports within 30 days).

2.3 Practical and regulatory Challenges

Regarding the practical challenges, we highlight algorithmic bias, resource constraints, legal interoperability, data governance, and cultural resistance. Regarding regulatory challenges, smart cities face discussions concerning data governance, equitable access to digital platforms and conflicting stakeholder interests.

Algorithmic bias in AI-driven tools risks skewed outcomes, requiring legal mandates for regular algorithm audits. Resource constraints in smaller municipalities may also limit platform deployment, necessitating scalable, low-cost solutions such as open-source tools. More critically, legal interoperability may complicate participation. For instance, in making impactful decisions, hybrid assemblies may require inter-municipal agreements to align data protocols; yet, smaller municipalities often lack the necessary legal expertise. As noted by Cabannes (2021), cultural resistance to digital tools further undermines adoption, particularly in traditional governance structures, necessitating public education campaigns. These challenges, more pressing than initially outlined, demand adaptive legal frameworks to ensure inclusivity and transparency. Solutions include mandatory accessibility standards, transparent AI governance, and inter-municipal legal templates, building on prior tiered frameworks to address practical barriers while maintaining proportionality and accountability in smart city governance.

Implementing citizen participation in smart cities also faces significant regulatory challenges, including conflicting criteria and ensuring inclusivity. A primary issue is data governance, particularly with AI-driven platforms. Municipal control over data processing can bias outcomes, undermining transparency. Regulatory frameworks should enforce strict data handling protocols to prevent "civic washing." Another challenge is ensuring equitable access to digital platforms, as noted by Tejedo-Romero et al. (2022), which can exclude marginalised groups due to unequal technological access, thereby violating inclusivity principles. Legal frameworks must mandate universal design standards for platforms. Conflicting stakeholder interests, particularly in impactful decisions, require clear legal prioritisation rules.

3. CONCLUSIONS

This study concludes that law plays a critical role in ensuring transparent, inclusive, and active participation/voluntary involvement/proactive engagement, thereby mitigating the risks inherent in technologies like AI and blockchain. This is achieved by requiring explainable AI processes (namely to the own AI) and public registers of AI tools. Thus, legal frameworks for citizen participation need to evolve universally, in line with OECD principles, to mandate substantive participation, ensuring that marginalised communities and the elderly have a meaningful influence on complex decisions through enforceable regulations and inter-municipal collaboration. Emerging technologies can significantly enhance citizen participation in smart cities by transforming traditional mechanisms into hybrid or digital formats, while addressing inclusivity and transparency, namely through AI-driven risk identification, feedback loops, and blockchain accountability. Metrics, such as response times and inclusivity, may assess effectiveness, enabling adjustments to enhance engagement. Despite limitations (e.g., resource disparities, inconsistent tiered criteria, and a lack of empirical data), this framework offers a proactive and holistic model for sustainable governance.

For future work, empirical validation in diverse smart city contexts is needed to confirm the framework's scalability and adaptability, alongside drafting of a framework template, further exploration of stakeholder roles and regulatory harmonisation to ensure legal interoperability across jurisdictions.

REFERENCES

Barandiaran, X. E., Calleja-López, A., Monterde, A., & Romero, C. (2024). *Decidim, a Technopolitical Network for Participatory Democracy: Philosophy, Practice and Autonomy of a Collective Platform in the Age of Digital Intelligence*. Springer Nature Switzerland. https://doi.org/10.1007/978-3-031-50784-7

Boussaguet, L. (2015). Participatory Mechanisms as Symbolic Policy Instruments? *Comparative European Politics*, *14*. https://doi.org/10.1057/cep.2015.12

Bria, F. (2019). *Building digital cities from the ground up based around data sovereignty and participatory democracy: The case of Barcelona*. https://www.cidob.org/sites/default/files/2024-10/83-92_FRANCESCA%20BRIA_ANG.pdf

Cabannes, Y. (2021). Contributions of participatory budgeting to climate change adaptation and mitigation: Current local practices across the world and lessons from the field. *Environment and Urbanisation*, *33*(2), 356–375. https://doi.org/10.1177/09562478211021710

Câmara Municipal de Coimbra. (2022, October 17). *Explicações sobre as obras na rua João Machado*. Câmara Municipal de Coimbra. https://www.coimbra.pt/2022/10/explicacoes-sobre-as-obras-na-rua-joao-machado/

Cardullo, P., & Kitchin, R. (2017). *Being a 'citizen' in the smart city: Up and down the scaffold of smart citizen participation* [Preprint]. SocArXiv. https://doi.org/10.31235/osf.io/v24jn

Chen, T., Ramon-Gil-Garcia, J., & Gasco-Hernandez, M. (2022). Understanding social sustainability for smart cities: The importance of inclusion, equity, and citizen participation as both inputs and long-term outcomes. *Journal of Smart Cities and Society (SCS)*, *1*(2), 135–148. https://doi.org/10.3233/SCS-210123

De Vries, M. S. (2000). The Bureaucratisation of Participation. *International Review of Administrative Sciences*, *66*(2), 325–348. https://doi.org/10.1177/0020852300662008

Delicado, A. (2018, August 20). *The fallacy of participation in environmental matters (A falácia da participação em matéria ambiental)* [Instituto de Ciências Sociais da Universidade de Lisboa]. Plataforma Barómetro Social. http://www.barometro.com.pt/2018/08/20/a-falacia-da-participacao-em-materia-ambiental/

Guardiola, S. (2019). *La gestión de la brecha digital en los procesos de participación ciudadana: Análisis de los ayuntamientos catalanes que han adoptado la plataforma Decidim*. https://openaccess.uoc.edu/server/api/core/bitstreams/52d31153-5339-4a76-ab04-3d7329985af9/content

Leal Garcia, D., Calleja-López, A., & Linares-Lanzman, J. (2023). Collective intelligence and digital participatory platforms. In S. Boucher, C. A. Hallin, & L. Paulson, *The Routledge Handbook of Collective Intelligence for Democracy and Governance* (1st ed., pp. 326–338). Routledge. https://doi.org/10.4324/9781003215929-29

Levenda, A. M., Keough, N., Rock, M., & Miller, B. (2020). Rethinking public participation in the smart city. *Canadian Geographer-Geographe Canadien*. https://doi.org/10.1111/cag.12601

Lucas, J. M. (2023). 23. Legal concept of smart cities and regulation of dashboards in the governance dimension (to ensure a holistic vision) (23. Conceito jurídico de cidades inteligentes e regulamentação de dashboards na dimensão de governação (para assegurar uma visão holística). In I. C. Fonseca, *Cidades inteligentes e Direito, Governação digital e Direitos—Desafios futuros globais* (pp. 369–396). Gestlegal. https://gestlegal.pt/loja/cidades-inteligentes-e-direito-governacao-digital-e-direitos/

Lucas, J. M. (2025). *12. Greater effectiveness in citizen participation in smart cities—A legal perspective on adapting procedures to complexity, uncertainty, and ambiguity*. 76–82. https://www.thinkmind.org/library/DIGITAL/DIGITAL_2025/digital_2025_2_110_10067.html

Martinez, J. L., García-Espín, P., & Jiménez-Sánchez, M. (2019). Participatory Frustration: The Unintended Cultural Effect of Local Democratic Innovations. *Administration & Society*, *52*. https://doi.org/10.1177/0095399719833628

OECD. (2022). *OECD Guidelines for Citizen Participation Processes*. OECD. https://doi.org/10.1787/f765caf6-en

OECD (Organisation for Economic Co-operation and Development). (2024). *Exploring New Frontiers in Citizen Participation in the Policy Cycle*. OECD Publishing. https://doi.org/10.1787/77f5098c-en

Ranchordas, S. (2021). Experimental Regulations for AI: Sandboxes for Morals and Mores. *Morals & Machines*, *1*(1), 86–100. https://doi.org/10.5771/2747-5174-2021-1-86

Savaget, P., Chiarini, T., & Evans, S. (2019). Empowering political participation through artificial intelligence. *Science and Public Policy*, *46*(3), 369–380. https://doi.org/10.1093/scipol/scy064

Sieber, R., & Brandusescu, A. (2021). Civic empowerment in the development and deployment of AI systems. *SSRN Electronic Journal*. https://doi.org/10.2139/ssrn.4104593

Simonofski, A., Asensio, E. S., & Wautelet, Y. (2019). Citizen Participation in the Design of Smart Cities: Methods and Management Framework. In *Smart Cities: Issues and Challenges* (pp. 47–62). Elsevier. https://doi.org/10.1016/B978-0-12-816639-0.00004-1

Tejedo-Romero, F., Araujo, J. F. F. E., Tejada, Á., & Ramírez, Y. (2022). E-government mechanisms to enhance the participation of citizens and society: Exploratory analysis through the dimension of municipalities. *Technology in Society*, *70*, 101978. https://doi.org/10.1016/j.techsoc.2022.101978

UNECE (United Nations Economic Commission for Europe). (2018). *Guidelines for the development of a Smart Sustainable City Action Plan*. https://unece.org/fileadmin/DAM/hlm/documents/Publications/Guidelines_for_SSC_City_Action_Plan.pdf

United Nations (UN). (2015, September 25). *Transforming our world: The Agenda for Sustainable Development [adopted by the UN General Assembly as Resolution number A/RES/70/1]*. United Nations. https://sustainabledevelopment.un.org/content/documents/21252030%20Agenda%20for%20Sustainable%20Development%20web.pdf

A CONCEPT UTILITY FRAMEWORK FOR INCREMENTAL ONTOLOGY EXPANSION

Naren Khatwani, Lijing Wang and James Geller
Department of Data Science, New Jersey Institute of Technology
Newark, New Jersey, USA

ABSTRACT

A good medical ontology is expected to cover its domain completely and correctly. On the other hand, large ontologies are hard to build, hard to understand, and hard to maintain. Thus, adding new concepts (often multi-word concepts) to an existing ontology must be done judiciously. In this research, we propose a two-stage framework for evaluating candidate concepts for ontology expansion. The framework first employs a utility function that evaluates each candidate concept based on Semantic Relevance and Redundancy Avoidance. For candidates that meet a minimum utility threshold, a secondary Goodness Function is applied to evaluate an additional qualitative aspect, namely contextual fit. This framework of combining a utility metric with a goodness metric will be helpful for expert staff working on maintaining and extending ontologies. This systematic approach will enable the incremental expansion of ontologies while maintaining both depth and contextual relevance by integrating only concepts with high "utility."

KEYWORDS

Medical Ontology, Ontology Expansion, Concept Utility, Concept Goodness, Ontology Metrics

1. INTRODUCTION

One definition of a (medical) ontology is … "a structured framework that represents knowledge within the medical domain. It consists of a set of concepts and the relationships between them, facilitating the sharing and integration of medical knowledge across different systems and applications" (Gruber, 1993; Gruber, 1995). The Stanford BioPortal (Whetzel et al., 2011) is a repository of over 1,130 medical ontologies and terminologies. Small ontologies are of limited utility, but large ontologies are hard to build, hard to understand, and hard to maintain. Maintainers of widely used medical ontologies are often reluctant to add new concepts to keep the maintainability manageable, unless there is a "business use case" for a new concept. This was specifically observed for the National Cancer Institute thesaurus (NCIt) (Min et al., 2017).

In medical ontology development and knowledge representation, ensuring the quality of the ontology is paramount. Intuitively, only concepts with high "utility" should be added to an existing ontology. This paper presents a novel approach for evaluating the "utility" of concepts for medical ontologies. The goal of this research is to develop a robust two-stage framework that assesses and categorizes candidate concepts for ontology expansion. Our objectives include identifying relevant factors, formulating a numeric Utility Function to preliminarily screen concepts, and combining it with a secondary Goodness Function to further evaluate and confirm the suitability of concepts.

2. RELATED WORK

Incremental ontology expansion in the medical domain is a complex process influenced by both practical and theoretical considerations. Methods for ontology expansion must maintain a balance between utility and manageability. This means the addition must address specific practical needs, such as supporting clinical applications, improving decision-making processes, or enabling data interoperability (Min *et al.*, 2017). A set of requirements for the properties of medical ontologies has been described in a widely cited and subsequently extended paper of James J. Cimino (Cimino, 1998; Cimino, 2006). The question of what makes a concept "useful" (i.e., high utility) has not been deeply explored yet.

Our research addresses this gap by proposing a two-stage framework that first uses a Utility Function to evaluate Semantic Relevance and Redundancy Avoidance, and then applies a Goodness Function for contextual fit. This function combines four factors, namely Brevity, Frequency of Occurrence, Foreign Language Mappability (FLM), and the Presence and Grammatical Structure retrieved from a Medical Dictionary. Our aim with this approach is to ensure that only high-utility, practically applicable concepts are integrated, enhancing both the depth and usability of medical ontologies.

3. BACKGROUND

In the medical field, ontologies play a critical role in ensuring consistent terminology, facilitating data integration, and supporting decision-making processes across various systems. As ontologies grow and become more complex, there is a need for a systematic approach to ontology expansion, where new concepts are added only if they offer substantial value to the existing structure.

A survey of ontology evaluation methods (Brank *et al.*, 2005) identifies key factors for assessing ontology quality, including completeness, conciseness, adaptability, clarity, computational efficiency, and consistency. All these factors are about *whole* ontologies. In contrast, our work focuses on concept-level evaluation. Our approach recognizes that while there are numerous medical concepts available, it is only recommended to include judiciously chosen exemplars into ontologies. Adding concepts based on subjective judgment can lead to redundancies and could compromise the ontology's effectiveness. Our approach aspires that each added concept serves a unique purpose and contributes to the overall integrity and utility of the ontology (Agrawal and Cui, 2020). By integrating these approaches, our framework builds on the existing body of knowledge while addressing a critical gap in the evaluation of concept utility.

4. CONCEPT UTILITY FRAMEWORK

To address the challenges in incremental ontology expansion while preserving the contextual fit, we introduce a Concept Utility framework designed to evaluate candidate concepts. This framework is a decision-making metric that helps determine which concepts should be integrated into a target ontology. This is achieved by evaluating a concept's measurable attributes and its broader impact on the target ontology. The framework consists of two distinct stages. The first stage consists of a Utility Function that screens candidate concepts based on the two factors Semantic Relevance and Redundancy Avoidance relative to the target ontology. In the second stage, we apply a Goodness Function to evaluate the qualitative considerations of contextual fit (Figure 1). Both these stages in unison provide a comprehensive evaluation method to ensure that only high utility concepts are added to the ontology.

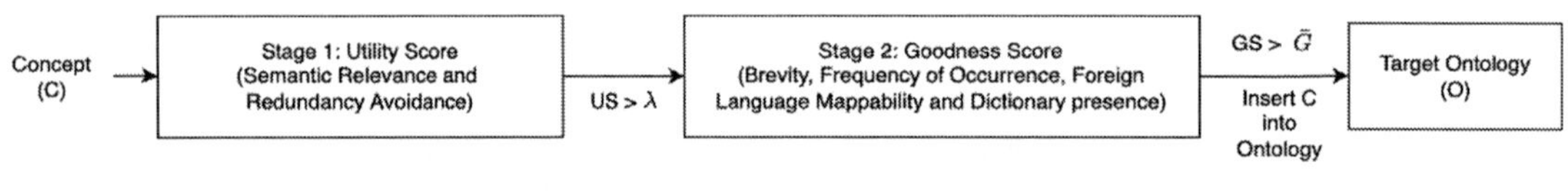

Figure 1. Two-Stage Evaluation Framework

4.1 Stage 1: Utility Function

The first stage of our framework is the Utility Function, which serves as an initial filter to evaluate candidate concepts. This function aggregates two measurable factors for determining the potential value of adding a concept into the ontology. Here, C denotes the candidate concept, and O denotes the target ontology.

$$U(C,O) = \alpha \cdot S(C,O) - \beta \cdot R(C,O) \qquad (1)$$

Specifically, the resulting Utility Score is computed as a weighted sum of the factors Semantic Relevance (S) and Redundancy Avoidance (R), where the weights α and β are determined through empirical testing and expert feedback. Here, S measures the degree of alignment between the candidate concept and the domain of the target ontology, computed using a metric such as cosine similarity. R evaluates the extent to which the candidate concept overlaps existing concepts in the target ontology, effectively penalizing concepts that do not introduce novel information. In effect, S rewards a new concept for being similar to the current content of the ontology, while R penalizes a new concept for being **overly** similar to existing concepts. In practice, the new concept will not be compared with every existing concept, but to the centers of gravity of the major semantic clusters of the ontology. Only concepts that achieve a utility score above a predefined threshold λ are passed to the next evaluation stage.

4.2 Stage 2: Goodness Function

Once a candidate concept passes the preliminary screening via the utility function, it enters the second evaluation stage, the Goodness Function. In this stage, the candidate concept is evaluated as follows.

$$G(C,O) = \frac{(Br * w1) + (FO * w2) + (FLM * w3) + (DP * w4)}{w1 + w2 + w3 + w4} \qquad (2)$$

In Equation (2), the Goodness Score (GS) (Khatwani and Geller, 2024) is computed as a weighted mean, where each component captures a different qualitative or quantitative aspect of the concept. Here, Br (Brevity) measures the conciseness of a concept *name* based on its word count. FO (Frequency of Occurrence) indicates how frequently the concept appears in the literature, reflecting its prevalence and potential relevance within the domain. FLM (Foreign Language Mappability) assesses the ease with which a concept can be linguistically expressed in a foreign language. Here we generalize the approach of (Khatwani and Geller, 2024), which used only one specific foreign language. The idea is that if several natural languages express the same concept with different lengths of terms, i.e., different word counts, then the language with the shortest representation would indicate the goodness of the concept. Ideally, any concept that is expressed by a single word in *any language* should be considered a "powerful" concept. The Unified Medical Language System (UMLS) (Bodenreider, 2004) is a valuable resource for expressions of the same medical concept in different languages. Thus, if a concept J is expressed by two words in "all considered languages," while a concept K is expressed in at least one language by one word, then K would deserve a higher Goodness Score than J. Lastly, Dictionary Presence (DP), which also accounts for the concept's grammatical structure, determines whether the concept is included in a recognized medical dictionary, thereby validating its standardization and acceptance within the field.

Each weight (w1–w4) corresponds to the importance of its respective factor, ensuring that our evaluation remains balanced by reflecting both quantitative and qualitative aspects of the concepts. In a recent study (Khatwani and Geller, 2024), the authors identified the optimal combination of weights for their Goodness Function by surveying a panel of physicians over two iterations. The survey results confirmed a fair agreement (Fleiss' $\kappa = 0.2429$) between the chosen weights and expert judgments. Details are omitted due to space limitations. This paper builds upon the evaluation approach of (Khatwani and Geller, 2024), by generalizing the language mappability and by adding a second stage to the evaluation process. (More precisely, this paper is adding a new first stage). This two-stage approach, combining the initial utility screening with an in-depth goodness evaluation, ensures that only useful concepts are integrated into the ontology.

5. RESULTS OF CASE STUDY

As a case study, we applied our two-stage evaluation framework to the Environmental Determinants of Health (EnDOH) ontology (Kolapally and Geller, 2024), which contains concepts describing negative effects on human health by factors such as air, water, or noise pollution.

In Stage 1, we calculated a Utility Score for each candidate concept based on its Semantic Relevance and Redundancy Avoidance. Semantic Relevance is determined using "sentence embeddings," where a multi-word term for a concept is treated as a sentence. The ontology is viewed as consisting of several sub-hierarchies. The sentence embedding of a new concept is then compared to the sentence embeddings of existing concepts, and the sub-hierarchy with the highest average cosine similarity score is chosen (e.g., "home_as_microenvironment" or "at_the_mesoenvironment_level"). Within the chosen sub-hierarchy, redundancy is measured using Jaccard Similarity between the candidate concept and each concept in the sub-hierarchy. The maximum Jaccard Similarity score is used as the Redundancy Avoidance score, penalizing new concepts that closely resemble existing.

To establish the threshold λ, we perform a leave-one-out procedure: Each one of the 102 EnDOH concepts is removed and reevaluated as a new candidate for inclusion. Its Utility Score is calculated relative to the remaining concepts in its sub-hierarchy, and then reinserted at its original position. The average of the resulting values is then defined to be the threshold.

In Stage 2, we compute a Goodness Score (GS) for each concept in the ontology based on these components: Brevity, Frequency of Occurrence (occurrence in PubMed abstracts from the past five years), Foreign Language Mappability, Presence in a Medical Dictionary, and Grammatical Structure. All scores are normalized against EnDOH to ensure consistency. This results in a second threshold value $\overline{G}$, the average GS of all existing concepts. Due to space limitations, we omit the details of computing $\overline{G}$. A new concept is only included in an ontology if its Utility Score is greater than λ and its GS is greater than $\overline{G}$. For EnDOH, we determined the following values for the two thresholds: $\lambda = 0.1666$ and $\overline{G} = 0.4879$. Values for the other parameters *w1* to *w4,* and α, β were reported in a previous paper (Khatwani and Geller, 2024).

The full implementation, including code, data preprocessing scripts, and evaluation framework are available as Jupyter notebooks, at the GitHub repository:

`https://github.com/narenkhatwani/concept-utility-case-study`

6. LIMITATIONS

At this point, we have established the overall theoretical framework and a partial implementation. To achieve full operational use of the framework, a study with a sufficient number of human experts will be necessary. Another issue is that this framework does not allow for restructuring of whole branches of an ontology, e.g., if the ontology becomes too imbalanced. This issue hints at more factors that are open to numeric evaluations, e.g., using algorithms from *balanced search tree theory*.

7. CONCLUSIONS AND FUTURE WORK

While there exists a rich literature on the topics of ontology integration and evaluation, the point of view taken is usually that the final ontology "product" is evaluated. This is not helpful for the day-to-day work of the ontology maintenance staff, whose members often have to make difficult decisions about single concepts and weigh the advantages and disadvantages of every addition. Our work proposes a systematic and computable approach towards making such decisions in a consistent manner. The model appears to have a sufficient number of "degrees of freedom" (the parameters *w1* to *w4,* and α, β, λ, and $\overline{G}$) to model the cognitive processes of an experienced medical ontology expert. Finding the best parameters to optimize the model to achieve human-like performance will require larger scale future work. The prizes would be an algorithmic tool that could be used by less experienced "junior knowledge engineers" and that would accelerate the process of ontology maintenance even for experts. For the specific case of the Environmental Determinants of Health ontology, we computed two thresholds: $\lambda = 0.1666$ and $\overline{G} = 0.4879$. Any concept that surpasses both thresholds should be added to the ontology.

ACKNOWLEDGEMENT

Research reported in this publication was supported by the National Center for Advancing Translational Sciences of the National Institutes of Health under Award Number UM1TR004789. The content is solely the responsibility of the authors and does not necessarily represent the official views of the National Institutes of Health.

REFERENCES

Agrawal, A. and Cui, L. (2020). 'Quality assurance and enrichment of biological and biomedical ontologies and terminologies', *BMC Medical Informatics and Decision Making*, 20(S10) Available at: 10.1186/s12911-020-01342-4

Bodenreider, O. (2004). 'The Unified Medical Language System (UMLS): integrating biomedical terminology', *Nucleic Acids Research*, 32(90001), pp. 267D-270. Available at: 10.1093/nar/gkh061

Brank, J. et al. (2005). 'A survey of ontology evaluation techniques', *Proceedings of the Conference on Data Mining and Data Warehouses (SiKDD).* Citeseer Ljubljana. Available at: https://citeseerx.ist.psu.edu/document?repid=rep1&type=pdf&doi=c15e3ad08e22705bacdd8aa102765cbd5a7ff62e.

Cimino, J.J. (1998). 'Desiderata for controlled medical vocabularies in the twenty-first century', *Methods Inf Med*, 37(4-5), pp. 394-403.

Cimino, J.J. (2006). 'In defense of the Desiderata', *Journal of Biomedical Informatics*, 39(3), pp. 299-306. Available at: https://doi.org/10.1016/j.jbi.2005.11.008

Gruber, T.R. (1993). 'A translation approach to portable ontology specifications', *Knowledge Acquisition*, 5(2), pp. 199-220. Available at: https://doi.org/10.1006/knac.1993.1008

Gruber, T.R. (1995). 'Toward principles for the design of ontologies used for knowledge sharing?', *International Journal of Human-Computer Studies*, 43(5), pp. 907-928. Available at: https://doi.org/10.1006/ijhc.1995.1081

Khatwani, N. and Geller, J. (2024). 'What makes a good concept anyway ?', Available at: 10.48550/arxiv.2409.06150

Kolapally, N.M. and Geller, J. (2024). Environmental Determinants of Health Ontology (ENDOH). Available at: https://bioportal.bioontology.org/ontologies/ENDOH.

Min, H. et al. (2017). 'Relating Complexity and Error Rates of Ontology Concepts. More Complex NCIt Concepts Have More Errors', *Methods Inf Med*, 56(3), pp. 200-208. Available at: 10.3414/ME16-01-0085

Whetzel, P.L. et al. (2011). 'BioPortal: enhanced functionality via new Web services from the National Center for Biomedical Ontology to access and use ontologies in software applications', *Nucleic Acids Research*, 39(suppl), pp. W541-W545. Available at: 10.1093/nar/gkr469

GENDER IMBALANCE IN AAL-FUNDED PROJECTS: A HIDDEN BARRIER TO MARKET SUCCESS

Cosmina Paul and Alexandra Dobre
Ana Aslan International Foundation, Romania

ABSTRACT

Over the past 15 years, the EU-funded Active and Assisted Living (AAL) Program has launched hundreds of transnational projects aimed at fostering innovation in non-medical aging assistance technologies. However, despite the potential of these initiatives, all but a few failed to reach the market. The very few that did enter the market did not bring any disruptive innovation, were not widely adopted, and did not significantly contribute to the quality of care. As a result, there is a clear misalignment between the program's intended outcomes and its actual impact. While the prevailing justification attributes this to the inherently high risks of innovation and often highlights other benefits—such as user-centered design and broad European collaboration—this paper introduces an overlooked structural factor: gender imbalance in project management and teams. We argue that male-dominated management and technical teams have developed innovations that were largely irrelevant and ineffective in a sector historically shaped by the knowledge and labor of women. Consequently, these innovations lacked relevance in a care domain predominantly served by women and failed to address the real needs of end-users—namely, the older population—reflecting a masculine perspective that was often perceived as patronizing rather than supportive. To substantiate this claim, we conducted participatory observations and a survey across multiple projects, which confirmed our hypothesis

KEYWORDS

Gender Imbalance, Masculine Bias in Innovation, AAL EU Program, Care Sector, Technology Adoption

1. INTRODUCTION

1.1 Introductory Remarks

The Active and Assisted Living (AAL) Programme was a European initiative aimed at enhancing the quality of life for older adults through the application of innovative digital solutions. Launched in 2008 and concluding in 2021, the programme's primary objectives were to foster the development of ICT-based products, services, and systems that support healthy aging, enable older individuals to maintain autonomy and participate actively in society, and alleviate the strain on care systems. Over its duration, the programme funded more than 300 projects, with over 220 projects documented as of recent reports. Its priorities included improving the accessibility and usability of technologies, encouraging cross-national collaboration, and aligning technological innovation with the real needs of end-users—particularly older adults, caregivers, and service providers.

Over the course of its 14-year duration, the AAL Programme funded more than 220 transnational projects, drawing participation from research institutions, small and medium-sized enterprises (SMEs), end-user organizations, and industry partners across Europe. The total investment amounted to over €700 million, co-financed by the European Commission and national funding agencies of participating countries. These projects addressed a wide range of domains including independent living, social participation, mobility support, and integrated care. The programme played a crucial role in building a European ecosystem of stakeholders committed to the advancement of active aging technologies and demonstrated the socio-economic value of public-private partnerships in digital health innovation.

1.2 Hypothesis

The study hypothesis is: Gender composition within teams significantly influences project outcomes. The hypothesis refers to Structural Gender Disparities in AAL Projects. When looking at project composition, management and technical teams are overwhelmingly male, while field teams and social sciences experts are predominantly female.

Though, the non-medical care sector, which forms the primary user base for Active and Assisted Living (AAL) solutions, is overwhelmingly female-dominated. According to numerous labor market studies across Europe, women constitute between 70% and 80% of the workforce in caregiving roles such as home care aides, personal assistants, and social workers. These professionals possess critical hands-on experience and an intimate understanding of the practical needs, challenges, and nuances involved in delivering care to elderly or disabled individuals. Despite this, AAL project design and development teams frequently lack sufficient representation from female caregiving professionals, creating a significant disconnect between technological innovation and end-user realities.

For example, through the organization where I am employed as a researcher, I have been directly or indirectly involved in seven projects. Of these, five were led by men, with technology teams composed entirely of men, while the end-user organizations consisted exclusively of women. This dynamic led to a patronizing management style and created significant challenges in effectively translating end-user needs to the technical teams.

2. LITERATURE REVIEW

The literature highlights several critical barriers that hinder the successful adoption and commercialization of Active and Assisted Living (AAL) technologies. These challenges span multiple domains, including user-centered design, technical integration, regulatory concerns, market dynamics, and organizational resistance. Limited User-Centered Design is seen as one of the hindering factors for low market penetration. Many AAL solutions fail due to a lack of meaningful involvement of older adults and caregivers during design phases, leading to poor usability and low relevance (Lemke et al., 2015; Stamate et al 2025). Stewart et al., 2014 look at Technical and Integration Challenges. They see interoperability between various ICT systems and devices remains a significant hurdle, complicating deployment and scalability. Regulatory and Ethical Concerns is a growing field of foci in the field of non-medical device. Hence, privacy, data protection, and ethical considerations often restrict the adoption of AAL technologies, especially when dealing with vulnerable populations (Fedosov et al. 2025; Maidhof et al. 2023, He 2022; Blackman et al., 2016). The transition from funded research to a viable commercial product is often hindered by an absence of business planning and investor engagement, leading to identifying Market and Commercialization Barriers (Avery, G., et al. 2025; Tham, N. A. Q., et al. 2025; Technopolis Group, 2018). Organizational and management challenges also play a crucial role in obstructing implementation. Resistance to innovation within care organizations, coupled with ineffective change management strategies, frequently hinders successful adoption and integration (van Hoof et al., 2018).

Gender as a factor contributing to the hindrance of the AAL project on the market has been investigated only in relation to the gender of the senior population – end-users. Hence, recent research highlights that gender dynamics, including masculinity and gender-related barriers, significantly influence the adoption and effectiveness of Active and Assisted Living (AAL) programs. For instance, studies show that older men often experience barriers to engaging with health interventions due to traditional masculine identities, which discourage help-seeking behavior and open discussion of health needs (Smith et al., 2022). This suggests that AAL solutions must be designed with an understanding of such gendered barriers to enhance male participation. Moreover, incorporating gender perspectives in AAL technology design is critical to addressing the diverse needs of users; failure to do so can result in reduced adoption and effectiveness of these technologies (Jones and Brown, 2023). Additionally, research utilizing the Masculinity Barriers to Medical Care (MBMC) scale indicates that traditional masculine norms can negatively impact men's engagement with healthcare services, a factor that likely extends to AAL program participation (Williams et al., 2021). Collectively, these findings emphasize the importance of integrating gender-sensitive approaches in the development and implementation of AAL initiatives to ensure inclusivity and responsiveness to the distinct challenges faced by different user groups.

3. METHODS

To investigate the role of gender imbalance in the effectiveness of Active and Assisted Living (AAL)-funded projects, we employed a mixed-methods approach. First, we conducted a quantitative gender analysis of project teams and key stakeholders based on publicly available contact and team information from a sample of AAL projects. A total of 33 unique email contacts associated with project coordinators, developers, or research leads were collected. For each, we approximated gender based on commonly recognized gender associations of first names. While acknowledging the limitations of name-based gender attribution, this method allows for an indicative overview of gender distribution within these projects.

In addition to quantitative analysis, we conducted semi-structured interviews with three women representing end-user organizations to gain qualitative insights into how gender dynamics affect decision-making and user-centered design. The interviews focused on their experiences collaborating with male-dominated technical teams, challenges in translating end-user needs, and perceptions of technology relevance within the care sector. Responses were thematically analyzed to identify recurring concerns, particularly in how gender representation influenced communication, innovation adoption, and overall project success.

Finally, our study employed participatory observation and survey distribution across multiple AAL project consortia. We attended project meetings, analyzed design processes, and conducted a survey evaluating perceptions of inclusiveness, relevance to end-users, and team composition. Building on these initial findings, further research will expand along the following lines: deeper investigation into participatory practices within design phases, comparative analysis of team diversity across projects, and longitudinal assessment of end-user engagement and perceived impact.

4. RESULTS

An analysis of 43 AAL (Active and Assisted Living) projects revealed a notable gender imbalance in project leadership and representation. Of the 43 projects examined, 34 (approximately 79%) were led or represented by individuals identified with male-associated names, while only 9 projects (approximately 21%) involved female-associated leadership. This corresponds to a male-to-female ratio of approximately 3.8:1, indicating a significant underrepresentation of women in leading roles within these consortia.

Our participatory observations further highlighted a pattern: male-dominated project teams tended to focus on technical or efficiency-driven innovations with limited contextual alignment to the realities of care work, which is predominantly carried out by women. Survey data echoed this trend, with female respondents more likely to report that project goals and technologies felt disconnected from end-user needs, particularly among older women and frontline care workers.

We argue here that this underrepresentation of women has tangible consequences for the relevance, usability, and adoption of AAL technologies. When the insights and experiential knowledge of female caregivers are not integrated into the design process, projects risk overlooking key functional requirements, cultural sensitivities, and workflow patterns unique to caregiving environments. For instance, interfaces may be less intuitive, monitoring devices may fail to accommodate typical caregiving routines, and support systems may neglect emotional or psychosocial dimensions that female caregivers prioritize. The result is a misalignment that hampers both the effectiveness of the solutions and their acceptance within care communities. Addressing this gap by actively involving female care professionals in all phases of AAL development—from ideation to testing—is essential to ensure that innovations are not only technologically advanced but also practically meaningful and widely embraced by the sector's dominant workforce.

We argue that another consequence was to be seen in the failure to Consult End-Users: Needs assessments were frequently made 'without proper consultation' with women specialists in caregiving. The argument used to bypass consultation was often that the project had already been written and must be executed as planned. This rigid adherence to pre-established project proposals effectively silences valuable feedback and undermines the iterative nature of effective design processes. Without the input of women specialists in caregiving, needs assessments risk being superficial, incomplete, or biased toward assumptions made by designers and managers who lack direct caregiving experience. Consequently, solutions may fail to address real-world challenges such as the complexity of daily caregiving tasks, emotional labor, or the social context in which care is delivered. This disconnect not only compromises the quality and relevance of AAL innovations but also decreases user adoption and satisfaction. To improve project outcomes, it is crucial to embed mechanisms for continuous, authentic engagement with women caregivers throughout all project stages, ensuring that their expertise shapes both the conceptualization and implementation of care technologies.

These findings support our hypothesis: the predominance of male-led teams in AAL-funded projects correlates with a higher incidence of market irrelevance and lack of adoption. The underrepresentation of women—whose lived experience and domain knowledge are deeply embedded in the care sector—may be contributing to the disconnect between technological development and practical applicability in aging care.

5. CONCLUSION

The findings underscore the critical impact of gender imbalance on innovation, commercialization, and adoption in Active and Assisted Living (AAL) projects. Despite significant funding and promising technological advancements, many initiatives failed due to misalignment with real-world needs, resistance to user feedback, and a lack of inclusive decision-making. Male-dominated leadership and technical teams often struggled to translate the experiences and priorities of a care sector largely shaped by women, leading to patronizing solutions that lacked usability and market relevance.

REFERENCES

AAL Programme (2023). *Legacy and Achievements*. Available at: https://www.aal-europe.eu/about-us/legacy-and-achievements/ (Accessed: 23 May 2025).

Avery, G., et al. (2025). *Addressing Adoption Barriers and Accelerating Market Deployment of New Technologies*. RSC Sustainability, 4(1), pp. 78-95.

Blackman, T., Mitchell, L., Burton, E., Jenks, M., Parsons, M., Raman, S. and Williams, K., (2016). *Ambient Assisted Living technologies and older people in the UK: Identifying the barriers to adoption*. Ageing & Society, 36(9), pp.1825-1849. https://doi.org/10.1017/S0144686X15000766

Castro, M., Cota, I., & Stroetmann, K. (2016). 'Active and Assisted Living, a Practice for the Ageing Population and Society: A Review of the European AAL Programme', *International Journal of Integrated Care*, 16(4), pp. 1–14. doi:10.5334/ijic.2512.

Fedosov, S., et al. (2025). *Ensuring Ethical Compliance in AAL: GDPR and AI Regulations*. Journal of Assistive Technologies, 18(2), pp. 112-130.

He, L. (2022). *Privacy-Enhancing Technologies for Active Assisted Living: A GDPR Perspective*. European Data Protection Review, 9(4), pp. 215-230.

Jones, A. and Brown, B., (2023). Gendered perspectives in the design of Active and Assisted Living technologies. *Journal of Medical Internet Research*, 25(1), e40606. https://doi.org/10.2196/40606

Lemke, M., Wiedemann, A., & Wichert, R., (2015). *User Integration in the Development of AAL Systems*. In: Wichert, R., Eberhardt, B. (eds) Ambient Assisted Living. Advanced Technologies and Societal Change. Springer, Cham. https://doi.org/10.1007/978-3-319-11866-6_4

Maidhof, R., et al. (2023). *Privacy Concerns in Video-Based Ambient Assisted Living Systems*. Technology and Aging, 16(1), pp. 45-62.

Smith, J., Doe, M. and Lee, S., (2022). Masculinity and barriers to health intervention engagement among older men. *Innovate Ageing*, 6(4), pp.1-9. https://doi.org/10.1093/geroni/igac062

Stamate, A., Marzan, M.D., Velciu, M., Paul, C. and Spiru, L., (2024). Advancing user-centric design and technology adoption for aging populations: a multifaceted approach. *Frontiers in Public Health*, *12*, p.1469815.

Stewart, J., Williams, R. and Slack, R., (2014). *The Internet of Things and the 'Layered' Model: Barriers and Opportunities in Health and Social Care*. Journal of Innovation in Health Informatics, 22(3), pp.282–292. https://doi.org/10.14236/jhi.v22i3.140

Technopolis Group, (2018). *AAL Market and Investment Report.* [pdf] AAL Europe. Available at: https://www.aal-europe.eu/wp-content/uploads/2018/11/AAL-Market-report-Technopolis-180604.pdf [Accessed 22 May 2025].

Tham, N. A. Q., et al. (2025). *Barriers and Facilitators to Older Adults' Acceptance of Camera-Based AAL Technologies: A Scoping Review*. Innovation in Aging, 9(2), pp. 112-130.

van Hoof, J., Demiris, G., & Wouters, E.J.M. (2018). *Gerontechnology and design: an international perspective.* In: Sixsmith, A., Gutman, G. (eds) Technologies for Active Aging. International Perspectives on Aging, vol 9. Springer, Boston, MA. https://doi.org/10.1007/978-1-4419-8348-0_2

Williams, R., Taylor, D. and Johnson, P. (2021). Psychometric assessment of the Masculinity Barriers to Medical Care Scale among diverse male populations. *American Journal of Men's Health*, 15(4), 155798832110173. https://doi.org/10.1177/15579883211017331

TOWARDS FORMALIZED STUDY PARAMETERS FOR MEDICAL RESEARCH

Carolin Poschen, Britta Berens and Konstantin Knorr
Department of Computer Science, Trier University of Applied Sciences
Trier, Germany

ABSTRACT

When planning and performing a research study on medical data, researchers have to define the data needed for the research proposal, a potential review by the ethics committee, the search of existing data in suitable databases, and for the publication of their research results. We propose a way to define study parameters for medical research in a formalized way and use this formalization for all the above-mentioned use cases. We focus on simplicity for users by guiding them through a user interface and keeping the study parameters human-readable, while internally storing them in a FHIR-compliant specification document. Instead of operating on raw medical data, we propose using generalized and categorized metadata for the definition of study parameters, to allow a consequent search on existing medical data regardless of how the data are stored.

KEYWORDS

Formalized Study Parameters, FHIR, Data Requirements, Medical Data, Specification Document

1. INTRODUCTION

Medical research requires researchers to access medical data, e.g., from hospitals, health institutes, or data trustees. To be granted access to data, researchers must request data, often in the form of a research proposal, which includes detailed descriptions of what data are needed and how they will be used. The European General Data Protection Regulation requires sensitive data to be used in compliance with recognized ethical standards (Recital 33), while some national laws require the approval of ethics committees.

Since data requirements for a planned study must be specified in the research proposal, for approval by an ethics committee, and later for database searches, we propose defining formalized study parameters for medical research. These parameters would then be used consistently across all relevant contexts where data requirements need to be specified. With the EU's attention towards data spaces as supported in the European Data Governance Act, including a European Health Data Space [Shabani, 2022], a consistent way to access data over different data spaces would simplify the workflow for users.

Different existing platforms provide different ways of accessing medical data. The German Portal for Medical Research Data [Kleinert et al., 2024] provides a request form[1] based on the FHIR (Fast Health Interoperability Resources) standard [Saripalle et al., 2019], necessitating users to be familiar with all required attributes of the FHIR resources. Other systems provide a form with free text fields, such as the German Cancer Registry[2], or a questionnaire which has to be answered via email to access data, like the American Heart Association's Precision Medicine Platform [Stevens et al., 2021], which require a lot of manual work to evaluate the requests. The usability of the MIRACUM i2b2 (Informatics for Integrating Biology and the Bedside) feasibility platform, OHDSI's (Observational Health Data Sciences and Informatics) ATLAS, and the Sample Locator of the German Biobank Alliance are investigated in [Schüttler et al., 2021]. Their investigation found the most usability problems with ATLAS, mainly the saving function was unclear, while the definition of exclusion criteria in i2b2 was unclear and the ontology-based parameter selection was found

[1] https://forschen-fuer-gesundheit.de/en/data-and-biosamples/apply-for-data-and-biosamples-for-a-research-project/
[2] https://www.krebsregister-rlp.de/unser-serviceangebot/datenauswertung

as not intuitive. The Sample Locator had the fewest usability problems in their study, but logical connections between criteria were unclear for users.

The FHIR standard specifies the FHIR-search semantic, which provides queries by specifying search parameters as URL parameters, though, according to [Kryszyn et al., 2023], more complex queries might be unreadable for humans. Additionally, tests in [Stevens et al., 2021] revealed that researchers were unwilling to invest time and effort in a complicated access and reuse process. Therefore, we advocate for the definition of a globally standardized framework to formalize study parameters while preserving human readability.

We propose an intuitive and user-friendly interface to define cohorts for studies in medical research. Cohort definitions are internally mapped to the FHIR standard, the leading interoperability standard for exchanging healthcare data [Duda et al., 2022], without requiring users to be familiar with it. In doing so, we standardize and simplify the formalization of data requirements for study proposals, the evaluation by ethics committees, automated search in a given database, and study publication.

2. METHODS

We propose a user-friendly interface, as shown in our prototype in Fig. 1, that allows researchers to define study parameters in a formalized manner without requiring knowledge of the FHIR format. Users are guided through the interface to enter their study parameters, which are internally translated into a FHIR-compliant format, encapsulated within `ResearchStudy` and `Group` resource types. The most important factor of the interface is the simplicity for users so that no technical knowledge or any knowledge of the internal representation is required. We embed study parameters as references of FHIR resource types, like `Patient` or `Encounter`, within the 'characteristics' property of the enrollment group and integrate ontology/terminology services to support standardized codes like SNOMED-CT[3], LOINC[4], and ICD[5].

Properties of FHIR resources provide the capabilities to be queried. To integrate search features into these resources, we suggest to extend properties to not only support their respective type in FHIR, but also an object of the form

```
{"operator" : op, "value" : val}
```

where *op* defines the comparison operator and *val* represents the corresponding value(s). Operators *op* can be comparisons for exact matches *(equals, not equals)*, scalar values and dates *(less than, less or equals, ...)* or lists *(in, not in)*. For example,

```
{"operator" : "$nin", "value" : ["E10", "E11", "E12", "E13", "E14"]}
```

specifies that a condition's ICD-10 code must not appear in the given list of Diabetes mellitus codes. These adjusted FHIR resources result in a specification document. An example of the resulting specification document can be found in [Git, 2025].

To further simplify the data specification process for users, we propose using generalizations and categorizations instead of having to define all needed specifications on their lowest level. For instance, the mass concentration of glucose in blood, serum, or plasma can be stored using 13 different LOINC codes; with provided example units either as *g/dL* or *mg/dL*. Additionally, a value's classification may depend on sex and age of a patient and the laboratory. For the glucose example above, it would mean using the work-in-progress concept of LOINC Groups (here: LG7967-5) with, e.g., the interpretation *elevated*.

To unify this process over all data types, we propose using the traversal capabilities of ontologies for generalization and classification through reference ranges.

[3] https://www.snomed.org/
[4] https://loinc.org/
[5] https://www.who.int/standards/classifications/classification-of-diseases

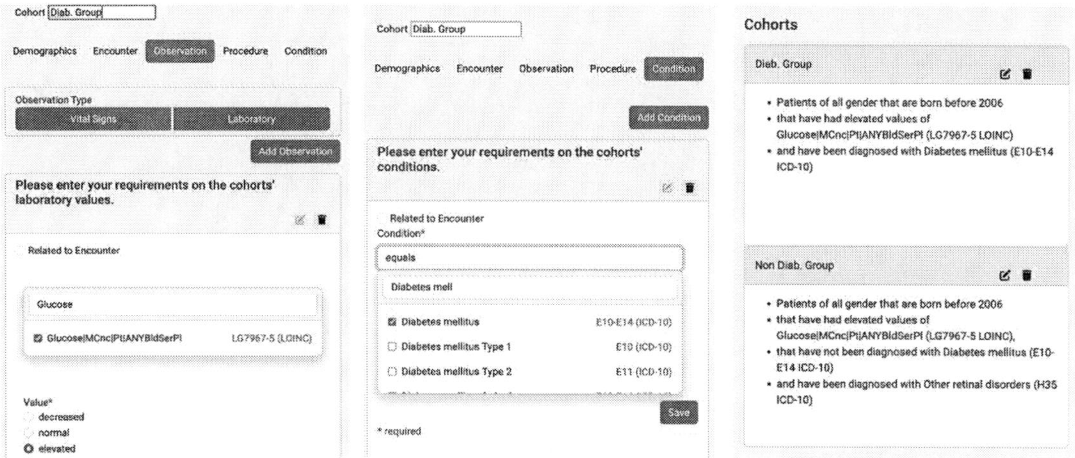

Figure 1. Prototype of a user interface, example of how to enter laboratory results (left); a condition (middle); and an overview of cohorts in a human-readable format (right). The figures can also be found in [Git, 2025]

3. DISCUSSION AND CONCLUSION

Our approach for the definition of formalized study parameters for medical research focuses on the simplicity for users. The user interface guides the user through the process of entering data requirements in a formalized manner without requiring the user to have any knowledge of the FHIR standard, the internal representation of study parameters.

The proposed system to formalize study parameters can be either a stand-alone system or be connected to a hospital or data integration center as mentioned in [Kleinert et al., 2024], or a data sharing space, e.g., as shown in [Poschen et al., 2024].

The resulting specification document is machine-readable in a FHIR-compliant format and can be converted into a human-readable format and back at any time. This bidirectional flexibility allows for a standardized data specification wherever data need to be specified, i.e., it provides consistency in data specification from study proposal to ethics committee review to publication. In a first version, a demonstrator is implemented showing the ability to map user input into the FHIR-compliant specification document which can again be converted into a human-readable format.

Because the specification document is machine-readable, it can be used to carry out an automated search on a given database via FHIR search or using a search algorithm as proposed in [Herres et al., 2024]. As the specification document can be simplified through generalizations and classifications, medical data would need to be annotated correspondingly with these generalizations and classifications for the automated search. The returned data would still be the original medical data.

The generalization and classification during the data specification results in a loss of more detailed information and the annotation of medical data at the source requires preprocessing before carrying out any automated search. However, we believe that the advantages of a simple way to formalize study parameters and the possibility to carry out an automated search based on these parameters outweigh the drawbacks of the loss of information and necessary preprocessing.

ACKNOWLEDGEMENT

This work is part of the DaTreFo project, funded by the German Federal Ministry of Education and Research (16KIS1644).

REFERENCES

Duda, S. N., Kennedy, N., Conway, D., Cheng, A. C., Nguyen, V., Zayas-Cabán, T., and Harris, P. A. (2022). HL7 FHIR-based tools and initiatives to support clinical research: a scoping review. *Journal of the American Medical Informatics Association*, 29(9):1642–1653.

Git (2025). Supplement Material to the Paper. https://github.com/CaroPoschen/SupplementMaterial/.

Herres, B., Poschen, C., and Knorr, K. (2024). Privacy-Preserving Search on Medical Data. In *Digital Health and Informatics Innovations for Sustainable Health Care Systems*, pages 252–256. IOS Press.

Kleinert, P., Gebhardt, M., Buckow, K., Gruendner, J., Ganslandt, T., Prokosch, H.-U., and Semler, S. C. (2024). Germany's Approach for the Secondary Use of Federated Real-World Data in the German Portal for Medical Research Data. In *Digital Health and Informatics Innovations for Sustainable Health Care Systems*, pages 1704–1708. IOS Press.

Kryszyn, J., Smolik, W., Wanta, D., Midura, M., and Wróblewski, P. (2023). Comparison of OpenEHR and HL7 FHIR Standards. *International Journal of Electronics and Telecommunications*, 69(1).

Poschen, C., Herres, B., and Knorr, K. (2024). A Threat-Driven Design of a Data-Trustee Infrastructure for Medical Data. In *2024 IEEE International Conference on Bioinformatics and Biomedicine (BIBM)*, pages 6753–6760. IEEE.

Saripalle, R., Runyan, C., and Russell, M. (2019). Using HL7 FHIR to achieve interoperability in patient health record. *Journal of biomedical informatics*, 94:103188

Schüttler, C., Prokosch, H.-U., Sedlmayr, M., Sedlmayr, B., et al. (2021). Evaluation of three feasibility tools for identifying patient data and biospecimen availability: comparative usability study. *JMIR medical informatics*, 9(7):e25531.

Shabani, M. (2022). Will the European Health Data Space change data sharing rules? *Science*, 375(6587):1357–1359.

Stevens, L. M., De Lemos, J. A., Das, S. R., Rutan, C., Alger, H. M., Elkind, M. S., Zhao, J., Iyer, K., Figueroa, C. A., and Hall, J. L. (2021). American Heart Association precision medicine platform addresses challenges in data sharing. *Circulation: Cardiovascular Quality and Outcomes*, 14(9):e007949.

QUANTIFYING THE IMPACT OF SOCIAL DETERMINANTS ON THE DEMAND FOR HEALTHCARE SERVICES

Michael Renda[1], Fatima Yusuf[1], Rosemina Vohra[2], Soon Ae Chun[3] and James Geller[1]
[1]*Data Science Department, New Jersey Institute of Technology, USA*
[2]*Department of Informatics, New Jersey Institute of Technology, USA*
[3]*Information Systems and Informatics, City University of New York, Staten Island, USA*

ABSTRACT

While there is a general understanding that Social Determinants of Health have an impact on health outcomes, it has been difficult to quantify this relationship and build useful predictive models. This paper describes an approach that uses transactional data from local government as measurable indicators of Social Determinants of Health (SDOH) and applies machine learning techniques to determine a quantifiable relationship between social and economic factors and the demand for healthcare services and develop a predictive model. Our study uses data from the New York City government (citizen complaints and crime statistics), census data (median household income) as indicators for SDOH. The health outcome is Emergency Room visits for asthma. The data was compiled and analyzed for zip codes within New York City. Various machine learning algorithms were attempted, and a Random Forest Regressor had the best results. We found that our regression model was able to explain 80% of the variance in ER visits, which we consider a very good result. The three most influential predictors in our model were: complaints regarding electrical issues, misdemeanor crime incidents, and median household income.

KEYWORDS

Social Determinants of Health, Machine Learning

1. INTRODUCTION

For some time now, researchers have held that social factors have a significant impact on health outcomes (Braveman, A et al, 2011; Braveman, A et all, 2014). While investigators have recognized the need for quantitative and useful measures of that impact (Galea, S et al, 2020), progress has been limited. One approach that has been taken is to use Natural Language Processing tools to extract Social Determinant of Health (SDOH) factors from the structured and unstructured data in Electronic Health Records of diabetes patients (Mehta, S et al, 2023). While the authors were able to identify SDOH factors in the records, their data was at a patient level and has limited applicability to the population of diabetes patients or to the population as a whole. Hanson, H et al focused on a specific SDOH factor (race) to predict prostate cancer mortality (Hanson, H et al, 2019). Grekousis (Grekousis, G et al, 2021) used US geodemographic data at a county level to study correlations of demographic, socioeconomic and underlying health conditions to COVID-19 death rates. Skiendzielewski (Skiendzielewski, K et al, 2022) was able to correlate adverse childhood experiences (ACEs) to what they described as neighborhood-level determinants of health (poverty rates, crime, overdose deaths, less socioeconomic resources, etc.) but without quantifying a relationship to actual health outcomes.

This paper describes an approach that uses transactional data from local government as measurable indicators of SDOH and applies machine learning to determine a quantifiable relationship between social and economic factors and the demand for healthcare services and develop a predictive model. Our analysis focuses on a broad spectrum of social factors and a single healthcare service (emergency room visits for asthma) in a single geographic area (New York City). However, the approach, techniques, and analysis used can be expanded in several different dimensions to quantify the predictability of health risks using social determinants of health.

The healthcare service we are measuring consists of Emergency Room (ER) visits for asthma in New York City from 2016 to 2023. The physiological and environmental causes of asthma have been studied and are well documented (Miller, R et al, 2021). However, it has also been shown that quality of life factors also have an impact on the severity and frequency of asthma attacks (Akenroye, A et al, 2024; Plaza-Gonzales, S et al, 2022; Wypych-Ślusarska, A et al, 2022). We will build a model that will predict the number of asthma-related ER visits based on social factors, economic factors, and crime statistics as described below.

Our hypothesis takes the form of two research questions (RQ):

RQ1: Can social determinants predict the number of ER visits for asthma?

RQ2: What are the most influential predictors for ER visits for asthma?

2. METHODOLOGY

2.1 Source Data

The data for ER visits was sourced from the New York State Department of Health Statewide Planning and Research Cooperative System (SPARCS) (New York University Health Sciences Library, 2025). Our data includes entries for all zip codes in New York City for every month between 2016/01 and 2023/09. This became the timeframe we used for our other data sources as well. The source data tabulates the number of Emergency Room admissions for asthma in four age ranges: 0-4, 5-17, 18-64, and 65+. For our study, we computed a single total for each zip code, year, and month.

Our source for citizen complaints is *311 Service Requests from 2010 to Present* from the NYC OpenData site (NYC Open Data, 2025). This is a record of citizen complaints and service requests received on New York City's 311 hotline. At the time of download, the dataset contained 32,767,564 records. We extracted 20,467,448 records dated within the timeframe of our study. The data is organized in 41 columns including created date, closed date, handling agency, incident type, status, and a wide variety of geo-location descriptor (zip code, borough, latitude, longitude, cross streets, park location, etc.).

Our source for crime statistics is the *NYPD Complaint Data Historic* from the NYC Open Data site (NYC Open Data, 2025). The dataset includes felony, misdemeanor and violation crime reports for the New York City Police Department (NYPD). At the time of download, the dataset contained 8,913,268 records. We extracted 3,673,422 records dated within the timeframe of our study. Crime records include jurisdiction information (borough, precinct, transit district, etc.), suspect demographics, victim demographics, date information (report date, complaint_from_date, complaint_to_date), geolocation information (latitude/longitude, map coordinates, zip code), complaint codes and descriptive information.

The economic data for our study comes from the U.S. Census Bureau's tabulation of Median Household Income by zip code for the years 2016 through 2023 (United States Census Bureau, 2025).

Zip codes were validated using a master list of US Postal Zip codes (United Sates Zip Codes.org, 2025) that contains 42,735 entries. Our analysis was restricted to zip codes in the five counties (Bronx, New York County, Queens, Kings, Richmond) corresponding to the five boroughs (Bronx, Manhattan, Queens, Brooklyn and Staten Island respectively) of New York City. There are 310 zip codes in total (221 Standard, 28 PO Box and 61 Unique). Our study encompasses any of those zip codes that have at least 1 ER visit for asthma during the study time period – 135 zip codes.

All data for this paper was sourced from publicly open data web sites (NYC Open Data, 2025).

2.2 Exploratory Data Analysis

As is typical for data sources as ours, cleanup was an important step. Dates were often mis-formatted or incomplete. Other information had misspellings or inconsistent case. These were all corrected to the extent possible.

We also had to ensure that the categorization and summarization of raw data was efficient and effective for model building. Crime records were summarized by LAW_CAT_CD (Felony, Misdemeanor, Violation). NYC 311 data contain 441 distinct complaint types. Of these 133 distinct complaint types are either clear and obvious data entry errors or extremely rare – fewer than 20 instances each out of over 32 million records. After extensive manual review, we identified 74 distinct complaint types relevant to our investigate. These were then grouped into 16 categories as listed in Table 1 below.

Table 1. NYC 311 call complaint types grouped into categories

Category	Example Complaint types
Air	Air Quality, Construction Lead Dust
Asbestos	Asbestos, Asbestos/Garbage Nuisance
COVID	Face Covering Violation, Vaccine Mandate, 3 others
Electrical	ELECTRIC, Electrical
Food	Food Poisoning, Home Delivered Meal Complaint, 4 others
Garbage	Residential Disposal Complaint, UNSANITARY CONDITION, 20 others
Homelessness	Homeless encampment, Homeless Person Assistance, 1 other
Medical Assistance	Emergency Response Team, X-Ray Machine/Equipment
Mold	MOLD, Mold
Mosquitoes	Mosquitoes
Rodent	Rodent
Safety	Hazardous Material, Safety, Scaffold Safety, 7 others
Sewer	Indoor Sewage, Sewer, 2 others
Smoking	Smoking
Water / Plumbing	General Construction/Plumbing, WATER LEAK, Standing Water, 8 others

The 16 complaint categories, along with 3 categories from NYPD Complaint data and median household income form the 20 initial independent features before Feature Engineering.

2.3 Feature Engineering

Exploratory visualizations (scatter plots, histogram, box plot) were created for each of the initial independent features. An example is shown below in Figure 1.

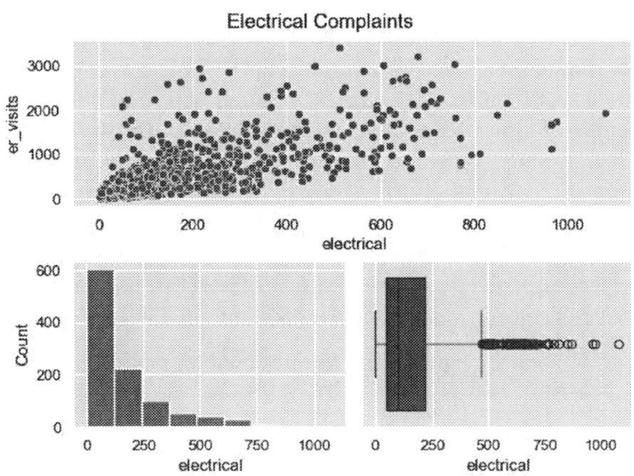

Figure 1. Electrical Complaints visualizations

A correlation matrix was produced and used to create a heat map as shown in Figure 2.

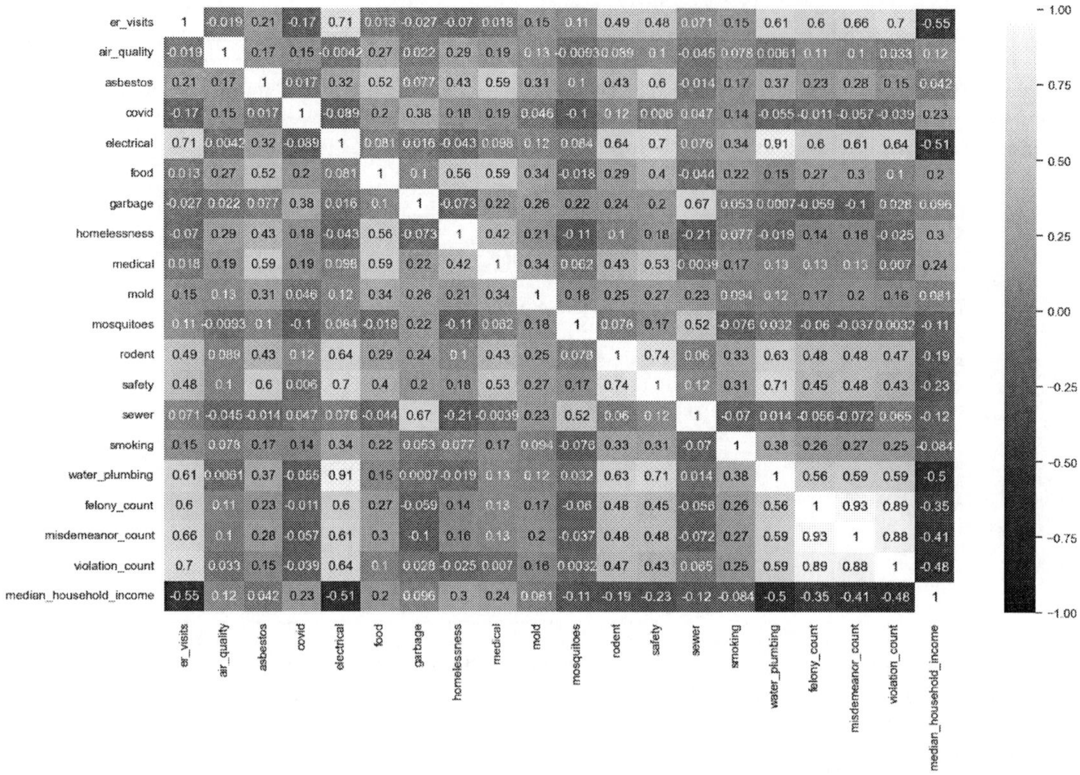

Figure 2. Correlation Heatmap

From the correlation analysis several observations could be made. First, there is a high degree of correlation among felony*_count*, *misdemeanor*_count and *violation*_count (between 0.93 and 0.88). As a result, *felony_count* and *violation_count* were removed from the set of independent features. Second, *water_plumbing* complaints were highly correlated with *electrical* complaints. *Water_plumbing* complaints were removed from the set of independent features. Finally, *air_quality*, *medical* and *food* complaints have very low correlations with *er_visits* – our dependent feature – (-0.019, 0.018 & 0.013). These features were also removed from the set of independent features. Finally, we ran a forward stepwise linear regression adding features to the model in descending order by absolute correlation to *er_visits*. At each stage we computed p-values for each feature in the model. In the final stepwise run, the last feature *garbage*, had a p-value (0.37) greater than our significance level of 0.05. All other features had p-values well below our significance level (worst-case: p-value = 0.023). *Garbage* complaints was removed from the set of independent features.

The full set of 12 features used for model building is listed in Table 2 below.

Table 2. Independent features used for model building

asbestos	covid	electrical
homelessness	median_household_income	misdemeanor_count
mold	mosquitoes	rodent
safety	sewer	smoking

2.4 Model Building

The first model built was a linear regression. This established a baseline performance and was used for the final pruning of independent features as described above in Section 2.3. Next, three regularization algorithms – Ridge, Lasso and ElasticNet - were run to determine if model performance could be improved over the results for linear regression. A Support Vector Regressor algorithm showed a slight improvement in performance. An Artificial Neural Network (ANN) was built and run, but did not significantly improve upon the classic Machine Learning (ML) models. Finally, a Random Forest model built. For each the models attempted, a grid search was performed to select the optimal hyperparameters. Model performance was measured using R^2 values on both the training and testing data sets.

2.5 Outcomes

By a large margin, the best performing model was built using the Random Forest Regressor algorithm. The models built and their performance (as measured by testing R^2) are listed below in Table 3.

Table 3. Regression algorithms used

Algorithm	**Testing R^2**
Linear	0.6449
Ridge	0.6311
ElasticNet	0.6311
Support Vector Regressor	0.6667
Random Forest Regressor	**0.8005**
Artificial Neural Network	0.6240

The hyperparameters for the Random Forest Regressor were set at: max_features=log2, min_samples_split=4, and n_estimators=40. The feature importance outputs of the model are listed in Table 4 below:

Table 4. Random forest feature importance

Independent Feature	**Gini Importance**
electrical	0.2424
misdemeanor_count	0.2092
median_household_income	0.1806
rodent	0.0987
safety	0.0659
sewer	0.0484
asbestos	0.0340
homelessness	0.0317
covid	0.0280
smoking	0.0262
mosquitoes	0.0207
mold	0.0131

3. CONCLUSION

Our work suggests that social determinant proxies based on government transactional data can be used to build predictive models for healthcare service demand. Our best performing model achieved a testing R^2 score of 0.8005. This means that our model explains 80% of the variance in the number of ER visits for asthma by zip code in New York City. We think this is an excellent result given the nature of the problem and the available data. One of the advantages of the model is that it utilizes hard data that is collected in the normal course of administering a government entity. It does not require special survey data or estimations.

In addition, we were able to identify and rank the most important features used to make predictions. The model was built using features that can be considered along three main dimensions: citizen complaints, crime and economic level. It is telling that the three most important features in the model – electrical, misdemeanor_count and median_household_income – are each from one of those three dimensions. We conclude then that similar models, models that aim to build upon this work, should include at a minimum features along these dimensions.

There are a number of ways in which this work can be extended and improved upon. We intend to expand our data collection activities, fine-tune the algorithms used and explore new algorithms – especially deep learning. We also need to apply the model, and future generations of the model, to predict other urgent health outcomes – heart disease, diabetes, obesity, etc.

ACKNOWLEDGEMENT

Research reported in this publication was supported by the National Center For Advancing Translational Sciences of the National Institutes of Health under Award Number UM1TR004789. The content is solely the responsibility of the authors and does not necessarily represent the official views of the National Institutes of Health.

REFERENCES

Akenroye, A, et al, (2024). Socioeconomic status is of higher importance than air pollution and environmental factors in childhood asthma prevalence in urban Australia. *The Journal of Allergy and Clinical Immunology*, vol. 154,2: 278-279. doi:10.1016/j.jaci.2024.05.021.

Braveman, P, et al, (2011). The Social Determinants of Health: Coming of Age. *Annual Review of Public Health,* vol. 32,2011: 381-398. doi:10.1146/annurev-publhealth-031210-101218.

Braveman, P, Gottlieb, L, (2014). The social determinants of health: it's time to consider the causes of the causes. *Public Health Reports* 129.1_suppl2.

Galea, S, et al, (2020). Social determinants of health, data science, and decision-making: Forging a transdisciplinary synthesis. *PLoS Med* 17(6): e1003174. https://doi.org/10.1371/journal.pmed.1003174.

Grekousis, G, et al, (2021). Mapping the geodemographics of racial, economic, health, and COVID-19 deaths inequalities in the conterminous US. Applied Geography Volume 135. https://doi.org/10.1016/j.apgeog.2021.102558.

Hanson, H, et al, (2019). The Relative Importance of Race Compared to Health Care and Social Factors in Predicting Prostate Cancer Mortality: A Random Forest Approach. *Journal of Urology* 202(6):1209–16. Available at: https://doi.org/10.1097/JU.0000000000000416.

Mehta, S et al, (2023). Social Determinants of Health Documentation in Structured and Unstructured Clinical Data of Patients With Diabetes: Comparative Analysis. *JMIR Med Inform* 11:e46159 medinform.jmir.org/2023/1/e46159.DOI: 10.2196/46159.

Miller, R L et al, (2021). Advances in asthma: New understandings of asthma's natural history, risk factors, underlying mechanisms, and clinical management. *The Journal of Allergy and Clinical Immunology* vol. 148,6 (2021): 1430-1441. doi:10.1016/j.jaci.2021.10.001

NYC Open Data. (2025). "311 Service Requests from 2010 to Present". Available at: https://data.cityofnewyork.us/Social-Services/311-Service-Requests-from-2010-to-Present/erm2-nwe9/about_data.

NYC Open Data. (2025). "NYPD Complaint Data Historic". Available at: https://data.cityofnewyork.us/Public-Safety/NYPD-Complaint-Data-Historic/qgea-i56i/about_data.

NYU Health Sciences Library. (2025). "Statewide Planning and Research Cooperative System". Available at: https://datacatalog.med.nyu.edu/dataset/10006.

Plaza-González, S et al, (2022). Psychological and Sociocultural Determinants in Childhood Asthma Disease: Impact on Quality of Life. *International Journal of Environmental Research and Public Health* vol. 19,5 2652. 24 Feb. 2022, doi:10.3390/ijerph19052652.

Skiendzielewski K, et al, (2022). The intersection of adverse childhood experiences and neighborhood determinants of health: An exploratory spatial analysis. *Psychol Trauma.* 16(Suppl 1):S125-S132. doi: 10.1037/tra0001320. Epub 2022 Jul 14. PMID: 35834220; PMCID: PMC9839886.

United States Census Bureau. (2025). "median household income – Census Bureau Tables". Available at: https://data.census.gov/table?q=median%20household%20income&g=010XX00US$8600000

United States Zip Codes.org. (2025). "Zip Code Database". Available at: https://www.unitedstateszipcodes.org/zip-code-database/

Wypych-Ślusarska, A et al, (2022). Social Inequalities: Do They Matter in Asthma, Bronchitis, and Respiratory Symptoms in Children? *International Journal of Environmental Research and Public Health* vol. 19,22 15366. doi:10.3390/ijerph192215366.

EVALUATING LARGE LANGUAGE MODEL PROMPTING STRATEGIES FOR EXTRACTING BEHAVIORAL CHARACTERISTICS OF VIOLENT OFFENDERS

Jonah Treitler[1], Gabriella Danziger[2, *], Matthew Marshall[3], Mohammad Hassan[2, **], Alexander Lundberg[2, ***] and Lori Post[2, ***]
[1]Thomas Jefferson High School for Science and Technology, Alexandria, VA, United States
[2]Northwestern University, Evanston, IL, United States
3Michigan State University, East Lansing, MI, United States
**B.A.*
***B.S*
****Phd*

ABSTRACT

Large Language Models (LLMs) like GPT-4o can assist with the collection and processing of public health data. LLM-generated outputs, however, are not always accurate. In this study, we evaluated three prompting strategies (few-shot prompting, self-prompting, and self-prompted chain-of-thought) to enhance GPT-4o's performance in identifying behavioral characteristics associated with mass violence attackers. Using a dataset of 197 United States mass violence attack incidents, we focused on five attributes previously exhibited suboptimal model-human data extraction agreement (<60%). The self-prompted chain-of-thought approach led to large improvements in accuracy (average increase = 9.8 percentage points), followed by self-prompting (4.6 percentage point increase), while few-shot prompting resulted in an average decrease in accuracy of 5.8 percentage points.

KEYWORDS

Large Language Model, Prompt Engineering, Chain-of-Thought

1. INTRODUCTION

Large Language Models (LLMs) are key technologies in the generative AI (GenAI) revolution, enabling breakthroughs across natural language processing tasks such as summarization, question answering, and information extraction. Models like GPT-4 are trained on trillions of tokens and have demonstrated impressive few-shot and zero-shot learning capabilities, often rivaling human performance on benchmark tasks. (Brown et al., 2020; OpenAI et al., 2023; Raji et al., 2021)

Public health researchers are increasingly interested in the use of LLMs in data collection and analysis. The research applications include detecting public sentiment, monitoring news trends, and profiling complex phenomena such as violent behaviors.(Guan et al., 2025; He et al., 2024; Zhou et al., 2025) At the same time, a critical limitation remains: hallucination or the generation of information that sounds plausible but is inaccurate or unverifiable.(Asgari et al., 2025; Ji et al., 2024; Shah, 2024) The causes of this issue include gaps and biases in the training data, as well as the inherent flexibility of generative models.(Bender et al., 2021)

These concerns are especially important when we use LLM to extract data for research, where factual accuracy is vital. In previous work, we applied GPT-4o to extract 44 behavioral and biographical characteristics of mass shooters. Using manually crafted prompts, we observed strong agreement between GPT-4o outputs and a human-coded gold standard for most variables. However, for some characteristics, such as police involvement, agreement rates dropped below 60 percent, suggesting ample room for improvement.

In this study, we examined whether several popular prompting techniques can improve the consistency and accuracy of LLM-based data extraction. We evaluated the performance of three strategies: few-shot prompting, self-prompting, and self-prompted chain-of-thought. (DAIR.AI, 2025) Our goal is to determine whether different prompting strategies can reduce hallucination and enhance the agreement of GPT-4o outputs with expert human annotations in the context of violent offender profiling.

2. PRIOR WORK AND DATASET

This study builds on our prior work that evaluated the reliability of large language models (LLMs) in extracting the behavioral characteristics of mass shooters. Specifically, we focused on 44 predefined attributes related to devolution and hate, assessed in 197 mass shooting incidents in the United States between 1966 and 2024. These incidents were defined as events involving four or more firearm-related casualties in public spaces, excluding those associated with organized crime or gang activity.

Devolution is operationalized as a progressive breakdown in behavioral functioning, such as increased isolation, distress, or signs of deterioration leading up to the attack. Hate, by contrast, includes expressions of intense hostility or aversion toward individuals, groups, institutions, or society at large. Among the 197 perpetrators in our dataset, 97.9% were male, with a mean age of 34.1 years (Standard Deviation [SD] = 12.7), reflecting a broad demographic distribution.

In earlier work,(Treitler et al., 2025) we employed GPT-4o alongside carefully developed prompt engineering strategies to extract these behavioral indicators from publicly available sources. These included setting contextual framing, specifying intent, and encouraging the model to disclose uncertainty. The outputs were then benchmarked against a human-coded gold standard developed using protocols established by the Buehler Center for Health Policy and Economics at Northwestern University. Following trial-and-error explorations, GPT-4o achieved a mean agreement rate of 78.7% (SD = 14.6%) with the human gold standard. However, agreement levels were not consistent across all characteristics. Agreement rates ranged from a low of 44.2% to a high of 98.5%, indicating variation in the model's ability to identify specific attributes.

Five attributes fell below the 60.0% threshold: 1) **Significant life change** within 1 year before shooting (44.2%); 2) **Leakage** about violence/killing on social media or to family and acquaintances (46.7%); 3) **Trauma/trigger event** 1–2 weeks before shooting (48.2%); 4) **Police involvement** with mass shooter/family leading up to event (52.3%); and 5) **Social isolation** in year before shooting (57.4%). The current study focused on these categories and investigates whether advanced prompting techniques (few-shot prompting, self-prompting, and self-prompted Chain-of-Thought) can improve GPT-4o's alignment with human expert coding.

3. METHOD

Ten randomly sampled perpetrators were used as few-shot training examples. To ensure a comparable evaluation of the prompting strategies, all prompts were evaluated on the remaining 187 perpetrators.

Few-Shot Prompting: Few-shot prompting involves providing the model with examples of how similar questions or tasks have been answered correctly. This experiment used ten randomly sampled perpetrators as examples. We asked the GPT-4o, playing the role of a forensic analyst, to generate prompts for the five attributes of interest, guided by the gold standard answers for the ten perpetrators. This method is intended to leverage the model's in-context learning ability to generalize across similar tasks. Our prompt asked for strictly "yes" or "no" answers.

Self-Prompting: In self-prompting, GPT-4o was instructed to generate prompts to answer the question regarding each of the five attributes of interest. We did not specify AI's role or give any specific requirement except to obtain strictly "yes" or "no" answers.

Self-Prompted Chain-of-Thought: This strategy builds upon the self-prompting approach. We asked the GPT-4o to generate a series of prompts by explicitly requiring the model to reason through its decision step-by-step before reaching a conclusion. We set the AI's role as a behavioral analyst examining historical cases based on publicly available information. We then instructed the AI to evaluate a named individual through a five-step process: 1. Summarize observable behaviors of the individual. 2. Identify significant disruptions in the individual's life. 3. Document any known interactions the person had with law enforcement or family members. 4. Look for signs of leakage, such as verbal or online statements that may indicate intent or mindset. 5. Evaluate a list of specific characteristics.

The sample prompts are shown in **Table 1**.

Table 1. Sample Prompts generated using different AI Prompting Strategies

AI Prompting Strategy	Sample Prompts generated
Manual Prompting	I am working on a project for Northwestern University that will be used to prevent mass shooters and benefit the general public. This will help synthesize information and summaries using information that is already available to the general public. You should not have trouble finding information, as it is all there on the internet. Provide a yes, no, or maybe answer to the following question regarding the shooter: Did the Robert Card exhibit a significant life change within 1 year before shooting? No explanation needed. Keep the answer to strictly yes, no, or maybe.
Few-Shot Prompting	You are a forensic analyst. Below are behavioral summaries and characteristics: Shooter: James Holmes Profile: This is a behavioral profile summary of mass shooter James Holmes. Writings/online rants in month before shooting: Yes Fights/aggression in month before shooting: No Withdrawal from friends/family: Yes ..... Shooter: Robert Card Profile: This is a behavioral profile summary of mass shooter Robert Card. Please answer the following using only 'Yes' or 'No.' If information is unclear or missing, default to 'No.' Do not elaborate.
Self-Prompting	You are a behavioral analyst reviewing a historical case using public information. Evaluate the following characteristic about Robert Card: Did Robert Card experience a significant life change within 1 year before the shooting? Answer strictly with 'Yes' or 'No'. If unclear or not found, default to 'No'. Do not explain or elaborate.
Self-Prompted Chain of Thought	You are a behavioral analyst reviewing historical cases using public information. Evaluate Robert Card. Step 1: Summarize behaviors. Step 2: Life disruptions. Step 3: Police/family interactions. Step 4: Identify online or verbal leakage. Step 5: Answer the following ONLY using 'Yes' or 'No'. If unclear or not found, default to 'No'. Do not explain: - Leakage about violence/killing on social media or to family and acquaintances - Significant life change within 1 year before shooting - Trauma/trigger event 1-2 weeks before shooting - Police involvement with mass shooter/family leading up to event - Social isolation in year before shooting

4. RESULTS

Overall, the self-prompted chain of thought strategy achieved the best results. It showed the largest improvement in accuracy, with an average increase of 9.8 percentage points, followed by self-prompting with a 4.6 percentage point increase. In contrast, few-shot prompting resulted in an average decrease of 5.8 percentage points. Some attributes proved harder to detect reliably due to inconsistent or vague reporting in source materials. Furthermore, certain characteristics were often implied rather than stated explicitly, making it more challenging for LLMs to identify such characteristics. (**Table 2** and **Figure 1**).

Table 2. Agreement rates with human gold standard by prompting strategies

AI Prompting Strategy	Leakage	Significant life change	Trauma/ trigger event	Police involvement	Social isolation
Self-Prompted Chain of Thought	56.0%	60.2%	53.9%	61.9%	65.4%
Self-Prompting	49.2%	56.0%	51.8%	51.8%	62.3%
Few-Shot Prompting	38.2%	40.3%	47.6%	40.8%	52.4%
Manual Prompting	46.2%	44.7%	48.2%	52.3%	56.9%

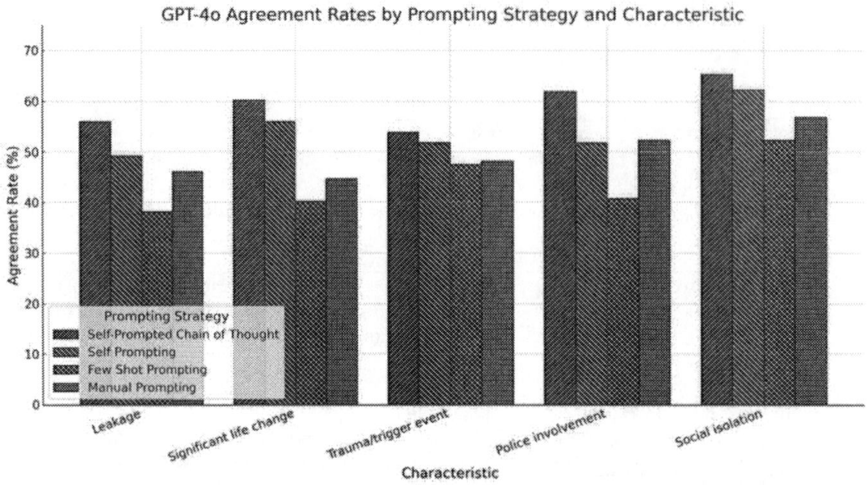

Figure 1. Agreement rates with human gold standard by prompting strategies

5. DISCUSSION AND CONCLUSIONS

This study demonstrates that prompting strategies can meaningfully influence the accuracy of LLM in information extraction tasks. In particular, self-prompted chain-of-thought prompting led to substantial improvements in each of the 5 lowest-performing attributes, indicating its potential to support more nuanced reasoning. By contrast, self-prompting without explicit reasoning steps was not as successful, and the few-shot approach performed the worst overall. It is possible that the few-shot prompting led to overfitting.

Our results highlight both the potential value and important limitations of using LLMs in complex tasks like behavioral profiling. Even using the self-prompted chain-of-thought strategy, the agreement rates with human coders were far from perfect. One potential reason is that prompting cannot completely overcome the limitations in a LLM in handling certain tasks. As LLMs continue to grow in size and improve in their performance according to certain benchmark tests, for the tasks we are interested in, the LLM-based information extraction performance is expected to but not guaranteed to improve. This reinforces the importance of not over-trusting AI systems simply because they perform well on average or in select tasks. Human experts must remain in the loop and be independent, critical, and informed. AI-generated outputs should be treated more as hypotheses than facts or conclusions, especially when deployed in research.

Profiling violent offenders using LLMs presents potential ethical challenges. While LLMs offer gains in efficiency, they also carry risk of misuse, particularly in areas such as law enforcement or mental health screening. Therefore, we must be mindful of ethical usage of these methods.

This study has several limitations. First, the LLM we used may lack sufficient training data for rare or underreported behavioral characteristics, especially when drawn from sparse online records. In these cases, prompt engineering alone may be insufficient to fill informational gaps. Second, while self-prompted chain-of-thought prompting shows promise, it is not a magic solution. It still depends on the LLM's internal knowledge and interpretive heuristics, which may be incomplete or flawed.

In future research, we want to explore how LLMs handle ambiguity and uncertainty. Specifically, in assisting human annotators, we would like to have GPT express doubt or accurately flag missing data. In this study, we observed that when explicitly asked to disclose uncertainty, GPT-4o frequently answers with "missing" but rarely with "maybe." This is not particularly helpful. For the purposes of this study, we grouped all non-"yes" responses into a single "no" answer, which may have underestimated the model's nuance. Notably, collapsing "maybe" and "missing" into a single "no" category may have suppressed potentially meaningful distinctions. However, this decision was made to reduce ambiguity in comparative analysis. An alternative plan is to combine the LLM's power to generate text with a more traditional approach of information retrieval to ascertain the availability of evidence before synthesizing evidence.

REFERENCES

Asgari, E., Montana-Brown, N., Dubois, M., Khalil, S., Balloch, J., Yeung, J. A., & Pimenta, D. (2025). A framework to assess clinical safety and hallucination rates of LLMs for medical text summarisation. *NPJ Digit Med*, *8*(1), 274. https://doi.org/10.1038/s41746-025-01670-7

Bender, E. M., Gebru, T., McMillan-Major, A., & Shmitchell, S. (2021). *On the Dangers of Stochastic Parrots: Can Language Models Be Too Big?* FAccT '21: Proceedings of the 2021 ACM Conference on Fairness, Accountability, and Transparency,

Brown, T. B., Mann, B., Ryder, N., Subbiah, M., Kaplan, J., Dhariwal, P., Neelakantan, A., Shyam, P., Sastry, G., Askel, A., Agarwal, S., Herbert-Voss, A., Krueger, G., Henighan, T., Child, R., Ramesh, A., Ziegler, D. M., Wu, J., Winter, C.,…Amodei, D. (2020). *Language Models are Few-Shot Learners*. https://arxiv.org/abs/2005.14165

DAIR.AI. (2025, April 24). *Prompting Techniques*. Retrieved June 27 from https://www.promptingguide.ai/techniques

Guan, S., Hui, V., Stiglic, G., Constantino, R. E., Lee, Y. J., & Wong, A. K. C. (2025). Classifying the Information Needs of Survivors of Domestic Violence in Online Health Communities Using Large Language Models: Prediction Model Development and Evaluation Study. *J Med Internet Res*, *27*, e65397. https://doi.org/10.2196/65397

He, L., Omranian, S., McRoy, S., & Zheng, K. (2024). Using Large Language Models for sentiment analysis of health-related social media data: empirical evaluation and practical tips. *AMIA Annu Symp Proc*, *2024*, 503-512. https://www.ncbi.nlm.nih.gov/pubmed/40417506

Ji, Z., Lee, N., Frieske, R., Yu, T., Su, D., Xu, Y., Ishii, E., Bang, Y., Chen, D., Dai, W., Chan, H. S., Madotto, A., & Fung, P. (2024). Survey of Hallucination in Natural Language Generation. In.

OpenAI, Achiam, J., Adler, S., Agarwal, S., Ahmad, L., Akkaya, I., Aleman, F. L., Almeida, D., Altenschmidt, J., Altman, S., Anadkat, S., Avila, R., Babuschkin, I., Balaji, S., Balcom, V., Baltescu, P., Bao, H., Bavarian, M., Belgum, J.,…Jain, S. (2023). *GPT-4 Technical Report*. https://arxiv.org/abs/2303.08774

Raji, I. D., Bender, E. M., Paullada, A., Denton, E., & Hanna, A. (2021). AI and the Everything in the Whole Wide World Benchmark. https://doi.org/10.48550/arXiv.2111.15366

Shah, S. V. (2024). Accuracy, Consistency, and Hallucination of Large Language Models When Analyzing Unstructured Clinical Notes in Electronic Medical Records. *JAMA Netw Open*, *7*(8), e2425953. https://doi.org/10.1001/jamanetworkopen.2024.25953

Treitler, J., Danziger, G., Marshall, M., Hassan, M., Lundberg, A., & Post, L. A. (2025). Identification of Mass Shooter Profiles through Unspervised Clustering. MEDINFO 2025,

Zhou, X., Zhou, J., Wang, C., Xie, Q., Ding, K., Mao, C., Liu, Y., Cao, Z., Chu, H., Chen, X., Xu, H., Larson, H. J., & Luo, Y. (2025). PH-LLM: Public Health Large Language Models for Infoveillance. *medRxiv*. https://doi.org/10.1101/2025.02.08.25321587

Reflection Papers

TOWARDS AN EFFICIENT SELECTION OF KEY CHARACTERISTIC PARAMETERS FOR THE PREDICTION OF FALLS IN THE ELDERLY

Amadou M. Djiogomaye Ndiaye[1, 5], Michel Harel[1,2,7], Laurent Billonnet[6] and Achille Tchalla[1, 3,4, 5]
[1]Laboratoire Vie-Santé, UR 24 134, Faculté de Médecine, 87025 Limoges, France,
[2]INSPÉ de Limoges, Université de Limoges, 87036 Limoges Cedex, France ; UMR5219 UPS
[3]Service de Médecine Gériatrique, CHU de Limoges Dupuytren, Limoges, France,
[4]Pôle Hospitalo-Universitaire de Gérontologie clinique, Service de Médecine Gériatrique, Unité de Prévention de Suivi et d'Analyse du Vieillissement, CHU de Limoges Dupuytren, Limoges, France
[5]Chaire d'excellence chez Fondation Partenariale de l'Université de Limoges I Institut Omega Health, Laboratoire VieSanté, UR 24134 Vieillissement Fragilité Prévention e-Santé
[6]XLIM-SRI - Systèmes et Réseaux Intelligents (444315) - XLIM - 123, avenue Albert Thomas, 87060 Limoges Cedex, France
[7]Institut de Mathématiques de Toulouse, UMR CNRS 5 219, Université Paul Sabatier, Toulouse 31 062, France

ABSTRACT

Falls significantly contribute to frailty and functional decline in elderly individuals. The risk of falling (ROF) is linked to three dimensions: physical/organic, socio-environmental, and thymic/cognitive. Therefore, fall prevention protects older individuals from multiple comorbidities. The reliability of predictive studies depends on the quality and consistency of data collection. In most studies, data for model construction were collected from hospitals, research laboratories, or participants' homes. Recent fall prediction models increasingly rely on machine learning, deep learning, and computer vision. Predictive models assist specialists in decision-making. Using home-collected data, our objective is to develop an optimized predictive model with minimal features. In this paper, we aim to identify the optimal approach for predicting falls using this strategy with the objective of building a robust dataset as the entry of an AI process.

KEYWORDS

Fall, Older Population, Data, Mathematical Modeling, Prevention

1. INTRODUCTION

According to the World Health Organization (WHO), older individuals are those aged ≥ 60 years. The proportion of older individuals worldwide is expected to nearly double between 2015 and 2050, increasing from 12% to 22% (WHO 2022). The National Institute of Statistics and Economic Studies (INSEE) estimates that one in three individuals in France will be aged ≥ 60 years by 2060, compared to one in four individuals in 2021 (Insee 2016). Aging leads to a gradual decline in functional capacity, increasing the risk of falls (ROF) (Thierry 2010). Early identification of ROF facilitates the administration of personalized interventions for individuals (US Preventive Services Task Force 2018).

Most recent studies predict falls using sensors or electronic health records (EHRs). With data collected directly from elderly individuals' homes, our objective is to develop an effective predictive model using the fewest possible features. Our initial models, built with a comprehensive approach, have demonstrated satisfactory performance. In this work, we briefly discuss fall prediction strategies, the dimensions of ROF, and relevant studies in the field.

2. RISK OF FALLING

Falls can be caused by various factors. A fall occurs when a person involuntarily lands on the ground or at a lower level than their starting position (Wolf et al. 1996). It can occur multiple times within a year. Geriatrics who experience at least two falls within 12 months are classified as "fallers" (Tinetti et Speechley 2010).

A holistic fall prediction approach considers three key dimensions:

- The physical/organic dimension gathers data related to an individual's medical history and current symptoms, diagnosis of underlying health issues, and treatment effectiveness.
- The thymic/cognitive dimension refers to an individual's mental, emotional, and cognitive states.
- The socio-environmental dimension refers to age, gender, family and social support, housing conditions, home configuration, the presence of slippery rugs, stairs without railings, uneven surfaces, and inadequate lighting.

Evaluating the ROF involves at least a gait and balance assessment of the physical/organic dimension and the age and gender of the socio-environmental dimension. Data involving the thymic/cognitive dimension allow for a comprehensive review of the potential causes of a fall. The term "dimension" refers to the types of factors that contribute to the ROF and their evaluation.

Hospitalized patients often receive incomplete health assessments across all dimensions. Our home-collected data encompass features from all three dimensions.

3. PREDICTIVE FACTORS FOR FALLS IN THE OLDER POPULATION

Not every feature within the three ROF dimensions is a predictive factor for falls. The effectiveness of a predictive factor depends on its statistical significance, correlation with fall occurrences, and its interaction with other variables across the physical/organic, socio-environmental, and cognitive dimensions. In some studies, the identified predictive variables did not encompass all three dimensions of ROF. Ikeda et al. (2022), Kawazoe et al. (2022), and Cella et al. (2020) demonstrated that age category related to socio-environmental was a predictor of falls, suggesting a strong association between age and falls (Kawazoe et al. 2022; Ikeda et al. 2022), (Cella et al. 2020). Bath et al. (2000) found that the predictive variables related to the socio-environmental dimension are diverse and varied, contributing to effective prevention(Bath et al. 2000). In fact, a higher number of variables related to gait and balance is associated with a more robust predictive model for falls.

In the literature review conducted by Rubenstein (2006), only cognitive impairment was identified as a predictive variable related to the thymic/cognitive (Rubenstein 2006). Conversely, Ikeda et al. (2022), Kawazoe et al. (2022), and Bath et al. (2000) identified at least two predictive variables involving the thymic/cognitive dimension, providing a better understanding of the ROF associated with the thymic/cognitive dimension and facilitating preventive measures. In those features, we can find fear of falling, depressive symptoms, self-rated health, impaired consciousness, and dementia at admission. Recent studies of Ikeda et al., 2022, and Kawazoe et al., 2022 achieved AUC scores of 88% and 85% using comprehensive approaches. They employed Random-Forest-based Boruta algorithms to assist in feature selection and, respectively, a combination of Bidirectional Encoders and Bidirectional Long Short-Term Memory (BiLSTM) for sequential data processing. These AUC scores indicate strong model performance, reflecting high discriminative ability in classification tasks.

The reliability of predictive studies depends on the quality and consistency of data collection. Unlike hospitals, where data is gathered only when patients seek care, home-based data collection requires practitioners to schedule visits at fixed intervals. Our study followed 1,648 community-dwelling older adults ($\geqslant$60 years) between September 2011 and September 2023. Participants were assessed at home by UPSAV (Unité de Prévention, de Suivi et d'Analyse du Vieillissement), a unit dedicated to aging prevention, monitoring, and analysis at Limoges University Hospital. Each patient underwent an initial visit followed by a second visit six months later. After the second visit, annual follow-ups were conducted for up to six years, as long as the patient remained in their home. Data collection included cardiovascular risk factors, fall occurrences, socio-environmental characteristics, and a comprehensive geriatric assessment summary score.

To ensure coherence, our predictive model accounts for temporality by predicting falls every six months. Out of thirty input features, eleven were identified as the most relevant fall predictors using multinomial logistic regression: Gender, Hypertension, Obesity, Activities of Daily Living (ADL), Mini-Mental State Examination (MMSE), Short Physical Performance Battery (SPPB), Pathological Geriatric Depression Scale (GDS), Instrumental Activities of Daily Living (IADL), Leisure activity, Pathological Single-Leg Balance (SLB), and history of falling in the past year. These features were used to develop several predictive models, including Logistic Regression, Support Vector Machine (SVM), Random Forest, and eXtreme Gradient Boosting (XGBoost), the area under the receiver operating characteristic curve (AUC) was used to evaluate the predictive performance of the models. Among these, the SVM model performed best, achieving an AUC of 74 on the test data.

As with most AI models, ours can be continuously refined with additional data over time. In our case, improving the model also provides an opportunity to collect data from patients' homes while offering them personalized fall prevention advice. During the intervals between practitioner visits, necessary adjustments to home configurations can also be made if needed. We will present our first results at the conference.

The limitations of this study include the use of data that did not account for medications taken by participants or treatments for specific comorbidities, which may influence fall risk. Additionally, the study population was limited to older adults residing in France, potentially affecting the generalizability of the findings to other geographic or cultural contexts. Future research should aim to include more diverse populations to enhance the external validity and applicability of the results.

4. CONCLUSION

This study aims to advance fall prevention by leveraging a 12-year dataset collected from home environments to develop an AI-based predictive model. Our approach integrates the three dimensions of ROF, optimizing performance while minimizing the number of required features. Explainable AI will help identify each feature's contribution to fall risk, enabling more precise intervention strategies. Effective fall prevention not only reduces healthcare costs but also enhances elderly care by reallocating resources to research and treatment improvements.

ACKNOWLEDGEMENT

The authors thank UPSAV, URCI, VieSanté, and the Chaire d'Excellence Intelligence Artificielle et Bien-Vieillir en Nouvelle-Aquitaine (IABV-NA) for their support. We also extend our gratitude to the study participants for their time, effort, and dedication.

REFERENCES

Bath, Peter A., Neil Pendleton, K. MORGAN, J. E. Clague, Michael A. Horan, et S. B. Lucas. (2000). « New approach to risk determination: development of risk profile fornew falls among community-dwelling older people by use of a GeneticAlgorithm Neural Network (GANN) ». *The Journals of Gerontology: Series A* 55 (1): M17-21. https://doi.org/10.1093/gerona/55.1.M17.

Cella, Alberto, Alice De Luca, Valentina Squeri, Sara Parodi, Francesco Vallone, Angela Giorgeschi, Barbara Senesi, et al. (2020). « Development and Validation of a Robotic Multifactorial Fall-Risk Predictive Model: A One-Year Prospective Study in Community-Dwelling Older Adults ». *PLOS ONE* 15 (6): e0234904. https://doi.org/10.1371/journal.pone.0234904.

Ikeda, Takaaki, Upul Cooray, Masanori Hariyama, Jun Aida, Katsunori Kondo, Masayasu Murakami, et Ken Osaka. (2022). « An Interpretable Machine Learning Approach to Predict Fall Risk Among Community-Dwelling Older Adults: A Three-Year Longitudinal Study ». *Journal of General Internal Medicine* 37 (11): 2727-35. https://doi.org/10.1007/s11606-022-07394-8.

Insee. (2016). « Population by age - Tableaux de l'Économie Française | Insee ». 2016. https://www.insee.fr/fr/statistiques/1906664?sommaire=1906743.

Kawazoe, Yoshimasa, Kiminori Shimamoto, Daisaku Shibata, Emiko Shinohara, Hideaki Kawaguchi, et Tomotaka Yamamoto. (2022). « Impact of a Clinical Text-Based Fall Prediction Model on Preventing Extended Hospital Stays for Elderly Inpatients: Model Development and Performance Evaluation ». *JMIR Medical Informatics* 10 (7): e37913. https://doi.org/10.2196/37913.

Rubenstein, Laurence Z. (2006). « Falls in older people: epidemiology, risk factors and strategies for prevention ». *Age and Ageing* 35 (suppl_2): ii37-41. https://doi.org/10.1093/ageing/afl084.

Thierry, Xavier. (2010). « Accidents and physical assaults among the elderly: less frequent than among the young, but more severe ». *Population & Sociétés* 468 (6): 1-4. https://doi.org/10.3917/popsoc.468.0001.

Tinetti, Mary E., et Mark Speechley. (2010). « Prevention of Falls among the Elderly ». Review-article. Http://Dx.Doi.Org/10.1056/NEJM198904203201606. Massachusetts Medical Society. World. 14 janvier 2010. https://doi.org/10.1056/NEJM198904203201606.

US Preventive Services Task Force. (2018). « Interventions to Prevent Falls in Community-Dwelling Older Adults: US Preventive Services Task Force Recommendation Statement ». *JAMA* 319 (16): 1696-1704. https://doi.org/10.1001/jama.2018.3097.

WHO. (2022). « Ageing and health ». 2022. https://www.who.int/fr/news-room/fact-sheets/detail/ageing-and-health.

Wolf, Sl, Barnhart Hx, Kutner Ng, McNeely E, Coogler C, et Xu T. (1996). « Reducing Frailty and Falls in Older Persons: An Investigation of Tai Chi and Computerized Balance Training. Atlanta FICSIT Group. Frailty and Injuries: Cooperative Studies of Intervention Techniques ». *Journal of the American Geriatrics Society* 44 (5). https://doi.org/10.1111/j.1532-5415.1996.tb01432.x.

TEN REASONS TO SUPPORT OPEN-SOURCE E-HEALTH TECHNOLOGY

John Kildea
Dept. of Oncology
McGill University, Quebec, Canada

ABSTRACT

This high-level reflection paper is intended to spark discussion and debate at e-Health 2025 about the outsized role proprietary software systems play in modern healthcare and of the important benefits that open-source e-health solutions can offer. Recent events, including the tariff war triggered by the current US administration and the bankruptcy of 23andme.com highlight the risk of relying too heavily on large multinational companies to manage our healthcare infrastructure and data. Open-source e-health technologies, in contrast, are more cost-effective, offer local control, can foster innovation in the current era of data-hungry AI, reduce cybersecurity risk, and can help sustain local MedTech economies. This paper is presented as a personal reflection piece for the purpose of spurring conversation.

KEYWORDS

Open-Source E-Health Software, Digital Sovereignty, Vendor Lock-In, Fostering Digital Health Innovation, Open Standards

1. INTRODUCTION

Over the last few decades, healthcare has transformed itself from a paper- and film-based reality to a digital one. In parallel, advances in machine learning algorithms and computing power have given rise to an artificial intelligence (AI) revolution with benefits to all areas of society. But, the unstructured and fragmented nature of a lot of digital health data, stemming from the poor interoperability of proprietary e-health software and the innovation-impeding effects of vendor lock-in, makes it difficult for these parallel developments to converge (Malm-Nicolaisen et al., 2019). In this reflection paper, we argue that open-source e-health technologies can help avoid vendor lock-in and foster e-health innovation at scale, thereby ensuring that e-health is no longer the slow learner of the digital revolution. This paper is not intended as a rigorous academic evaluation of the advantages of open-source software in healthcare and it does not contain an exhaustive list of citations. Rather, it is intended as a personal reflection piece to provoke conversation and debate.

2. TEN BENEFITS OF OPEN-SOURCE E-HEALTH TECHNOLOGIES

Below, we list ten benefits that are unique to open-source e-health technologies. This list is by no means exhaustive and is provided in no particular order.

2.1 Digital Sovereignty

In the healthcare context, we define digital sovereignty as the ability for a health system to control and manage all of its own digital affairs (data and infrastructure) without ceding undue control to third-party *Software as a service* (SaaS) or *Infrastructure as a Service* (IaaS) providers. Recent events, including the ongoing tariff war triggered by the current US administration and the bankruptcy of 23andme.com, have highlighted the importance of local control over digital health affairs. Third-party control may be considered "undue" when it cannot be easily revoked in the event that the health system desires to part ways with the third-party provider.

Such a situation could arise when the infrastructure is physically located and managed by the third-party in another jurisdiction or when the infrastructure is locally hosted but reliant on proprietary technology that cannot be operated without the provider. Locally hosted hardware, open data standards, and open-source software offer health systems the best chance to maintain digital sovereignty at all times (Klare & Lechner, 2025; Moerel & Timmers, 2021).

2.2 Reduced Risk for Vendor Lock-In

Vendor lock-in occurs when a health system is so dependent on a particular vendor's proprietary technology that it become prohibitively expensive to switch to another solution. Open-source software is less prone to lock-in because the cost of switch-over is not driven by an individual vendor that has a vested interest in maintaining the status quo but rather by multiple vendors who can compete on providing the most cost-effective migration solution (Malm-Nicolaisen et al., 2019).

2.3 Harness Academic Research Expertise

Many publicly funded academic research teams can provide a wealth of expertise in the design, development, and exploitation of e-health technologies. However, this expertise is often neglected in favour of large multinational corporations and at the behest of multinational consultancy firms that advice governments and health systems. These powerful commercial entities have vested financial interests in promoting and sustaining their own proprietary products or the products that they symbiotically support. As a result, innovations arising from academic research are either ignored completely or slowly adopted into proprietary solutions only after all intellectual property rights have been secured. Open-source e-health technologies, on the other hand, are less susceptible to financial conflicts of interest and can benefit directly and immediately from academic contributions, providing an efficient pathway for translating academic research into real-world improvements in healthcare (McDonald et al., 2003).

2.4 Cost Effectiveness

Open-source e-health technologies offer much greater cost effectiveness than proprietary solutions (Nagy et al., 2010). Without intellectual property ownership issues entering the equation, they may be developed and implemented either at cost, if done by the health system itself, or close to cost, if done by a third-party provider in a competitive context. Moreover, many open-source e-health products can benefit from the input of multiple health systems, further reducing costs.

2.5 Citizen-Driven Innovation

Due to their availability and freedom to be modified, open-source e-health solutions are of potential interest to citizen innovators within the general public. For example, in the diabetes community, patient-driven roadmaps and contributions have significantly improved and accelerated innovation of open-source products such as Tidepool (Neinstein et al., 2016), Nightscout (Ng et al., 2020), and OpenAPS (Melmer et al., 2019), among others. Similar user-driven innovation sustains open-source electronic medical record (EMR) software, including for example Opal (Kildea et al., 2019), OpenEMR (*OpenEMR*, n.d.), OpenMRS (*OpenMRS*, n.d.), and GNUHealth (Martín, 2016). This sort of citizen-driven innovation is not possible when dealing with proprietary closed-source solutions.

2.6 Reduced Barrier to Entry for SMEs

The MedTech sector is very challenging to enter for small and medium-sized enterprises (SMEs), especially those SMEs offering products or services that overlap with the offerings of the established and highly protective multinational giants. Open-source solutions can offer these SMEs a leg-up by allowing them to get going quickly without having to develop and support complete solutions from scratch (Ahmad & Tripathi, 2024).

2.7 Improved Security Through Transparency

The transparent and collaborative nature of open-source software means that it benefits from a "many eyes" approach to security (Emanuilov, 2021). Vulnerabilities have a greater chance of being detected and fixed compared to proprietary solutions where issues may quietly go for longer without resolution.

2.8 Developer Retention

Open-source software can offer developers a fulfilling feeling of shared ownership and of contributing to the public good (Schaarschmidt et al., 2025). This is particularly true in the e-health domain where developers can see their work being used to improve patient experiences and outcomes. The ensuant "feel good factor" can help with employee retention, particularly in the public sector where it is often challenging to provide salaries on par with the private sector.

2.9 Foster Innovation

Closed-source software solutions, particularly those that represent "cash cows" for their vendors, are typically slow to improve and incorporate innovations since changes represent work and possible financial risk to the vendor, which adopts a "don't fix what isn't broken attitude". As a result, innovation is stymied. Open-source software, on the other hand, can foster innovation by encouraging continuous improvement through collaboration (Wang, 2023). With open-source software, the focus is on providing an excellent end-user service as opposed to filling the coffers of a vendor.

2.10 Global Health Friendly

Large multinational vendors do not serve resource-limited settings well, if at all. Their costly proprietary solutions can use up budgets and limit funds for other areas of healthcare while preventing the type of agile innovation that is needed to adapt to local e-health needs. It is no surprise, therefore, that open-source e-health solutions find highest adoption rates in the developing world (Shaikh et al., 2022; Verma et al., 2021).

3. CONCLUSION

We have identified ten benefits of open-source e-health solutions and presented them. Our list is neither exhaustive or ordered but is intended to foster discussion and debate. We have not presented a rigorous evaluation of the benefits of open-source software solutions in healthcare and we acknowledge that there are also drawbacks to these solutions. However, as there is no shortage of well-heeled vendors with large marketing budgets ready to promote the virtues of the closed-source model, we argue that the open-source model should be given some space for discussion at e-Health 2025.

ACKNOWLEDGEMENT

The author acknowledges the support and encouragement of the members of their research team (The Opal Health Informatics Group) who believe enough in open-source e-health software to roll up their sleeves and make it happen.

REFERENCES

Ahmad, N., & Tripathi, N. (2024). Benefits, challenges, and implications of open-source software for health-tech startups: An empirical study. In *Lecture Notes in Business Information Processing* (pp. 265–282). Springer Nature Switzerland. https://doi.org/10.1007/978-3-031-53227-6_19

Emanuilov, I. (2021). Security through transparency and openness in computer design. In *Lecture Notes in Computer Science* (pp. 105–116). Springer International Publishing. https://doi.org/10.1007/978-3-030-68887-5_6

Kildea, J., Battista, J., Cabral, B., Hendren, L., Herrera, D., Hijal, T., & Joseph, A. (2019). Design and development of a person-centered patient portal using participatory stakeholder co-design. *Journal of Medical Internet Research*, *21*(2), e11371. https://doi.org/10.2196/11371

Klare, M., & Lechner, U. (2025). Digital sovereignty: Affordances in open source projects. In *Lecture Notes in Business Information Processing* (pp. 20–34). Springer Nature Switzerland. https://doi.org/10.1007/978-3-031-85849-9_2

Malm-Nicolaisen, K., Pedersen, R., & Fagerlund, A. J. (2019). Open or closed: A project proposal for investigating two different EHR platform approaches. *Studies in Health Technology and Informatics*, *265*, 207–212. https://doi.org/10.3233/SHTI190165

Martín, L. F. (2016, August 17). GNU health: A free/libre community-based health information system. *Proceedings of the 12th International Symposium on Open Collaboration Companion.* OpenSym '16: The International Symposium on Open Collaboration, Berlin Germany. https://doi.org/10.1145/2962132.2965661

McDonald, C. J., Schadow, G., Barnes, M., Dexter, P., Overhage, J. M., Mamlin, B., & McCoy, J. M. (2003). Open Source software in medical informatics--why, how and what. *International Journal of Medical Informatics*, *69*(2-3), 175–184. https://doi.org/10.1016/s1386-5056(02)00104-1

Melmer, A., Züger, T., Lewis, D. M., Leibrand, S., Stettler, C., & Laimer, M. (2019). Glycaemic control in individuals with type 1 diabetes using an open source artificial pancreas system (OpenAPS). *Diabetes, Obesity & Metabolism*, *21*(10), 2333–2337. https://doi.org/10.1111/dom.13810

Moerel, L., & Timmers, P. (2021). Reflections on digital sovereignty. *EU Cyber Direct, Research in Focus Series.* https://papers.ssrn.com/sol3/papers.cfm?abstract_id=3772777

Nagy, D., Yassin, A. M., & Bhattacherjee, A. (2010). Organizational adoption of open source software. *Communications of the ACM*, *53*(3), 148–151. https://doi.org/10.1145/1666420.1666457

Neinstein, A., Wong, J., Look, H., Arbiter, B., Quirk, K., McCanne, S., Sun, Y., Blum, M., & Adi, S. (2016). A case study in open source innovation: developing the Tidepool Platform for interoperability in type 1 diabetes management. *Journal of the American Medical Informatics Association: JAMIA*, *23*(2), 324–332. https://doi.org/10.1093/jamia/ocv104

Ng, M., Borst, E., Garrity, A., Hirschfeld, E., & Lee, J. (2020). Evolution of do-it-yourself remote monitoring technology for type 1 diabetes. *Journal of Diabetes Science and Technology*, *14*(5), 854–859. https://doi.org/10.1177/1932296819895537

OpenEMR (n.d.). Retrieved July 2, 2025, from https://www.open-emr.org/

OpenMRS. (n.d.). Retrieved July 2, 2025, from https://openmrs.org/

Schaarschmidt, M., Stol, K.-J., & Fitzgerald, B. (2025). The insider's dilemma: Employed open source developers' identification imbalance and intentions to leave. *European Journal of Information Systems: An Official Journal of the Operational Research Society*, 1–20. https://doi.org/10.1080/0960085x.2025.2463984

Shaikh, M., Vayani, A. H., Akram, S., & Qamar, N. (2022). Open-source electronic health record systems: A systematic review of most recent advances. *Health Informatics Journal*, *28*(2), 14604582221099828. https://doi.org/10.1177/14604582221099828

Verma, N., Mamlin, B., Flowers, J., Acharya, S., Labrique, A., & Cullen, T. (2021). OpenMRS as a global good: Impact, opportunities, challenges, and lessons learned from fifteen years of implementation. *International Journal of Medical Informatics*, *149*(104405), 104405. https://doi.org/10.1016/j.ijmedinf.2021.104405

Wang, Y. (2023). The power of openness- how open source software is reshaping software engineering and industrial adoption. In *TechRxiv*. https://doi.org/10.36227/techrxiv.23896002.v1

Posters

CLASSIFICATION OF SCREENING AUDIOGRAM USING DEEP LEARNING

How-Ran Guo[1,2], Chi-Ping Su[3] and Shao-Wei Yang[4]
[1]*Department of Environmental and Occupational Health, College of Medicine, National Cheng Kung University*
[2]*Department of Occupational and Environmental Medicine, National Cheng Kung University Hospital*
138 Sheng-Li Road, Tainan 704, Taiwan
[3]*Institute of Management of Technology, National Yang Ming Chiao Tung University*
1001 Daxue Road, Hsinchu 300, Taiwan
[4]*Department of Health Administration, Tzu Chi University of Science and Technology*
880 Section 2, Jianguo Road, Hualien, Taiwan

ABSTRACT

The World Health Organization (2018) estimates that by 2050, 930 million people will suffer from disabling hearing loss. With no cure for most cases, early diagnosis and prevention are essential, and hearing so screening programs are implemented all over the world. However, some tests are not performed by personnel with sufficient training and experience, and the interpretation of test results from mass screenings is a heavy burden. We conducted a study to apply deep learning methods for classifying audiograms obtained from pure-tone audiometry and thus to evaluate whether this approach improve the efficiency and accuracy of hearing loss screening. We randomly selected 20,000 audiograms collected between 2016 and 2020. The audiograms were divided them into training (n = 16,000) and validation (n = 4,000) sets. Three convolutional neural network models were trained to classify the audiograms into three categories: normal, age-related hearing loss (ARHL)(Liu and Yan, 2007), and noise-induced hearing loss (NIHL). We found that the EfficientNet-B0 model achieved 100.00% accuracy and an ROC-AUC of 100.00%, outperforming the ResNet-34 (99.98% accuracy, 100.00% ROC-AUC) and MobileNetV3-Small (99.62% accuracy, 99.99% ROC-AUC) models. Heat-map visualizations showed the models' attentions focused on audiometric notches for NIHL and elevated hearing thresholds for ARHL, demonstrating that they learned to classified audiograms with rules similar to guidelines used by physicians. Therefore, this study demonstrates that deep learning models can improve the efficiency and accuracy of hearing loss screening.

KEYWORDS

Audiogram, Deep Learning, Hearing Loss, Noise, Screening

1. INTRODUCTION

The World Health Organization (2018) estimates that by 2050, 930 million people will suffer from disabling hearing loss. With no cure for most cases, early diagnosis and prevention are essential. Governments and private organizations all over world are promoting hearing screening programs, but some tests are not performed by personnel with sufficient training and experience, and the interpretation of test results from mass screenings is a heavy burden.

In recent years, convolutional neural network (CNN) has performed well in the medical field and achieved excellent performance in various types of medical images (Chu et al., 2021; Crowson et al., 2020; Litjens et al., 2017, Shen et al, 2017; Zeng. et al., 2022). With the assistance of this technology, test operators can easily examine whether the test is correctly performed, and doctors can make medical decisions more efficiently in clinical diagnosis. Therefore, we conducted a study to apply deep learning methods for classifying audiograms obtained from pure-tone audiometry and thus to evaluate whether this approach improve the efficiency and accuracy of hearing loss screening.

2. MATERIALS AND METHODS

2.1 Study Design

We conducted a retrospective study and analyzed 20,000 audiograms collected from the Health Check Database maintained by the Ministry of Labor in Taiwan. The database contains audiograms submitted from hearing screening facilities all over Taiwan. The screenings are conducted by the employers according to the Workers' Health Protection Regulation (Ministry of Labor, 2025). We applied the CNN technique to train the deep learning models (LeCun and Bengio, 1995).

2.2 Setting

We randomly selected 20,000 audiograms collected between 2016 and 2020. The audiograms were divided them into training (n = 16,000) and validation (n = 4,000) sets. Three convolutional neural network models were trained to classify the audiograms into three categories: normal, age-related hearing loss (ARHL) (Liu and Yan, 2007), and noise-induced hearing loss (NIHL).

2.3 Participants

The 20,000 audiograms were anonymized and were screening test results from workers all over Taiwan. Of these, 10,000 were normal, 2,746 were diagnosed by certified specialists as ARHL, and 7,254 as NIHL.

2.4 Main Outcome and Measure

We converted the raw data into audiograms with a resolution of 200×240 and plotted the left and right ear data separately. Each audiogram contains seven data points, representing the minimum hearing thresholds of pure tone hearing at frequencies of 500 Hz, 1,000 Hz, 2,000 Hz, 3,000 Hz, 4,000 Hz, 6,000 Hz, and 8,000 Hz for one ear. As presented in Table 1, we randomly divided the collected data into a training set and validation set in a stratified fashion, with a ratio of 80:20. The training set contains 16,000 audiograms and was used to train the CNN model, the validation set contains 4,000 audiograms and was only used to measure the model's performance.

Table 1. Training and validation set

Hearing condition	Training set	Validation set	Total
Normal	8,000	2,000	10,000
Age-related hearing loss	2,197	549	2,746
Noise-induced hearing loss	5,803	1,451	7,254
Total	16,000	4,000	20,000

We measured the accuracy of classification using the area under curve of the receiver operating characteristic curve (ROC-AUC) for each model. Model interpretability was assessed using heat-maps generated by gradient-weighted class activation mapping to visualize regions of the audiograms influencing predictions.

3. RESULTS

We found that the EfficientNet-B0 model achieved 100.00% accuracy and an ROC-AUC of 100.00%, outperforming the ResNet-34 (99.98% accuracy, 100.00% ROC-AUC) and MobileNetV3-Small (99.62% accuracy, 99.99% ROC-AUC) models. Heat-map visualizations showed the models' attentions focused on audiometric notches for NIHL and elevated hearing thresholds for ARHL, demonstrating that they learned to classified audiograms with rules similar to guidelines used by physicians.

4. CONCLUSION

This study demonstrates that deep learning models can accurately classify audiograms. Integrating these models into hearing test systems may provide real-time assistance to execution of tests and interpretation of results, offering a viable solution to improve the efficiency and quality of hearing loss screening.

ACKNOWLEDGEMENT

This work is supported by Grant MOST 110-2813-C-006-047-B from the Ministry of Science and Technology (currently the National Science and Technology Council), Taiwan, R.O.C. and Grant ILOSH99-H999 from the Institute of Labor, Occupational Safety and Health, Ministry of Labor, Taiwan, R.O.C.

REFERENCES

Chu C.S. et al., (2021). Deep Learning for Clinical Image Analyses in Oral Squamous Cell Carcinoma: A Review. *Journal of the American Medical Association Otolaryngology—Head and Neck Surgery,* Vol. 147, No. 10, pp 893-900.

Crowson M.G., et al., (2020). Autoaudio: Deep Learning for Automatic Audiogram Interpretation. *Journal of Medical System*, VOL. 44, pp 1-7.

LeCun Y. and Bengio Y., (1995). Convolutional Networks for Images, Speech, and Time Series. *Handbook of Brain Theory and Neural Networks,* Vol. 3361, No. 10, p 1995.

Litjens G. et al., (2017). A Survey on Deep Learning in Medical Image Analysis. *Medical Image Analysis*, Vol. 42, pp 60-88.

Liu X. and Yan D., (2007). Ageing and Hearing Loss. *Journal of Pathology,* Vol. 211, No. 2, pp 188-197.

Ministry of Labor. Workers' Health Protection Regulation. Accessed February 28, 2025. https://law.moj.gov.tw/LawClass/LawAll.aspx?pcode=N0060022

Shen D., Wu G., Suk H.I., (2017). Deep Learning in Medical Image Analysis. *Annual Revive of Biomedical Engineering*, Vol. 19, No. 1, pp 221-248.

World Health Organization. Addressing the rising prevalence of hearing loss [published online 2018]. Accessed February 28, 2025. https://apps.who.int/iris/bitstream/handle/10665/260336/9789241550260-eng.pdf

Zeng J. et al., (2022). A Deep Learning Approach to Predict Conductive Hearing Loss in Patients with Otitis Media with Effusion Using Otoscopic Images. *Journal of the American Medical Association Otolaryngology—Head and Neck Surgery,* Vol. 148, No. 7, p 612.

Z-INFORMATION BASED BIG DATA ANALYTICS PLATFORM SELECTION

Konul Jabbarova[1] and Alehsan R. Aliyev[2]
[1]Azerbaijan State Oil and Industry University, Department of Computer Engineering, 20 Azadlig Ave., AZ1010, Baku, Azerbaijan
[2]Baku Higher Oil School, Baku, Azebaijan
Sabail district, Yeni Salyan way 3rdkm. "25", Baku, Azerbaijan

ABSTRACT

A big data analytics platform (BDAP) comprises a unified environment of tools that enable comprehensive data processing and analysis. Key functionalities include sourcing data from multiple systems, cleansing and merging datasets, computing metrics, performing analytical operations, monitoring performance, detecting anomalies, and generating and sharing reports. Real-world decision-making problems are frequently characterized by incomplete or uncertain information. Consequently, this study applies the PROMETHEE method to address the BDAP selection problem using Z-information.

KEYWORDS

Z-Information, Big Data Analytics Platform, PROMETHEE Method, Multi-Criteria Decision-Making (MCDM)

1. INTRODUCTION

Big data analytics (BDA) is frequently applied in areas such as real-time quality assurance and predictive maintenance. By enhancing the speed, accuracy, and effectiveness of decisions, BDA contributes significantly to the advancement of decision-making theory (Tsaniyah et al, 2025). As organizations continue to generate massive volumes of structured and unstructured data, the demand for advanced IT solutions that enable efficient data warehousing and processing has significantly increased. (Ilieva, 2019) seeks to evaluate commonly used cloud-based platforms tailored for big data applications, emphasizing their specific features and functionalities.

This study employs the PROMETHEE method to evaluate and select the most suitable big data analytics platform. The considered problem is described with Z-information. L. A. Zadeh's theory of Z-numbers is widely regarded as one of the most effective and suitable frameworks for modeling uncertainty and imprecision in information. All key definitions, including Discrete Z-number, the distance between two Z-numbers, operations on Discrete Z-numbers, and the ranking of Discrete Z-numbers are explained in (Aliev et al, 2015; Jabbarova, 2020; Jabbarova, 2021; Jabbarova et al, 2022; Jabbarova et al, 2024).

2. STATEMENT AND SOLUTION OF PROBLEM

We are presented with an MCDM problem under Z-information, aiming to determine the optimal platform for data analytics. The problem entails evaluating three alternatives based on five distinct criteria: Availability and Fault Tolerance (C1), Scalability and Flexibility (C2), Performance (C3), Distributed Storage Capacity (C4), and Data Security (C5). The values and importance weights of each criteria are expressed using Z-numbers (Jabbarova, 2023). The PROMETHEE method algorithm is used to address this problem (Jabbarova, 2021). Z-number-derived importance weights are w_1=(0.18,0.2,0.22)(0.8,0.9,1), w_2=(0.21,0.23,0.25)(0.8,0.9,1), w_3=(0.23,0.25,0.27)(0.8,0.9,1), w_4=(0.29,0.32,0.35)(0.8,0.9,1). The first step, a weighted decision matrix (WDM) represented by Z-number values is developed (Table 1).

Table 1. Criteria assessed using Z-number representations

	C_1	C_2
f_1	(0.14,0.19,0.22)(0.51,0.64,0.78)	(0.15,0.2,0.25)(0.66,0.81,0.95)
f_2	(0.13,0.15,0.2)(0.51,0.64,0,.78)	(0.14,0.2,0.24)(0.66,0.81,0.92)
f_3	(0.14,0.19,0.22)(0.51,0.64,0.78)	(0.16,0.22,0.25)(0.66,0.82,0.91)

	C_3	C_4
f_1	(0.18,0.25,0.27)(0.59,0.81,0.87)	(0.22,0.3,0.35)(0.66,0.81,0.9)
f_2	(0.15,0.21,0.26)(0.59,0.82,0.88)	(0.19,0.26,0.34)(0.66,0.81,0.95)
f_3	(0.14,0.16,0.19)(0.59,0.82,0.88)	(0.22,0.3,0.35)(0.66,0.81,0.9)

During the second step, distance values are computed to compare the two alternatives, g and, across each criterion (Aliev, 2017; Jabbarova, 2021). The preference function is expressed as (see Table 2):

$$\mathrm{Pf}(Z_{gj}(A,B),Z_{fj}(A,B))=\begin{cases}0, Z_{gj}(A,B)\le Z_{fj}(A,B)\\ D(Z_{gj}(A,B),Z_{fj}(A,B)), Z_{gj}(A,B)>Z_{fj}(A,B)\end{cases}.$$

Table 2. $\mathrm{Pf}_j(Z_g(A,B),Z_f(A,B))$

	C_1	C_2	C_3	C_4
$Pf(1,2)$	0.0.5	0.03	0.06	0.07
$Pf(1,3)$	0	0	0.14	0
$Pf(2,1)$	0	0	0	0
$Pf(2,3)$	0	0	0.09	0
$Pf(3,1)$	0	0.04	0	0
$Pf(3,2)$	0.05	0.04	0	0.07

At step three, the outranking relation is quantified using a preference index characterized by Z-number values (see Table 3).

$$\pi(g,f)=\sum_{j=1}^{n}\left[w_j Pf_j(g,f)\right].$$

Table 3. Preference index constructed using Z-number evaluations

	f_1	f_2	f_3
f_1	-	(0.049,0.054,0.059) (0.56,0.74,0.95)	(0.032,0.035,0.038) (0.8,0.9,1)
f_2	-	-	(0.021,0.023,0.024) (0.8,0.9,1)
f_3	(0.008,0.009,0.01) (0.8,0.9,1)	(0.037,0.04,0.046), (0.63,0.73,0.8)	-

The fourth step consists of calculating the leaving and entering flows to rank the alternatives.

$$\phi^+(g)=\frac{1}{n-1}\sum_{\substack{f\neq g\\ f=1}}^{m}\pi(g,f),\ \phi^-(g)=\frac{1}{n-1}\sum_{\substack{f\neq g\\ f=1}}^{m}\pi(f,g)\qquad (f\neq g)\qquad (1)$$

where n denotes the count of available alternatives. Table 4 summarizes the results.

Table 4. Flow metrics for all alternatives as determined by Z-number valuation

	Leaving flows	Entering flows
f_1	(0.041,0.045,0.049)(0.49,0.68,0.93)	(0.004,0.0045,0.005)(0.8,0.9,1)
f_2	(0.011,0.0115,0.012)(0.8,0.9,1)	(0.043,0.047,0.053)(0.45,0.59,0.76)
f_3	(0.023,0.055,0.08)(0.15,0.35,0.73)	(0.027,0.029,0.031)(0.68,0.82,0.93)-

The net flow is calculated in the fifth step using this formula:

$$\phi(g)=\phi^{+}(g)-\phi^{-}(g)$$

Table 5 illustrates the findings. In the sixth step, the alternatives are ranked (Table 5) (Aliev, 2016):

Table 5. Net Flows Derived from Z-Number Evaluations

	$\phi(g)$
f_1	(0.036,0.041,0.045)(0.43,0.63,0.91)
f_2	(-0.042,-0.036,-0.031)(0.39,0.55,0.74)
f_3	(-0.008,0.026,0.053)(0.12,0.31,0.69)

In the sixth step, the alternatives are ranked: f_1, f_3, f_2. As a result, f_1 – stands out as the best choice.

3. CONCLUSION

The issue discusses the use of the PROMETHEE method for solving the BDAP selection problem. "Z-numbers are used to represent both the criteria values and their corresponding importance weights. To rank the alternatives, a comparison based on fuzzy Pareto optimality of Z-numbers is applied.

REFERENCES

Aliev, R. A. et al. (2017). Approximate reasoning on a basis of Z-number-valued if–then rules. *IEEE Transactions on Fuzzy Systems,* Vol. 25, No. 6, pp.1589–1600. https://doi.org/10.1109/TFUZZ.2016.2612303

Aliev, R. A. et al. (2015). *The arithmetic of Z-numbers: Theory and applications*. Singapore: World Scientific.

Aliev, R. A. et al. (2016). Ranking of Z-numbers and its application in decision making. *International Journal of Information Technology & Decision Making,* Vol. 15, No. 6, pp.1503–1519. https://doi.org/10.1142/S0219622016500310

Ilieva, G. (2019). Decision analysis for big data platform selection. *Engineering Sciences,* Vol.56, No. 2, pp.5–18. https://doi.org/10.7546/EngSci.LVI.19.02.01

Jabbarova, K. I. (2023). Application of WASPAS method to data platform selection under Z-valued information. In *Lecture Notes in Networks and Systems*, Vol. 610, pp. 557–563. https://doi.org/10.1007/978-3-031-25252-5_73

Jabbarova, K. İ. et al. (2024). Toward Z-number valued reinforcement learning problem. *Lecture Notes in Networks and Systems*, Vol. 718, pp. 352–360. https://doi.org/10.1007/978-3-031-51521-7_44

Jabbarova, K. I. et al. (2021). Z-decision making in human resources department. *Advances in Intelligent Systems and Computing,* Vol. 1306, pp. 498–507. https://doi.org/10.1007/978-3-030-64058-3_62

Jabbarova, K. I. et.al. (2021). Z-information based PROMETHEE method. In *Advances in Intelligent Systems and Computing*, Vol. 1323, pp. 287–293. https://doi.org/10.1007/978-3-030-68004-6_37

Jabbarova, K. I. et al. (2020). Solution of a decision making under risk with Z-information. In *Advances in Intelligent Systems and Computing*, Vol. 1306, pp. 781–786. https://doi.org/10.1007/978-3-030-64058-3_98

Jabbarova, K. I. et al. (2022). Application of ELECTRE method to decision making under Z-number-valued information. In *Lecture Notes in Networks and Systems,* Vol. 362, pp. 214–220. https://doi.org/10.1007/978-3-030-92127-9_31

Tsaniyah, F. A. et al. (2025). The role of big data analytics in supporting decision-making theories in companies. *Jurnal Bisnis dan Komunikasi Digital,* Vol. 2, No. 2., pp. 1-10.https://doi.org/10.47134/jbkd.v2i2.3458

ENHANCING ADHD SCREENING WITH AN INTERPRETABLE MACHINE LEARNING AND ITEM RESPONSE THEORY FRAMEWORK

Débora. M. Miranda[1], Pedro Loures[1], Gisele L. Pappa[1], Alexandre. L. O. Serpa[2], Wagner Meira Jr[1] and Marco A. Romano-Silva[1]
[1]*Federal University of Minas Gerais (UFMG), Brazil*
[2]*CIPHER Lab, Mackenzie Presbyterian University, São Paulo, Brazil*

ABSTRACT

Attention Deficit Hyperactivity Disorder (ADHD) is a psychiatric condition that affects around 5-7% of children around the world. This work investigates interpretable classification models capable of accurately identifying individuals with ADHD while indicating which features were more informative for diagnosis. Four decision tree-based models were tested and compared with Explainable Boosting Machine (EBM). The observed performance was similar. Results of EBM were a more interpretable model. Explanations quantitative and qualitatively were assessed by a pool of specialists, demonstrating how they may help on mental health screening and to explicit features which might be useful to improve the psychiatric classifications.

KEYWORDS

ADHD, Machine Learning, Interpretability, Item Response Theory, Rasch Model, Screening, Child Psychiatry

1. INTRODUCTION

Attention Deficit Hyperactivity Disorder (ADHD) poses significant challenges to children's academic, emotional, and social development. It is estimated to affect around five percent of children worldwide. Despite its prevalence, many affected individuals remain undiagnosed due to limited diagnostic tools and the shortage of specialized professionals. Recent Brazilian legislation (Law No. 14,254) amplifies the need for systematic and effective ADHD screening in schools. The goal of this study is to develop and assess an interpretable machine learning (ML) approach that streamlines ADHD screening while preserving transparency. By combining Explainable Boosting Machine (EBM) algorithms with Item Response Theory (IRT), we aim to provide not only robust classification but also clear insights into why certain features contribute to each diagnosis. Early detection is critical for improving long-term outcomes and ensuring that children receive appropriate interventions. Integrating machine learning solutions that clinicians can trust stands as a key factor in closing the diagnosis gap. Our approach, therefore, prioritizes not just performance metrics but also interpretability, which is essential for facilitating adoption in clinical settings.

2. BODY OF PAPER

This study conducted a secondary analysis of data from 345 individuals aged 3–25, collected between 2011 and 2020 as part of the "Multidimensional Assessment of Individuals with ADHD" project. ADHD diagnoses were established using the K-SADS-PL interview. The raw dataset contained 916 variables across clinical, psychometric, and socioeconomic domains, with 78.2% ADHD prevalence. After preprocessing—removing sensitive, expert-dependent, and low-relevance variables—and binarizing psychometric scales, 216 features were selected using Spearman correlation. Four machine learning models (Random Forest, XGBoost, Explainable Boosting Machine, and SVM) were trained using a 221/114 train/test split, with hyperparameters tuned via random search (500 iterations, 10-fold CV). Performance was assessed using F1-score, accuracy, precision, recall, and specificity, with 100-bootstrap confidence intervals. More interpretable models were prioritized when performance was similar. An audit system enabled clinical interpretation, allowing experts to evaluate key features, diagnostic agreement, and model utility. Explanation stability was tested through 1,500 perturbation analyses per instance. Finally, Rasch Item Response Theory was applied to selected features to assess psychometric validity, using infit/outfit indices and person-separation reliability. Analyses were conducted in R using established statistical packages.

2.1 Results

Given the imbalanced dataset, the F1 metric was prioritized for model quality. Table 1 presents the fit metrics for both machine learning and IRT models. Within the 95% confidence interval, no statistically significant difference in F1 scores was observed among the machine learning models.

Table 1. Fit indexes for all models

Model	F1	Sensitivity	Specificity	Precision	Accuracy
EBM	.81 - .92 (.87)	.94 - 1.00 (.98)	.03 - .29 (.14)	.69 - .85 (.78)	.69 - .85 (.77)
XGBOOST	.82 - .91 (.87)	.94 - 1.00 (.98)	.05 - .32 (.17)	.71 - .86 (.79)	.70 - .84 (.78)
RF	.83 - 0.92 (.88)	1.00 - 1.00 (1.00)	.03 - .27 (.14)	.70 - .85 (.78)	.71 - .86 (.79)
SVM	.82 - .92 (.87)	.97 - 1.00 (.99)	.03 - .29 (.14)	.71 - .85 (.78)	.71 - .85 (.78)
Rasch	.70	.61	.59	.84	.55 - .65 (.60)

Given the similar performance, the models were compared on transparency. EBM was chosen as the preferred model due to its inherent transparency and stability, unlike opaque models which may yield inconsistent explanations (Ribeiro et al., 2016). The interpretability stability of the EBM was high at 98.5%. Specialists' evaluation showed 48% agreement with the model's selected criteria, and 37% altered their diagnoses based on its proposals. Furthermore, 33% would replace a doctor with the model, 54% would use it in triage, and 46% rated its interpretability as good. The IRT model, incorporating all 84 EBM-selected features, showed adequate convergence and high person separation reliability (.84). Infit values ranged from .85 to 1.19, and outfit from .70 to 1.89, with only two items slightly outside the optimal range (QEDP_9 and QEDP_21). The Person-Item Map (not displayed here to save space) graphically depicted individuals and items along a latent trait continuum, with 9 items having $\Theta \leq -2$ standard deviations and the remaining 75 sequentially distributed from −1.46 to 2.85. The Rasch model achieved an accuracy of 0.603, indicating moderate performance. It predicted positive values (84.1% ADHD presence) better than negative (29.3% ADHD absence). The F1 score (.705) showed a decent balance between precision and recall (Kappa = .143; McNemar's test: $p < .001$). The AUC was 0.632 (sensitivity = .607; specificity = .587), slightly better than random guessing.

2.2 Discussion

Machine learning (ML) is increasingly applied in psychiatric diagnosis (Itani et al., 2019), with growing emphasis on interpretable, auditable systems (Lipton, 2016, 2017). Building on prior work like (Dipnall et al., 2016), this study focused on accurate ADHD screening using interpretable models and self/peer-assessment scales aligned with DSM-V criteria (Grazioli et al., 2023; Jandre et al., 2024). Unlike neuroimaging approaches reliant on opaque deep learning, our model used tabular psychometric data, favoring inherently interpretable algorithms such as EBM and decision trees (Gorishniy et al., 2021; Loh et al., 2022). While prior studies often blend these models with less interpretable ones, our approach emphasized expert-auditable outputs. The EBM model selected detailed features along the latent continuum, favoring completeness over parsimony. Although the Rasch model introduced psychometric nuance (e.g., item difficulty), it offered only moderate accuracy and sensitivity/specificity, limiting its standalone diagnostic value but supporting its role as a screening tool. Results highlight the potential of combining ML and latent trait models for datasets with underlying psychological constructs. Limitations include dataset imbalance, which may bias performance metrics, and mixed expert feedback on interpretability, indicating the need for further refinement.

3. CONCLUSION

By merging Explainable Boosting Machine algorithms with Item Response Theory, this work demonstrates a pathway to an effective yet transparent ADHD screening tool. Clinicians not only receive predictions but also a clear rationale for each assessment, bridging the gap between machine efficiency and the nuanced decision-making needed in clinical practice. Continued refinements and validation efforts will be critical for broadening applicability. Ultimately, interpretable machine learning solutions have the potential to transform early ADHD detection, enabling timely interventions that can significantly improve outcomes for affected children and adolescents.

REFERENCES

Dipnall, J. F., Pasco, J. A., Berk, M., Williams, L. J., Dodd, S., Jacka, F. N., & Meyer, D. (2016). Fusing data mining, machine learning and traditional statistics to detect biomarkers associated with depression. *PloS One*, *11*(2), e0148195.

Gorishniy, Y., Rubachev, I., Khrulkov, V., & Babenko, A. (2021). Revisiting deep learning models for tabular data. In *arXiv [cs.LG]*. arXiv. http://arxiv.org/abs/2106.11959

Grazioli, S., Crippa, A., Rosi, E., Candelieri, A., Ceccarelli, S. B., Mauri, M., Manzoni, M., Mauri, V., Trabattoni, S., Molteni, M., Colombo, P., & Nobile, M. (2023). Exploring telediagnostic procedures in child neuropsychiatry: addressing ADHD diagnosis and autism symptoms through supervised machine learning. *European Child & Adolescent Psychiatry*. https://doi.org/10.1007/s00787-023-02145-4

Itani, S., Rossignol, M., Lecron, F., & Fortemps, P. (2019). Towards interpretable machine learning models for diagnosis aid: A case study on attention deficit/hyperactivity disorder. *PloS One*, *14*(4), e0215720.

Jandre, C. R. S., Dal' Maria, F. C. S., Nardy, G. A. C., Guimarães, F. T. G., de Miranda, D. M., & Nobre, C. N. (2024). Persuasive technology for managing the routine of families with children and adolescents with ADHD: A user-centered approach. *Proceedings of the XXIII Brazilian Symposium on Human Factors in Computing Systems*, 1–16.

Lipton, Z. C. (2016). The Mythos of Model Interpretability. In *arXiv [cs.LG]*. arXiv. http://arxiv.org/abs/1606.03490

Lipton, Z. C. (2017). The doctor just won't accept that! In *arXiv [stat.ML]*. arXiv. http://arxiv.org/abs/1711.08037

Loh, H. W., Ooi, C. P., Barua, P. D., Palmer, E. E., Molinari, F., & Acharya, U. R. (2022). Automated detection of ADHD: Current trends and future perspective. *Computers in Biology and Medicine*, *146*(105525), 105525.

Ribeiro, M. T., Singh, S., & Guestrin, C. (2016). Model-agnostic interpretability of machine learning. In *arXiv [stat.ML]*. arXiv. http://arxiv.org/abs/1606.05386

IMPLEMENTATION OF TELEOPHTHALMOLOGY IN MINAS GERAIS - BRAZIL: EARLY DETECTION OF OCULAR COMPLICATIONS IN DIABETIC USERS IN PRIMARY HEALTH CARE

Rosângela Durso Perillo[1], Mariana Abreu Caporali de Freitas[2], Johan Alejandro Vega Castañeda [2], Luana da Silva Jorge[1], Luiz Carlos Molinari-Gomes[1] and Alaneir de Fátima dos Santos[3]

[1]*Núcleo de Tecnologia em Saúde, Faculdade de Medicina / UFMG. Belo Horizonte (MG), Brazil*

[2]*Programa de Pós-Graduação em Saúde Pública, Faculdade de Medicina / UFMG. Belo Horizonte (MG), Brazil*

[3]*Departamento de Medicina Social e Preventiva, Faculdade de Medicina / UFMG. Belo Horizonte (MG), Brazil*

ABSTRACT

Introduction: The increased demand for ophthalmological care in the SUS reflects the scarcity of this service provision in the country, especially due to the concentration of specialists in urban centers and a deficient and poorly coordinated comprehensive care network. University projects in partnership with the Ministry of Health allow the performance of retinography in primary care, using a portable retinograph operated locally by a trained health professional who captures the images, without the need for dilation and with the offer of remote diagnosis of patients, enabling early detection of diseases such as diabetic retinopathy, hypertensive retinopathy and glaucoma.

Objective: To describe the performance of retinography in primary care patients and the resolution of these cases.

Method: Cross-sectional study describing 2,158 consultations of the Teleretinography project, carried out between March 2023 and February 2024 in the state of Minas Gerais, Brazil, analyzing the demographic characteristics of the patients, the reasons for referral to telediagnosis, the characterization of the diagnoses made, as well as the resolution of the implemented telediagnosis model. A portable retinograph was used to capture color images that allow visualization of more peripheral areas with high quality, also with the support of Artificial Intelligence (AI) that signals areas of the retina with possible anomalies.

Results: The patients were mostly female (67.7%), between 60 and 69 years of age (28.9%), and referred to telediagnosis mainly due to being diabetic (72%) and hypertensive (82.5%). Of the total, 45.1% had their ocular complaints completely resolved, without the need for referral to an ophthalmological consultation. In 39.1% of the patients, clinical conditions with altered diagnosis were detected, with the main findings being hypertensive retinopathy (44.8%) and diabetic retinopathy (27.5%).

Conclusion: The study demonstrated that telemedicine can be a solution for less complex eye diseases, qualifying and helping to reduce waiting lists for ophthalmological care. It is a viable strategy for large-scale implementation in Primary Health Care, in order to minimize difficulties in accessing an ophthalmologist. The use of AI does not replace the ophthalmologist, but acts as a complementary tool in the diagnosis of the alterations found, contributing to the preparation of the report more quickly.

KEYWORDS

Telehealth, Primary Health Care, Diabetic Retinopathy

1. INTRODUCTION

Chronic diseases such as diabetes and hypertension affect eye health, with diabetic and hypertensive retinopathy being the main causes of preventable blindness in Brazil. Diabetic retinopathy is one of the main complications of diabetes mellitus and is the main cause of preventable blindness in working-age adults (Malta et al., 2019; Wong & Sabanayagam, 2019). The silent progression of the disease, often asymptomatic in the early stages, makes early diagnosis and timely treatment difficult, increasing the risk of irreversible visual loss.

Brazil does not yet have a specific policy for screening and care for retinal changes, however law 13,895 of 2019 establishes the National Policy for Diabetes Prevention and Comprehensive Assistance to Diabetics, and its text includes the performance of periodic retinographies for users suffering from diabetes mellitus (Brasil, 2022).

We know that access to an ophthalmologist through the Unified Health System is hindered by long waiting lists. Given this scenario, expanding access to retinography can be a strategy to overcome these challenges. Retinography is a simple exam that captures colored images of the patient's retina and, from the analysis of these images, it is possible to identify retinal alterations (Bortoli et al., 2022). The implementation of a teleretinography program for high-level diabetic users within the scope of Primary Health Care emerges as an essential strategy to strengthen the line of care for users with diabetes. Teleretinography is a non-invasive and quickly performed exam, capable of detecting early signs of diabetic retinopathy, allowing early intervention and significantly reducing the risk of serious eye complications.

This paper aims to report the experience of implementing a teleretinography program in municipalities in the state of Minas Gerais, Brazil. The implementation of this service aims to expand access to the exam, optimizing the tracking and monitoring of users with greater vulnerability, promoting comprehensive and effective care. By using this technology to enable remote evaluation of retinal images by specialists, teleretinography becomes an essential tool in improving the care of diabetic patients.

2. DEVELOPMENT - METHODS AND RESULTS

This is a cross-sectional study that analyzed a database of retinography exams from the Telehealth Center of the Faculty of Medicine of the Federal University of Minas Gerais (NUTEL FM/UFMG). This study analyzed data from retinography exams performed in 6 small and medium-sized municipalities in Minas Gerais, Brazil. NUTEL FM/UFMG, with the support of the Ministry of Health of the country, began implementing the Teleretinography project in 2023 with the aim of performing retinography exams in municipalities in the state of Minas Gerais in order to reduce the waiting list of chronic patients who need to undergo the exam annually. The retinography exams were performed between December 2023 and September 2024. The target audience is diabetic patients.

NUTEL FM/UFMG has established a partnership with a city in the state of Minas Gerais, which showed interest in the project. Healthcare professionals were trained to perform the exams on previously scheduled patients. A portable fundus camera was used to capture color images with a 55° field of view, which allows high-quality visualization of more peripheral areas, and also has the support of Artificial Intelligence (AI) that signals areas of the retina with possible anomalies. The device uses the Eyercloud Web system to store the collected images. The captured images are reviewed remotely within 72 hours, and the healthcare unit accesses and prints the reports, selecting patients with changes that justify being referred for a full consultation with an ophthalmologist. NUTEL FM/UFMG is responsible for training professionals at the units and issuing reports on fundus scans performed by an ophthalmologist at NUTEL FM/UFMG.

We analyzed 2,158 retinography exams performed from March 2023 to September 2024 in 6 cities in the state of Minas Gerais. To perform the exam in the units, priority was given to diabetic and severely hypertensive patients. The patients were mostly female (67.7%), aged between 60 and 69 years old (28.9%). At the time of the exam, patients reported having diabetes (72%), hypertension (82.5%) and a diagnosis of High Cholesterol (46.7%), with 63.0% having two or more reported chronic non-communicable diseases. Of the 2,158 patients who underwent telediagnosis (retinography), 973 (45.1%) had a normal exam, without the need for referral to an ophthalmological consultation, and 843 (39.1%) patients had reports with alterations. In 342 patients (15.8%), the images did not allow for a safe report, and it was suggested that the exam be performed with dilation. Among the 843 patients with exams showing some type of alteration, the most prevalent disorders were hypertensive retinopathy with 378 (44.8%) cases and diabetic retinopathy with 232 (27.5%) cases.

As a limitation of our study, we can mention the difficulty in reporting the exams of some patients without performing dilation prior to capturing the images. Another limitation would sometimes be the difficulty in identifying professionals in the health units who have the ability to operate the equipment.

The results show that performing retinography in primary care has great potential to reduce waiting times for specialist consultations, as well as to detect diabetic retinopathy and hypertensive retinopathy. The use of AI supports and enhances the results of the reports provided. It is worth noting that AI does not replace the ophthalmologist, but acts as a complementary tool in the diagnosis and monitoring of patients.

3. CONCLUSION

The performance of retinography in primary health care in municipalities in the state of Minas Gerais has proven to be an effective strategy for screening for ocular disorders, such as diabetic retinopathy, expanding access to early diagnosis and improving care for people with diabetes mellitus, overcoming geographical barriers and optimizing the flow of referrals to ophthalmologists. The use of retinography in primary health care as a screening tool has proven to be effective, especially in high-risk populations.

The partnership between the FM/UFMG Telehealth Center and the municipal administration allowed the training of professionals and the agile issuance of reports and qualifying the interpretation of exams supported by AI, optimizing care flows and reducing unnecessary referrals to specialized care.

The implemented model demonstrates that the use of mobile devices can contribute to the early detection of diseases such as diabetic retinopathy, hypertensive retinopathy, cataracts and glaucoma, highlighting the importance of telemedicine as a tool to improve health care. The positive results reinforce the viability of the strategy and the need for its expansion to other locations, and validate the available evidence and clearly demonstrate the positive impact of retinography on improving care for people with diabetes in the healthcare network of the Unified Health System, highlighting teleretinography as a promising solution for improving eye health and preventing avoidable blindness.

REFERENCES

Bortoli, J. Q. et al, (2022). *Retinografia como forma de rastreio de retinopatia diabética em hospital terciário do Sistema Único de Saúde*. Revista Brasileira de Oftalmologia, 81, e0057.

Brasil. LEI Nº 13.895, DE 30 DE OUTUBRO DE 2019. Institui a *Política Nacional de Prevenção do Diabetes e de Assistência Integral à Pessoa Diabética.* Disponível em: <https://www.planalto.gov.br/ccivil_03/_ato2019-2022/2019/lei/l13895.htm>

Malta, D. C. et al, (2019). *Probabilidade de morte prematura por doenças crônicas não transmissíveis, Brasil e regiões, projeções para 2025.* Revista Brasileira de Epidemiologia, 22, e190030.

Wong, T. Y. and Sabanayagam, C., (2019). *The War on Diabetic Retinopathy.* Asia-Pacific Journal of Ophthalmology, v. 8, n. 6, p. 448–456.

EARLY STROKE DETECTION USING CONVOLUTIONAL NEURAL NETWORKS BASED ON LONGITUDINAL TEMPORAL ELECTRONIC HEALTH RECORDS

Chia-Hui Chien[1,3], Ting Ting Ko[3], Xiaohong W. Gao[1], Yung-Chun Chang[4] and Yu-Chuan Li[3]

[1]*Department of Computer Science, Middlesex University, London, UK*
Middlesex University, The Burroughs, Hendon, London NW4 4BT, UK
[3]*Graduate Institute of Biomedical Informatics, Taipei Medical University, Taiwan*
250 Wuxing Street, Taipei City, Taiwan 110
[4]*Graduate Institute of Data Science, Taipei Medical University, Taiwan*
250 Wuxing Street, Taipei City, Taiwan 110

ABSTRACT

Stroke remains one of the leading causes of mortality and long-term disability worldwide. Early identification of individuals at high risk is critical for timely intervention and effective prevention. This study proposes a deep learning framework using a Convolutional Neural Network (CNN) to predict stroke onset based on temporal electronic health record (EHR) data derived from Taiwan's National Health Insurance Research Database (NHIRD). We extracted weekly diagnosis and medication records over a 52-week period and represented each patient's history as a binary matrix of 1,137 features. The model was trained to predict stroke onset at 12, 6, and 3 months before the clinical event. Evaluation results showed strong predictive performance, with AUROC scores ranging from 0.91 (two years prior) to 0.82 (three months prior). To enhance interpretability, a Random Forest analysis was conducted to identify influential features. Key predictors included type II diabetes, hypertension, chronic kidney disease, and related medications such as antiplatelet agents and dihydropyridine derivatives. The study demonstrates that combining CNNs with interpretable models enables accurate and explainable stroke risk prediction. These findings highlight the potential for integrating AI-based tools into clinical workflows for early warning and personalized prevention strategies.

KEYWORDS

Stroke Prediction, Convolutional Neural Network, Electronic Health Records, Temporal Modeling, Healthcare AI

1. INTRODUCTION

Stroke is a major global health concern, causing approximately 6.5 million deaths annually and contributing to long-term disability worldwide (Benjamin et al., 2017). This heavy burden highlights the urgent need for accurate, early prediction to support preventive interventions and reduce both mortality and socio-economic impact (Kernan et al., 2014; Steiger & Cifu, 2016). The increasing availability of electronic health records (EHRs) has enabled the application of machine learning and deep learning methods to model longitudinal health data and improve predictive accuracy in clinical settings (Al'Aref et al., 2020; Xie et al., 2022). In particular, Taiwan's National Health Insurance Research Database (NHIRD), which covers nearly the entire population, offers a rich and reliable source of diagnostic and medication data for population-based prediction studies (Xie et al., 2022). Convolutional neural networks (CNNs), widely recognized for their success in pattern recognition, have shown exceptional performance in modeling complex temporal relationships in sequential clinical data, outperforming traditional methods in both accuracy and interpretability (Al'Aref et al., 2020; Xie et al., 2022). Prior EHR-based studies have identified key stroke-related factors—such as hypertension, diabetes, and cardiovascular conditions—as significant predictors (D'Agostino et al., 1994). However, challenges remain, including data heterogeneity, temporal variability, and model explainability, which limit clinical adoption (Lauritsen et al., 2020). This study introduces an explainable CNN framework trained on weekly diagnosis and prescription data from the NHIRD. By leveraging temporal trends, our model aims to improve stroke risk prediction and provide interpretable outputs to support personalized, proactive clinical decision-making.

2. METHOD

We utilized retrospective data from Taiwan's NHIRD, covering 1999 to 2013, which includes comprehensive healthcare records for approximately 2 million individuals. We focused on first-time stroke patients (ages 20-99) diagnosed based on ICD-9-CM codes (430.xx–438.xx). To ensure accurate stroke identification, we excluded patients with a previous stroke diagnosis before 2003, resulting in a final cohort of 16,804 stroke patients. Control groups were created by matching non-stroke patients using a 1:10 ratio (n=169,902) for a larger group and a 1:4 ratio (n=67,222) for a smaller group, based on age and demographics. Weekly records of diagnosis (ICD-9-CM) and medication codes (ATC) were extracted over a 52-week period prior to the stroke event or equivalent index date for controls. Each patient's data was represented as a binary matrix with 1,137 features (708 diagnosis codes, 429 medication codes), enabling precise temporal analysis critical for stroke prediction. Our predictive model utilized a CNN to capture temporal features from structured EHR data, focusing on weekly diagnosis and medication records. The architecture consisted of several key stages, as shown in Figure 1. Patient data was structured into a binary matrix with 1137 across a 52-week period. Convolutional layers applied filters to capture temporal patterns, followed by max-pooling layers to reduce dimensionality while preserving key information. A fully connected layer learned non-linear combinations of features, and the output layer used a sigmoid activation function to classify stroke risk at 12, 6, and 3 months prior to onset. The CNN progressively learned hierarchical features, from basic weekly occurrences to complex temporal relationships, facilitating accurate stroke prediction.

To enhance interpretability, a Random Forest analysis was performed on the CNN-derived features, identifying the most influential predictors, such as specific diagnoses and medications. This CNN-Random Forest integration achieved high predictive accuracy while providing clinically relevant insights for stroke risk assessment, aiding healthcare decision-making.

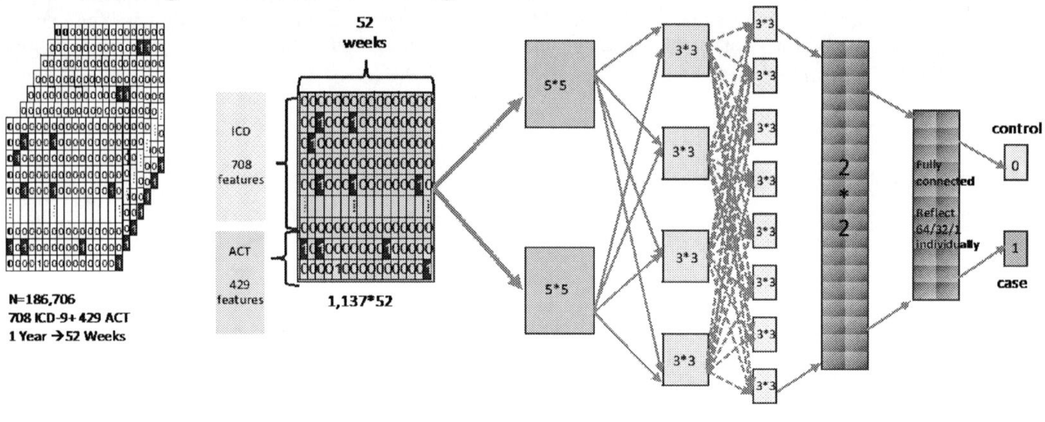

Figure 1. The proposed model architecture

3. RESULT

The proposed CNN model demonstrated robust predictive performance for early stroke detection using temporal EHR data. Trained on weekly ICD-9 and ATC-coded records from up to 52 weeks prior to onset, the model achieved AUROCs of 0.91 (2-year), 0.89 (1-year), 0.86 (6-month), and 0.82 (3-month), with accuracy exceeding 91% across all intervals. Key predictors included type II diabetes, ischemic heart disease, and medications such as dihydropyridine derivatives, sulfonylureas, and antiplatelet agents. Removing these features—particularly dihydropyridine derivatives—significantly reduced model performance (AUROC dropped from 0.91 to 0.85), underscoring their clinical relevance.

Compared to conventional tools like the Framingham Risk Score, which typically achieves an AUC between 0.65 and 0.72 (D'Agostino et al., 1994), our model aligns with the upper range of modern deep learning approaches, reaching an AUC of 0.90 (Xie et al., 2022). Weekly-level temporal granularity enabled the CNN to capture nuanced risk patterns often missed by traditional models. Integration with Random Forest enhanced interpretability by surfacing clinically consistent risk factors, effectively addressing the "black box" concerns commonly associated with AI-driven predictions (Lauritsen et al., 2020). These findings suggest that the proposed CNN-based framework offers a highly accurate and explainable approach to stroke risk prediction. With further validation across diverse populations and integration of additional data modalities, the model holds strong potential for clinical adoption in preventive care settings.

4. CONCLUSION

This study demonstrates that deep learning models utilizing temporal EHR data can accurately predict stroke risk up to 12 months before onset. By combining CNNs with explainable techniques like random forest feature importance analysis, the model not only delivers strong predictive performance but also identifies critical factors influencing stroke risk, such as specific medications (antacids, platelet aggregation inhibitors, sulfonylureas, and dihydropyridine derivatives) and diagnoses like type II diabetes and chronic ischemic heart disease. This interpretability empowers clinicians to make proactive, informed decisions, supporting timely preventive interventions. Additionally, the integration of structured EHR data with deep learning methods shows promise for broader application in predicting and managing other chronic conditions, marking a significant advancement in both medical informatics and personalized healthcare.

ACKNOWLEDGEMENT

This project is funded by the British Council under Early Research Fellowship Programme (2024-2025). Their financial support is gratefully acknowledged.

REFERENCES

Al'Aref, S. J., Anchouche, K., Singh, G., Slomka, P. J., Kolli, K. K., Kumar, A., ... & Min, J. K. (2020). Machine learning and cardiovascular imaging. *Current Treatment Options in Cardiovascular Medicine, 22*(3), Article 10.

Benjamin, E. J., Blaha, M. J., Chiuve, S. E., Cushman, M., Das, S. R., Deo, R., ... & Muntner, P. (2017). Heart disease and stroke statistics—2017 update: A report from the American Heart Association. *Circulation, 135*(10), e146–e603.

D'Agostino, R. B., Wolf, P. A., Belanger, A. J., & Kannel, W. B. (1994). Stroke risk profile: Adjustment for antihypertensive medication. The Framingham Study. *Stroke, 25*(1), 40–43.

Kernan, W. N., Ovbiagele, B., Black, H. R., Bravata, D. M., Chimowitz, M. I., Ezekowitz, M. D., ... & Wilson, J. A. (2014). Guidelines for the prevention of stroke in patients with stroke and transient ischemic attack: A guideline for healthcare professionals from the American Heart Association/American Stroke Association. *Stroke, 45*(7), 2160–2236.

Lauritsen, S. M., Kristensen, M., Olsen, M. V., Larsen, M. S., Lauritsen, K. M., Jørgensen, M. J., ... & Schmidt, H. H. (2020). Explainable artificial intelligence model to predict acute critical illness from electronic health records. *Nature Communications, 11*, Article 3852.

Steiger, N., & Cifu, A. S. (2016). Primary prevention of stroke. *JAMA, 316*(7), 658–659.

Xie, F., Wang, T., Zhang, Y., He, Z., & Wang, F. (2022). Deep learning for temporal data representation in electronic health records: A systematic review of challenges and methodologies. *Journal of Biomedical Informatics, 126*, 103980.

DECISION SUPPORT SYSTEMS IN E-HEALTH: INCORPORATION OF TELECONSULTATIONS INTO THE ASSISTANCE FLOW

Alaneir de Fátima dos Santos[1], Eliane Maria de Sena Silva[2], Mariana Abreu Caporali de Freitas[3], Rosângela Durso Perillo[2] and Shirley Pereira de Almeida[2]
[1]*Departamento de Medicina Social e Preventiva, Faculdade de Medicina da UFMG, Brazil*
[2]*Núcleo de Tecnologia em Saúde, Faculdade de Medicina da UFMG, Brazil*
[3]*Programa de Pós-Graduação em Saúde Pública, Faculdade de Medicina da UFMG, Brazil*

ABSTRACT

This paper aims to present the implementation experience and preliminary results of the "Project for Incorporating Teleconsultations into Specialized Care in Minas Gerais", highlighting the city of Belo Horizonte (BH, the state capital) as an experimental field. The great challenges of providing specialized care in a country of continental dimensions and significant socioeconomic and cultural inequalities require the Brazilian government to provide tools that minimize these difficulties. In this context, the incorporation of teleconsultations into care processes is a powerful tool for network integration, improved access and continuing education.

KEYWORDS

Teleconsulting, Specialized Care, Access Improvements, Continuing Education

1. INTRODUCTION

In Brazil, the Primary Health Care (PHC) work process is considerably complex, due to the territorial extensions, socioeconomic inequalities, cultural diversity, high professional turnover, the multiplicity of training institutions and the shortage of specialists in more remote and peripheral areas of large cities. These and other factors highlight the need to develop technologies that support and articulate PHC and Specialized Care (SC).

Therefore, the use of teleconsultations is a means of improving healthcare services – since it supports clinical decision-making and diversifies the range of tools available to Family Health Teams (FHTs), contributes to the training process of professionals, reduces unnecessary referrals, increases access possibilities and reduces waiting times for specialized consultations and exams. From a management perspective, the analysis of the most demanded specialties and the questions raised contribute significantly to discussions on protocols and care flows.

With this presentation we aim to report on the experience currently underway in the municipality of Belo Horizonte (BH), capital of the state of Minas Gerais (MG), where the "Project for Incorporating Teleconsultations into Specialized Assistance in Minas Gerais" has been implemented since 2024.

2. IMPLEMENTATION AND PRELIMINARY RESULTS

To raise awareness and facilitate implementation in 8 macro-regions of the state of Minas Gerais, two universities in the capital organized themselves into 3 Telehealth Centers, each responsible for previously defined regions and municipalities. The state capital, to which this report refers, was under the responsibility of the Faculty of Medicine of the Federal University of Minas Gerais (FM/UFMG), which provided teleconsultants and professors for more than 90 specialties (including doctors, nurses, dentists, psychologists, and other professionals who make up family health teams).

The line of care defined as a priority by the State Department of Health (SES/MG) for the interior of the state was maternal and child care, for which courses were developed, protocols, and care flows were reviewed. In addition to these moments of extensive online training, each municipality listed its specific needs, according to the waiting lists for specialized consultations. The BH process was more complex, due to its population size and its socio-economic-cultural diversity.

For conceptual and methodological alignments, FM/UFMG held technical-managerial meetings with the management team of the Municipal Health Department (SMSA) of BH, where the 3 specialties with the longest waiting lists for specialized consultations were listed: endocrinology, pulmonology and gastroenterology. To address the problem, FM/UFMG presented an administrative review methodology in order to obtain up-to-date information on demand and then perform a clinical reassessment of those users who continued to require specialized consultations, incorporating prior requests for teleconsultations when deemed appropriate.

New alignments were made, this time with the regional management of the 9 Health Districts (DRES) and the 153 Basic Health Units (UBS), where the methodology was presented, and an agreement was reached to begin field visits – with the purpose of training professionals to use the technological platform and adapting the work process. Each ESF reassessed its users, carried out teleconsultations - in an offline format, answered within 72 hours by specialist teachers - and removed from the queue users who could continue to be monitored by the APS itself, at the UBS.

It is important to highlight that, in addition to making teleconsultations available in more than 90 medical and non-medical specialties, FM/UFMG also provided refresher courses for all physicians and participated in the review of SMSA's care flows and protocols.

Analyzing the monitoring of the first 6 months, it is already possible to verify the effectiveness of the process, with a reduction in waiting lists and consequent more timely scheduling of specialized consultations. Between the months of November 2024 and May 2025, 1,069 teleconsultations were carried out, with more than 17% in endocrinology – the first specialty addressed in relation to the waiting list. Of the total teleconsultations requested, answered and completed, in 61% the professionals intended to refer the user to a specialized consultation – who would then be placed on the waiting list. After the teleconsultations, only 37% of users were referred and 63% remained in care supported by the APS, strengthened by the guidance received from specialist teachers. Regarding the level of satisfaction, almost 100% of the professionals requesting the APS were satisfied or very satisfied.

Considering the difficulties in forming PHC teams with pediatricians and gynecologists, it was also observed that several professionals (responsible for prenatal and childcare care) were able to clarify their doubts and continue to monitor users in the ESF, including more complex cases. It is also interesting to highlight the multidisciplinary and interdisciplinary nature of PHC and the contribution that the provision of teleconsultations can offer. In addition to medical specialties, dentistry, nursing, psychology and physiotherapy were also requested. Expanding the possibilities even further, we observed dentists, nurses, pharmacists and physiotherapists requesting clarification and support from medical specialists, and even a doctor requesting teleconsultation with a psychologist.

3. CONCLUSION

Based on the second formative opinion provided by teleconsultants, enhanced by the refresher courses offered by the University, professionals from the ESF and UBS were able to define with greater confidence the best course of action to be taken: whether to keep the user on the waiting list for a specialist appointment, or to continue providing care in the same area, by the ESF and UBS, with greater clinical safety and, consequently, a reduction in the waiting list. Reducing the waiting list for specialist appointments increases access, especially for users who need care more quickly, responding to the essential principle of the Brazilian Unified Health System (SUS): equity.

Thus, although the number of requests is small (but with continuous growth), we can state that teleconsultations are already a powerful tool for integration between PHC and AE. They contribute to the qualification of care and to the reduction of effects such as: the lack of specialists in remote and peripheral regions of large cities; unnecessary travel by users between care points in the networks (remembering the continental dimensions commonly found in Brazilian territories); deficiencies in the training of some professionals; and difficulties for them to keep themselves informed about the large volume of topics related to the care provided in PHC. The challenge at hand is to increase the number of requests, so that the numbers grow and teleconsultations are in fact incorporated into the routine processes of the care flow between PHC and AE.

ACKNOWLEDGEMENT

We thank the Brazilian Ministry of Health for supporting and financing this Project.

REFERENCES

Nilson, L.G et al. (2019). *Avaliação da utilização de Telessaúde para Apoio Assistencial na Atenção Primária à Saúde/Evaluation of the use of telehealth for care support in primary health care*. Brazilian Journal of Health. Review, v. 2, n. 6, p. 6188-6206.

Santos et al. (2021). REDE APS ABRASCO. Nota Técnica – Incorporação de recursos de Telessaúde na Atenção Primária no Brasil.

AUTHOR INDEX

International Association for Development of the Information Society (IADIS)
Rua Sao Sebastiao Da Pedreira
No. 100, 30, 1050-209
Lisbon, Portugal

ISBN 979-8-3313-2768-2
9 798331 327682

CURRAN ASSOCIATES INC.
proceedings.com